Optical Fibre Lasers and Amplifiers

Edited by

P.W. FRANCE
British Telecom Research Laboratories
Ipswich

Blackie
Glasgow and London

Published in the USA and Canada by
CRC Press, Inc.
Boca Raton, Florida

Blackie and Son Ltd
Bishopbriggs, Glasgow G64 2NZ
and
7 Leicester Place, London WC2H 7BP

Published in the USA and Canada by
CRC Press, Inc.
2000 Corporate Blvd, N.W., Boca Raton, FL 33431

© 1991 Blackie and Son Ltd
First published 1991

*All rights reserved
No part of this publication may be reproduced, stored in a retrieval system,
or transmitted, in any form or by any means—graphic,
electronic or mechanical, including photocopying,
recording, taping—without the written permission of the Publishers*

British Library Cataloguing in Publication Data

Optical fibre lasers and amplifiers.
I. France, P.W.
621.36

ISBN 0-216-93157-6

Library of Congress Cataloging-in-Publication Data

Fibre lasers and amplifiers / edited by P.W. France.
  p.   cm.
 Includes index.
 ISBN 0-8493-7716-1
  1. Rare earth lasers. 2. Fiber optics. 3. Optical
communications.   I. France, P.W.
TA1677.F53 1991
621.36'6—dc20   91-2870
                                                          CIP

Phototypesetting by Thomson Press (India) Limited, New Delhi
Printed in Scotland by M. & A. Thomson Litho, East Kilbride

Preface

One of the most exciting new developments in the field of optics has been the rediscovery of the fibre laser amplifier. This device is incredibly simple, yet promises to revolutionise the field of optical communications: it is probably the most significant development since the monomode optical fibre. Merely by using a short length of optical fibre doped with a small amount of a rare earth element and 'pumping' this fibre with light from a semiconducting diode, gains of greater than 1000 can be achieved. This simple amplifier can be spliced into a length of normal optical fibre and can greatly increase the transmission distance. This device is going to have a major impact on the topology of optical communication systems and will allow information technology to be accessed by a greater portion of the population.

The work presented in this book reviews the development of optical fibre lasers and amplifiers over the past few years. The book is a compilation of chapters written by several contributors, all of whom are internationally acknowledged experts. Professor Elias Snitzer, now at Rutgers University, New Jersey, is regarded as the grandfather of the whole field and was one of the early pioneers of laser development in the mid sixties. In Chapter 1 he presents an overview and a historical introduction to the field. Chapter 2 is theoretical and introduces the concepts of energy levels and of lasing. Chapter 3 then discusses the fabrication of rare-earth doped optical fibres, whilst Chapter 4 covers the spectroscopy of rare-earth dopants in these fibres. Chapter 5 is written by several contributors, all of whom are involved in making components used to assemble fibre laser amplifiers, and it emphasises the practical aspects of the device. Chapter 6 then moves on to the operation of erbium amplifiers in optical systems, and a comparison is made here with other alternative amplifying devices. Chapter 7 reviews work into fibre laser oscillators made in silica fibre, while Chapter 8 covers similar work using fluoride glass fibres, an alternative glass host: this is a novel material with some interesting lasing properties which include amplifiers at $1.3\,\mu m$ using neodymium. Chapter 9 describes Q-switching and mode-locking in fibre lasers, and finally Chapter 10 examines future prospects for these devices and discusses potential developments such as distributed fibre amplifiers, soliton propagation, and the general use of fibre lasers in optical sensors.

This book is aimed at scientists and system engineers in industry and the academic world who might use or be involved with optical fibre lasers and amplifiers. It is the first comprehensive book to be published in this emerging

field and will be of interest to all people who are eager to hear of new and exciting developments in advanced optics.

The contributors wish to thank their respective establishments for support during the writing of this work, and to acknowledge the community of scientists and engineers who have helped to advance and revolutionise this field.

P.W.F

Contributors

Mr B.J. Ainslie	Head of Group, British Telecom Research Laboratories, Ipswich IP5 7RE
Mr J.R. Armitage	Executive Engineer, British Telecom Research Laboratories, Ipswich IP5 7RE
Mr M.C. Brierley	Executive Engineer, British Telecom Research Laboratories, Ipswich IP5 7RE
Dr D.M. Cooper	Head of Group, British Telecom Research Laboratories, Ipswich IP5 7RE
Ms S.P. Craig-Ryan	Executive Engineer, British Telecom Research Laboratories, Ipswich IP5 7RE
Professor A.I. Ferguson	Department of Physics and Applied Physics, Strathclyde University, Glasgow G4 0NG
Dr T. Finegan	Executive Engineer, British Telecom Research Laboratories, Ipswich IP5 7RE
Dr P.W. France	Head of Group, British Telecom Research Laboratories, Ipswich IP5 7RE
Professor D.C. Hanna	Department of Physics, Southampton University, Southampton SO9 5NH
Dr N. Langford	Postdoctoral Research Fellow, Strathclyde University, Glasgow G4 0NG
Dr C.A. Millar	Head of Group, British Telecom Research Laboratories, Ipswich IP5 7RE
Mr C.J. Rowe	Executive Engineer, British Telecom Research Laboratories, Ipswich IP5 7RE
Professor E. Snitzer	College of Engineering, PO Box 909, Piscataway, New Jersey 08855
Dr D.M. Spirit	Executive Engineer, British Telecom Research Laboratories, Ipswich IP5 7RE

CONTRIBUTORS

Dr A.C. Tropper — Lecturer, Department of Physics, Southampton University, Southampton SO9 5NH

Mr I.J. Wilkinson — Assistant Executive Engineer, British Telecom Research Laboratories, Ipswich IP5 7RE

Dr R. Wyatt — Head of Group, British Telecom Research Laboratories, Ipswich IP5 7RE

Contents

1 Perspective and overview — 1
E. SNITZER

1.1 Introduction — 1
1.2 Brief history of fibre lasers — 1
1.3 Early work on glass lasers — 3
1.4. Ligand field and ion interaction — 7
1.5 Fibre applications — 10
References — 12

2 Introduction of glass fibre lasers and amplifiers — 14
J.R. ARMITAGE

2.1 Introduction — 14
2.2 Comparison of bulk and fibre lasers — 15
    2.2.1 Threshold pump power requirements — 15
    2.2.2 Other advantages of fibre lasers — 17
2.3 Energy levels of the rare-earth ions — 18
    2.3.1 'Free-ion' energy levels — 18
    2.3.2 Rare-earth ions in a crystal field — 19
    2.3.3 Rare-earth ions in glasses — 25
    2.3.4 Judd–Ofelt theory — 26
2.4 Population dynamics in optically pumped rare-earth ion-doped media — 28
    2.4.1 Radiative and non-radiative decay processes — 28
    2.4.2 Which transitions will lase? — 30
    2.4.3 Multi-ion effects in rare-earth ion-doped glasses — 32
    2.4.4 Comparison of rare-earth ion-doped silica and fluorozirconate glasses — 36
2.5 Modelling of fibre laser oscillators and amplifiers — 38
    2.5.1 Introduction — 38
    2.5.2 Rate equation analysis of the population dynamics — 39
    2.5.3 Propagation equations with gain and loss — 42
    2.5.4 Conclusions — 48
References — 48

3 Glass structure and fabrication techniques — 50
S.P. CRAIG-RYAN and B.J. AINSLIE

3.1 Introduction — 50
3.2 Doped glass structure and properties — 50
    3.2.1 Glass properties — 50
    3.2.2 Bonding in glasses — 53
    3.2.3 Rare-earth compounds — 56
    3.2.4 Rare earth in glass — 57
    3.2.5 Glass composition — 58

CONTENTS

3.3 Fabrication methods — 58
   3.3.1 Modified chemical vapour deposition (MCVD) — 61
   3.3.2 Outside vapour deposition (OVD) — 63
   3.3.3 Vapour axial deposition (VAD) — 65
3.4 Results — 67
   3.4.1 Rare-earth ions in silica-based host glasses — 67
   3.4.2 Fibre properties — 74
3.5 Conclusion — 77
References — 77

4 Spectroscopy of rare-earth doped fibres — 79
R. WYATT

4.1 Introduction — 79
4.2 Absorption and emission cross-sections — 79
4.3 Excited-state absorption — 86
   4.3.1 Excited-state absorption at the pump wavelength — 86
   4.3.2 Excited-state absorption at the signal wavelength — 92
4.4 Fluorescence measurements — 96
   4.4.1 Neodymium — 97
   4.4.2 Erbium — 100
4.5 Conclusions — 104
References — 105

5 Components for fibre amplifiers and lasers — 106
D.M. COOPER, T. FINEGAN, C.J. ROWE and I.J. WILKINSON

5.1 Introduction — 106
5.2 Pump sources — 107
   5.2.1 Ti:sapphire lasers — 107
   5.2.2 0.98 μm InGaAs–GaAs lasers — 108
   5.2.3 1.48 μm InGaAsP–InP lasers — 108
5.3 Pump/signal multiplexing components — 112
   5.3.1 Fibre couplers — 112
   5.3.2 Fused-fibre wavelength division multiplexers — 115
   5.3.3 Pump rejection filters — 119
   5.3.4 Dielectric filter WDMs — 122
   5.3.5 Grating-based WDMs — 123
5.4 Laser cavity elements — 125
   5.4.1 Introduction — 125
   5.4.2 Dielectric stack mirrors — 125
   5.4.3 Loop reflectors — 126
   5.4.4 Ring cavities — 129
5.5 Packaged fibre amplifiers — 130
   5.5.1 Component assembly — 130
   5.5.2 Amplifier performance — 131
References — 131

6 Silica fibre amplifiers and systems — 134
D.M. SPIRIT

6.1 Introduction — 134
6.2 Applications of optical amplifiers — 135
6.3 Rare-earth doped fibre amplifiers — 137
6.4 Non-linear fibre amplifiers — 144
   6.4.1 Raman amplification — 144
   6.4.2 Brillouin amplification — 148

CONTENTS

6.5	Semiconductor laser amplifiers	149
6.6	Comparison between fibre and semiconductor laser optical amplifiers	150
	6.6.1 Amplifier characteristics	150
	6.6.2 Suitability for specific applications	152
6.7	Advanced system demonstrations	154
	6.7.1 Long-span high-capacity transmission	154
	6.7.2 Pre-amplifier receiver	156
	6.7.3 Multichannel systems	156
	6.7.4 Non-linear transmission	157
	6.7.5 Routeing switch	158
	6.7.6 OTDR enhancement	158
6.8	Conclusions	158
References		159

7 Silica fibre laser oscillators — 161
D.C. HANNA and A.C. TROPPER

7.1	Introduction	161
7.2	Tuneable operation	165
7.3	Single longitudinal mode operation	168
7.4	High power operation	169
7.5	$1.5\,\mu m$ Er fibre lasers	172
7.6	$1.3\,\mu m$ Nd fibre lasers	175
7.7	Superfluorescent operation (ASE operation)	176
7.8	Visible operation	177
7.9	Laser transitions with low quantum efficiency	179
References		181

8 Fluoride fibre lasers and amplifiers — 183
P.W. FRANCE and M.C. BRIERLEY

8.1	Introduction	183
8.2	Fluoride glasses as laser host materials	184
8.3	Spectroscopy of rare earths in fluoride glasses	187
	8.3.1 Absorption and emission spectra	187
	8.3.2 Detailed spectroscopy of Nd	194
	8.3.3 Detailed spectroscopy of Er	196
8.4	Fabrication technique	197
	8.4.1 Glass melting and preform fabrication	198
	8.4.2 Fibre drawing	199
	8.4.3 Monomode fibre	199
8.5	Fluoride fibre lasers	200
	8.5.1 Neodymium-doped fluoride fibre lasers	200
	8.5.2 Holmium-doped fluoride fibre lasers	202
	8.5.3 Thulium-doped fluoride fibre lasers	202
	8.5.4 Erbium-doped fluoride fibre lasers	204
	8.5.5 Systems experiments	204
8.6	Up-conversion lasers	205
8.7	Fluoride fibre amplifiers	206
8.8	Other systems	208
References		210

9 Q-switched and mode-locked fibre lasers — 212
N. LANGFORD and A.I. FERGUSON

9.1	Introduction	212
9.2	Mode-locking of fibre lasers	213

	9.2.1	Introduction	213
	9.2.2	Loss modulation	214
	9.2.3	Phase modulation	215
	9.2.4	Optical non-linearities in fibres	215
	9.2.5	Mode-locking of doped fibres with bulk modulators	218
	9.2.6	Mode-locking of doped fibre lasers with integrated devices	221
9.3	Q-switching of optical fibre lasers		223
	9.3.1	Introduction	223
	9.3.2	Q-switched and Q-switched mode-locked operation of fibre lasers with bulk devices	225
	9.3.3	Q-switched and Q-switched mode-locked fibre lasers with integrated optic devices	226
9.4	Conclusions		226
References			227

10 Future directions 229

C.A. MILLAR

10.1	Introduction		229
10.2	Fibre amplifiers – the impact on optical communications		231
	10.2.1	Fibre power amplifiers	232
	10.2.2	Fibre amplifiers as signal repeaters	240
	10.2.3	Fibre amplifiers at 1320 nm wavelength	241
	10.2.4	Fibre pre-amplifiers	242
	10.2.5	Distributed fibre amplifiers	243
10.3	Fibre lasers		246
	10.3.1	Short-wavelength fibre lasers	246
	10.3.2	Long-wavelength fibre lasers	248
	10.3.3	Fibre lasers for short pulse sources	249
10.4	Fibre amplifiers at different wavelength regions		250
10.5	Active optical fibre sensors		252
10.6	Conclusions		254
References			254

Index 257

1 Perspective and overview

E. SNITZER

1.1 Introduction

Fibre lasers are the most recent major activity in glass lasers. These devices offer promise for significant components for use in telecommunications and for sensors. In this introduction the various eras of glass laser technology are briefly reviewed, including a description of some of the early fibre laser work. Some of the current activities are then described. Finally, some projections are made concerning possible applications.

The first glass laser was based on trivalent neodymium in an alkali–alkaline earth silicate that emitted at 1.06 μm [1]. It was in the form of what would now be described as a multimode fibre with a core index of refraction of 1.54 and a cladding index of 1.52. The core diameter was 0.3 mm. The motivation for this configuration was in part to take advantage of the increase in brightness at the core by the light-concentrating effect of the clear cladding. Also, the early laser glasses at American Optical Company were made in 0.5 kg melts and in those days no methods had been worked out for making good optical quality glass in such small melts. Hence, with a lower refractive index cladding, the structure was more forgiving of stria. Since then laser emission has been made to occur in most of the rare earths and in thousands of different glass compositions. However, only a handful of these have been commercially significant. There are a number of books and review articles that cover glass lasers in general [2–6]. The other chapters of this book have detailed descriptions of fibre lasers, hence a minimum of references will be cited in this chapter.

1.2 Brief history of fibre lasers

The early fibre lasers were side pumped with flashlamps. The first reported activity demonstrated single-mode operation in fibres that were small enough to exhibit one or a few modes of operation in the device [7]. Later a neodymium glass fibre was used as a power amplifier for a He–Ne laser emitting at 1.0621 μm [8]. The amplified light preserved the coherence inherent to the He–Ne laser. The experiment that was performed used a single-mode fibre with a 15 μm diameter core and gave 40 dB of gain to increase the He–Ne output from 230 μW to 0.6 W. The output from the amplifier had a coherence length in excess of 12.2 m as measured by a Michelson interferometer. In a second experiment, an $InAs_{0.17}P_{0.83}$ laser diode operating at 1.063 μm had 4 μW of its output focused into the same fibre amplifier for 47 dB of gain to produce 0.2 W of output [9].

In another experiment, the fibre laser was used as the pre-amplifier in a detection system. In such a device the signal S is amplified by a gain, G. In addition, the spontaneous emission due to the inversion associated with the laser constitutes an input noise, N, which experiences a net gain of $G - 1$. The output is given by $SG + N(G - 1)$. It can be shown that the spontaneous emission noise, N, when expressed as photons emitted per second, is given by the product of the optical bandwidth (Δv) multiplied by the number of propagating modes. For low noise, the cross-section can be reduced to single-mode propagation. In addition to limiting propagation to a single mode, narrowing the optical bandwidth would further reduce the noise. As an alternative strategy, a post-electronic detector can be used which can discriminate against the DC spontaneous emission. Under the latter conditions, the noise is the fluctuations in the number of photons arriving in a given time interval. The motivation for a fibre laser pre-amplifier is to amplify the signal to the point where the noise of the input impedance in the post-electronic detector is not limiting, but instead the limitation is associated with the spontaneous emission noise. This, in principle, would allow the detection of only a few tens of photons in a pulse with bit error rates of 10^9 or less. In 1969, such a detector was used to detect a Q-switched neodymium glass laser pulse that had a linewidth of 10 nm and a detectivity of 4×10^3 photons in the pulse – a value that is not very exciting today, but in 1969 was an achievement.

The next major development was the CW fibre laser [10]. The initiation of this phase of glass laser work resulted from the recognition that with laser diodes to provide high-brightness pumps at 0.8 μm, where neodymium is strongly absorbing, and with the development of low-loss glass fibres of predominately fused silica as the host, in small diameters there is not a serious problem with heat loading and CW operation is readily obtained. Although the end-pumped CW fibre laser was demonstrated in 1973 [10], it was not until the publication twelve years later by Mears *et al.* [11] that a widespread activity in fibre lasers began, which included the extension of the host from fused silica to heavy metal fluorides [12] as well as the investigation of many other rare-earth transitions.

Many fibre lasers have now been fabricated with different rare-earth ions and in a variety of low-loss silica and fluoride glass hosts. Thus far, two general categories of application have been identified. One is the optical amplifier. The greatest attention has been given to erbium operating at 1.55 μm, but there is also potential for ZBLAN doped with neodymium to operate at 1.33 μm or Tm for operation at 1.48 μm. The other application is for an incoherent source that could be used in an interferometric type gyroscope or a relatively incoherent, but bright, source for use in a printer. In the latter application the incoherence is desirable to avoid speckle patterns at the image plane. In both these applications research initially was done using laser diodes. The much greater familiarity on the part of the communications companies with laser diodes has led them to prefer such devices for the optical amplifier, because they are electronic devices that are readily modulated light sources. Similarly, for the bright incoherent source for the interferometric fibre gyro, the superluminescent laser diode was selected as the first choice. It was only when it could be demonstrated that fibres could compete, and perhaps give considerably better performance in these two applications, that fibre lasers began to assume the status of a serious electro-optic component.

1.3 Early work on glass lasers

In the early 1960s glass laser research was to a large extent concerned with establishing which ions could be made to lase and in which transitions this was possible. There was a hunt for new host materials, both glasses and crystals. The laser materials were in the form of fibres, rods, slabs, discs and a variety of oscillator–amplifier configurations [13]. Most of the glass work carried out then was with flashlamp pumping. While glass was made to lase CW, the much lower threshold for neodymium in a crystal such as YAG and the better thermal conductivity of the crystal made it the preferred material for CW operation.

Most of the attention was given to neodymium operating at $1.06\,\mu m$. This was because it contained a large number of absorption bands in the visible and near infrared which facilitated efficient operation. For example, the slope efficiency above threshold for long-pulse operation, of the order of $0.1–1$ ms, commonly was in excess of 5%. While neodymium could be made to emit at 0.9 and $1.4\,\mu m$ and laser emission was obtained from other rare earths such as ytterbium, holmium, erbium, praseodymium, samarium and thulium, the high efficiency and ease of operation of neodymium at $1.06\,\mu m$ made this glass laser of greatest interest.

The broad fluorescent linewidth was used to good advantage in Q-switched operation in which the energy is stored first as inversion in the excited states of the ion with the laser cavity in a relatively low Q state. The Q is then switched to a high value. The inversion which had been stored is substantially higher than the steady-state value corresponding to the high Q value to which the system had been switched, which resulted in the emission of most of the stored energy in a single large pulse of typically 30 ns duration. While crystals such as Nd:YAG can be Q-switched, its higher grain cross-section for Nd depleted the inversion due to amplified spontaneous emission which led to less energy in the Q-switched pulse.

A large part of the work in the early 1960s on neodymium glass lasers was concerned with the influence of different glass hosts on the gain cross-section. Attention was also given to 'athermal glasses', that is, glasses that could tolerate a relatively high temperature gradient from centre to edge without degrading the beam quality [2]. Self Q-switching glasses were also developed, in which the ultraviolet light from the flashlamp produced short-lived colour centres that could be saturated at the onset of laser action, with the result that a Q-switched pulse or train of Q-switched pulses could be generated [14].

Q-switching configurations were among the important configurations pursued with glass lasers. This could be done by rotating one of the end mirrors of the cavity. The rotation was synchronised with the pump so that a high population inversion was first obtained and then the mirror brought into place for the production of a Q-switched pulse. Alternative arrangements consisted of electro-optic or acousto-optic devices placed within the laser cavity, and which were synchronised with the flashlamp pump to give a high Q after the inversion was obtained. Another method for Q-switching consisted of using a saturable absorber, usually located between one end reflector and the laser rod. The absorption of the saturable absorber at low light intensity is chosen so that in combination with the end reflector it gives a low effective reflection but not so low that laser action cannot be initiated. When the system does lase, the absorber saturates and gives an effective high reflectivity which results in a Q-switched pulse. If the saturable absorber is located close to one end reflector and is also of relatively short length, say 1 mm or less, it was found that the Q-switched pulse had an interesting structure within it. It now consisted of a train of pulses, each several orders of magnitude shorter and brighter than the overall Q-switched pulse and with a separation time between individual pulses equal to the round trip time in the cavity. This behaviour results from the fact that the effective reflectivity is associated with the light level at the location of the saturable absorber. As the inversion within the cavity increases, and with the build up of the light level at the saturable absorber, preference is given to those fluctuations of the various cavity modes which coherently reinforce to give a high intensity at the location of the saturable absorber. This so-called mode locking can be thought of as resulting from the various possible modes that can exist in the cavity coherently reinforcing each other to give a well-defined single pulse which now propagates back and forth between the end reflectors of the cavity or around the circuit in which the cavity is configured. In the case of self-Q-switching glass the saturable absorber typically extends over the whole, or a significant fraction, of the laser rod. Under these conditions there is no tendency to produce mode locking. In a general sense, mode locking can be viewed as a consequence of some time-dependent change occurring in the cavity such that the lowest threshold for operation occurs for a short-duration pulse which propagates around the cavity. It is also possible to obtain CW mode locking – that is, where there is continuously generated a series of single pulses or pulse trains which are separated by the round trip time within the cavity. In later chapters of this book methods of CW mode locking as well as Q-switched and mode-locked Q-switched operation of fibre lasers are described.

In the mid-1960s there emerged several well-defined applications for glass lasers. These were based primarily on neodymium operating at $1.06\,\mu m$ because of its relatively high efficiency when pumped with a flashlamp. They were used mainly for welding and cutting in long-pulsed and Q-switched operations.

Work was also undertaken on high peak power Q-switched oscillator–amplifier combinations for the study of high-temperature plasmas. In the later 1960s the high-temperature plasma work tended to dominate the research activity because of the large amount of glass used in these large systems. A primary focus for the host material was the requirement that it be capable of withstanding the high-intensity laser light that was generated. These systems were in the form of Q-switched or mode-locked Q-switched pulses which could have a duration of a few nanoseconds down to a fraction of a nanosecond. Light would be generated from several amplifier trains that were synchronised by being driven from a single oscillator. These large systems are used for research on controlled thermonuclear reactors (CTR), also called inertial confinement fusion (ICF). The largest such system, at the Lawrence Livermore Laboratory, is the Nova Laser which is capable of delivering 100 kJ in 10 beams onto a target pellet containing deuterium or tritium. The short-duration pulses are furthermore shaped in such a way as to be capable of producing an implosion of the material to give a sufficiently high temperature over an appropriate time duration to initiate a thermonuclear reaction which converts the isotopes of hydrogen to helium, thereby releasing substantial amounts of energy. A practical power plant is believed to require pulse energies perhaps 100 times or more greater than the presently available Nova system.

With such high flux densities, a major concern is laser damage caused by inclusions within the glass, surface damage, or self-focusing damage. The latter is associated with the non-linear index of refraction of the material, i.e. the index of refraction n can be written as $n_0 + n_2 E^2$, where n_0 is the index of refraction at low light levels and E is the electric field associated with the laser light. The inevitable small variation in light intensity across the aperture of the beam can lead to a catastrophic collapse of the light in a cross-section of a few millimetres into a filament of $1\,\mu m$ or less in which the light intensity is so high that multiphoton absorption can occur which melts the glass locally. The rapid heating and cooling leaves a fossil track which can degrade the beam quality for subsequent laser use. If the track extends to the surface it can lead to surface pitting. Hence one of the main requirements for large glass lasers was a low value for the non-linear index n_2.

Although glass is capable of storing a large amount of energy in an inverted population to give high Q-switched pulses, the early alkali–alkaline earth silicates gave gain cross-sections that were less than could optimally be used. A typical value for a potassium–barium silicate was 1.5×10^{-20} cm^2. By going to alkali phosphates the peak gain cross-section could be increased to 4 or 5×10^{-20} cm^2. Under the same pumping conditions this led to higher energy within the pulse. Furthermore, there were less problems with self-focusing because the n_2 value could be reduced to less than 10^{-13} e.s.u. as contrasted with n_2 values for silicates which were higher typically by 50% or more.

For some fibre and integrated optic applications there is an interest in materials of high values of n_2 because these can be the basis for devices in which light pulses can switch other pulses in optical circuits [15]. For example, in a twin-core fibre there is a resonant coupling of the light between the two cores. Efficient cross-talk requires the velocity of propagation in each of the cores to be identical. If a series of high-intensity and low-intensity pulses are coupled into one core, the high-intensity pulses could produce a sufficiently large index change to prevent the resonant coupling to the second core, but the low-intensity pulses would couple to the second core, thereby producing a pulse height sorter. Alternatively, a separate high-intensity pulse could be used to switch out preselected pulses from a pulse train. Work on these light-switching-light devices is continuing but practical components appear to require sensitivities about two orders of magnitude greater than those presently available.

Erbium emission at 1.54 μm received some early attention in the mid-1960s and continues to be a commercially significant laser because of its use as an eye-safe laser for rangefinder applications [16,17]. This is a three-level system which is not easily pumped by a flashlamp. However, by the addition of ytterbium in relatively high concentrations (10–15 wt% of the oxide as compared with 0.1–0.3 wt% of Er_2O_3) sensitised fluorescence can be obtained by energy transfer from the Yb to Er. Such energy transfer has also been utilised by the addition of a small amount of Nd or Cr to a Yb–Er laser.

Along with the development of various rare-earth doped glass lasers there has been a substantial effort in various rare-earth doped crystals. The study of selected rare-earth ions in these two categories of hosts has been synergistic in extending our understanding of processes affecting rare earths. They also complemented each other by providing a wider range of properties that could be tailored for different applications. This is especially evident for neodymium in glasses versus neodymium in YAG. Where CW operation or high-pulse repetition rates are important, YAG is preferred as a host. On the other hand, for high-energy, long-pulsed operation or where high-energy storage for Q-switching is important, the glass hosts are preferred. However, there are many applications which fall in the grey area between these two, where the choice made is based primarily on personal preference. For example, neodymium in YAG for rangefinders and materials processing have tended to be more popular in the US than elsewhere. Europe and the Soviet Union have made more extensive use of glass lasers for these applications.

1.4 Ligand field and ion interaction

Along with the extensive experimental work on materials development and device application, there has been the development of a significant body of theoretical analysis of the processes operative in various hosts. A ligand field analysis of the different rare earths has been done for a number of the crystalline hosts. This is more difficult to do with the glass hosts because of the considerably lower symmetry expected and because of the possibility of multiple sites which unfortunately cannot be easily inferred, as would be the case for a crystalline host. Nevertheless in the case of trivalent ytterbium, which has a relatively small ionic radius, an analysis was made assuming a distorted octahedron which gave reasonable agreement with the absorption and fluorescent spectrum for the transition between the $^2F_{7/2}$ and $^2F_{5/2}$ energy levels [18].

The observed spectra in the visible and infrared for the various rare-earth ions involve transitions for the same f-electron configuration but different values for the total spin S, orbital angular momentum L and total angular momentum J. The coulomb interaction between electrons and the spin–orbit coupling are comparable, with the result that only J is a good quantum number, i.e. the rare earths are intermediate between L–S and j–j coupling. Independently, but more or less simultaneously, Judd [19] and Ofelt [20] each calculated the total electric dipole transition probabilities between the various term values for ligand fields that lacked a centre of symmetry. Since electric dipole transitions depend on matrix elements involving the electric dipole vector (**er**) for transitions between the same f-electron configuration, the transition probability would be zero. Only to the extent that the ligand field admixes a small amount of 5d- or 5g-electron configuration into the f-electron configuration is there a finite electric dipole transition probability. The resulting probabilities are considerably less than for allowed dipole transitions. Typical rare-earth transitions give fluorescent radiative lifetimes of the order of 0.1–10 ms as compared with an allowed electric dipole transition of 10^{-8} to 10^{-9} s for spontaneous emission. This parity forbidden electric dipole transition is comparable to the magnetic dipole transition in some of the rare earths. A measure of the strength of interaction with the ligand field is indicated by the splitting associated with the sublevels for a term with a given J value. Whereas the transition metal ions with partially filled 3d orbitals have ligand field splittings of the order of several thousand inverse centimetres, for rare earths the sublevel splitting for a given J value is only a few hundred inverse centimetres. This latter value is also the typical linewidth for absorption and fluorescence in glasses. The total width for glasses and crystals is about the same, but unlike crystals where there can exist one or a few well-defined sites which give clearly recognised splitting into sublevels, in glasses there is much less structure in the absorption and fluorescent lines and the widths for any portion of the lines are a significant fraction of the total linewidth.

The Judd–Ofelt parameters are calculated on the assumptions that the sublevels within a term are uniformly occupied and that the separation between term values is small compared with the energy separation to the 5d- or 5g-electron configurations that are admixed by the ligand field. The result is a transition probability for absorption or induced emission from initial state J to final state J', given by

$$p = a[(n^2 + 2)^2/n](2J + 1)^{-1}S(J,J') \qquad (1.1)$$

where a has only a slight host dependence due to the frequency of the transition. The term containing n is a local field correction for the index of refraction of the host material. $S(J,J)$ is the line strength, which can be written as the sum of three terms

$$S(J,J) = \Omega_2 U_2 + \Omega_4 U_4 + \Omega_6 U_6. \qquad (1.2)$$

U_2, U_4 and U_6 are calculated quantities, tabulations of which are available [21], and Ω_2, Ω_4, and Ω_6 are the Judd–Ofelt parameters, determined by a least-squares fit to the absorption spectra of a rare earth–host combination. This powerful technique was first applied to rare earths in glass by Krupke [22]. An extensive table of computed values of the Ω_i's for Nd in a variety of glass hosts has been published [23]. The calculated values for various rare earths in representative hosts and some interpretation of the physical significance of the Ω_i's is also available [24]. It is remarkable that all the transitions for a given ion within the f-electron manifold can be described by only three parameters with a precision that is typically about 10%. The only exception to this is accounted for by the so-called hypersensitive transitions where the transition probabilities tend to be much larger. Often the omission of these hypersensitive transitions leads to a more accurate set of Judd–Ofelt parameters. For example, neglecting the praseodymium hypersensitive transition $^3H_4 \rightarrow {^3P_2}$ gives a more accurate set of Judd–Ofelt parameters for Pr.

Another important consideration in evaluating the performance of various rare earths in different hosts relates to the non-radiative transition probability to go from one level to the next level below. The non-radiative process is mediated by interaction with the phonon spectrum of the host. If the energy level separation is small the non-radiative probability will be quite large and phonon emission would take place very rapidly. For energy separations that are a few thousand inverse centimetres, the radiative process could effectively compete with the non-radiative process. For a large enough separation virtually no quenching occurs from the upper state. The decisive parameter for the host is the energy of the highest frequency phonon. Lane et al. [25] have shown a good correlation between the phonon spectra for borate, silicate, phosphate and tellurite glasses and the non-radiative transition probabilities. The low-energy phonon spectrum for heavy metal fluorides permits these hosts for rare earths to have fluorescence and laser action originating from energy levels that are quenched in silicates. This non-radiative process has a major bearing on the levels at which efficient fluorescent emission is obtained. It also has a considerable influence on up-conversion fibre pumping schemes such as those recently reported for thulium [26] and holmium [27] in heavy metal fluoride hosts. Here there is some population in an excited state from which additional pumping can result in populating a higher energy level, which is then capable of giving fluorescence and laser emission at a wavelength shorter than the pump wavelength.

The ability to populate an upper level relative to some lower state to which a radiative transition occurs is, of course, a necessary condition for laser action. However, it is possible that from that upper level there can be a transition to a still higher level whose cross-section is greater. Such an excited-state absorption (ESA) is what has prevented neodymium from lasing at the peak of the fluorescence band at 1.32 μm for the $^4F_{3/2} \to {}^4I_{13/2}$ transition in silica. The ESA band was to the edge of the $^2G_{7/2}$ state in SiO_2 co-doped with Nd_2O_3 and Al_2O_3 with the result that laser emission occurred at 1.4 μm. The addition of 13 mol% P_2O_5 to SiO_2 allowed laser action to occur at 1.363 μm [29]. In ZBLAN hosts the lines were sufficiently narrowed and shifted, and the transition probabilities changed so that the $^4F_{3/2} \to {}^4I_{13/2}$ transition could give gain at 1.3 μm [28]. The fluoride hosts generally offer promise for lasers at wavelengths longer than those available with the silicates [30].

Energy transfer between neighbouring ions is another important consideration. This occurs for Yb sensitisation of erbium fluorescence. It can be the basis for populating Yb after first absorbing pump photons in Nd. This process is also important in concentration quenching effects if the Nd concentration is much above 10^{20} ions/cc in SiO_2 co-doped with Al_2O_3. Quenching of the $^4F_{3/2}$ state can result from an interaction of one Nd ion in this state with another Nd in the ground state, leading to both ions ending up in the $^4I_{15/2}$ state, from which rapid quenching occurs down to the ground state by phonon emission. Energy transfer can also be used to advantage in other ways. For example, if thulium is excited to the $^3H_4$ state interaction with an unexcited thulium ion can lead to both ions populating the $^3F_4$ level in the so-called two for one population scheme. Another example of ion interaction effects is the apparent decrease in quantum efficiency for erbium in SiO_2 as a laser amplifier at 1.5 μm where it appears that the two erbium ions in the excited state can interact such that one of them makes a non-radiative transition to the ground state while the other is up-converted to the $^4I_{9/2}$ level from which it non-radiatively cascades back down to the $^4I_{13/12}$ state, with the result that one excited erbium has been quenched. The investigation of the spectroscopic properties of rare earths in glasses have led to many other interesting observations. These include fluorescent line narrowing, anomalous linewidths and persistent hole burning at low temperatures, with novel two-level system modelling to describe their behaviour [31].

Another category of phenomena relates to the subject of defects created in the glass in the drawing process itself which is a quenching process from a high temperature. These defects can furthermore be changed by exposure to ultraviolet light, as was indicated earlier in self-Q-switching silicate glasses. A very interesting defect is associated with the observation that in single-mode silica fibre, which also contains germania, there can develop a defect that leads to second harmonic generation of light when illuminated with mode-locked Q-switched light from a Nd:YAG laser [32, 33]. Another defect, which is related to the presence of germania, is the photorefractive effect in which there is created a grating as a result of standing waves established within the germania silica core [37].

In the case of erbium operated as an optical amplifier, the light to be amplified is incident from a single-mode fibre and coupled out of the single-mode erbium amplifier into another single-mode fibre. The erbium amplifier will be discussed in greater detail later; suffice it to state here that because of the requirement of two single-mode couplers a dispersive single-mode coupler is probably the preferred method of pumping. However, for four-level laser oscillators or super-luminescent sources, double-clad configuration can profitably be used with multimode pumps [34]. In the latter configuration the laser ion is contained in a single-mode core, surrounded by a first cladding of lower refractive index but of considerably larger area than the core. This first cladding is in turn surrounded by a second cladding of lower refractive index. The pump light is coupled into the first cladding and as it propagates down the fibre is eventually absorbed by the core. Configurations of this type have been used with neodymium that permitted pumping with relatively incoherent multiple stripe laser diodes with conversion efficiencies for pump light to single-mode laser light output in excess of 50%.

1.5 Fibre applications

In the past two years a considerable amount of work has been reported on Er fibre amplifiers. As a light amplifier it appears to offer significant competition to the electro-optic repeater in which the signal is detected and then relaunched into a further length of fibre. Work has progressed to such practical questions as Er concentration and index of refraction profiles to give best performance for dispersion shifted fibre. Later chapters in this book discuss these questions; suffice it here to say that practical pumps exist at a number of wavelengths, but particularly attractive are the efficient laser diode pumps, free of excited-state absorption, at $0.98\,\mu m$ and $1.47\,\mu m$ [35]. The latter is particularly interesting because the pump and signal wavelengths are so close to one another that both can be tightly coupled to the fibre core in single mode. This is an important consideration because erbium is a three-level system in the $1.55\,\mu m$ region at room temperature. Gains of 3 to 5 dB/mW of absorbed power with total gains in excess of 35 dB and saturation outputs of 18 dB m have been reported. We tend to think of fibres as low-power devices, but recently an output of 1 W at 1.535 μm from an 8.2 μm Er-doped germano-silicate core single-mode fibre, with 28% conversion efficiency from a 0.532 μm Nd:YAG pump, was described [36].

Because the Er amplifier at 1.55 μm is at longer wavelengths than the zero group dispersion point of silica, impressive bright soliton propagation experiments have been reported [38, 39].

With an optical amplifier a number of interesting topologies for systems can be considered. For example, in a broadcast mode, the signal can be split into many outputs to go to multiple stations without running out of signal, by first amplifying the signal, or possibly amplifying the signal at strategic places in a star coupler. Another approach is to use extensive wavelength division multiplexing (WDM). Unlike the electro-optic repeater which requires a separate repeater module for each channel, a single amplifier could boost all the wavelengths. Of course, attention has to be given to accomplishing this without cross-talk between channels.

In discussing the possibilities that optical amplifiers provide, it becomes clear that a number of other components are necessary. A reliable compact pump, and probably one or more laser diodes, are required. These and the signal have to be coupled into the amplifier, preferably by a coupler that is dispersive for high coupling efficiency at both wavelengths. The amplifier can give gain in both directions of propagation. Are Faraday isolators needed, and if so where should they be placed? For WDM the erbium laser could be the oscillator, but frequency selective elements would be required in the cavity to maintain carrier wavelength stability; modulators would also be required. In short, the full realisation of the potential of the Er amplifier requires the development of a number of other components and considerable skill in their integration.

In addition to Er at 1.55 μm there is the possibility of using thulium in ZBLAN at 1.48 μm for a long-wavelength communications amplifier. The ZBLAN host is more difficult to work with and the efficiency is likely to be less because of the need to take special measures to empty the terminal state, but it does have the advantage of being a four-level laser system.

Another ZBLAN fibre that has been considered for an amplifier is Nd doped. This provides a four-level laser at 1.33 μm, which is at the shorter communications window where much of the lightwave communication operates.

Another important application that is emerging for fibre lasers is the superluminescent source. The feedback for these devices can be reduced sufficiently so that little or no spectral structure exists in the laser output. Both Er at 1.55 μm and Nd at 1.06 μm are potential candidates for this application.

In the 1960s there was a narrowing of attention for laser glasses to Nd and Er in a handful of hosts – Nd because it gave efficient emission for high-brightness lasers at 1.06 µm, and Er for its use as an eye-safe laser rangefinder at 1.54 µm. While there now appears to be a narrowing of attention for fibre lasers to these same ions, but for different reasons, the number of fibre laser systems of commercial interest is likely to proliferate rather than remain confined to these two ions. The variety of hosts that are now possible can provide special lasers for medical and sensor applications. With the development of the various auxiliary components necessary to operate an instrument or system, the comfort level of dealing with fibres will probably lead to increasing uses for these devices. The non-linear phenomena associated with selected damage in germania silica fibres, fibre gratings and couplers of various sorts are all now available which would make it possible to construct quite complex but reliable optical systems for communications, sensors and medical instruments. Add to this the integrated optic devices, and the range of components available competes with microwave hardware. And the remarkable feature of the evolution of lightwave technology is that it allows greater amounts of information to be handled with less space and cost, and enables novel measurements and medical procedures to be performed.

References

1. E. Snitzer, *Phys. Rev. Lett.* **7** (1961) 444.
2. E. Snitzer and C.G. Young, 'Glass lasers', in *Advances in Lasers* (A. Levine, ed.), vol. 2, Dekker, New York (1968).
3. K. Patek, *Glass Lasers*, Butterworth, London (1970).
4. M.J. Weber 'Optical materials for neodymium fusion lasers', in *Critical Materials Problems in Energy Production* (C. Stein, ed.), Academic Press, New York (1976).
5. D.C. Brown, *High-Peak-Power Nd-Glass Laser Systems*, Springer, Berlin (1981).
6. S.E. Stokowski, 'Glass lasers' in *Handbook of Laser Science and Technology*, vol I, *Lasers and Masers*, CRC Press, Boca Raton, FL (1982), p. 215.
7. C.J. Koester and E. Snitzer, *Appl. Opt.* **3** (1964) 1182.
8. G.C. Holst, E. Snitzer and R. Wallace, *IEEE J. Quant. Electron.* **QE-5** (1969) 342.
9. B. Ross and E. Snitzer, *IEEE J. Quant. Electron.* **QE-6** (1972) 361.
10. J. Stone and C.A. Burrus, *Appl. Phys. Lett.* **23** (1973) 388.
11. R.J. Mears, L. Rekie, S.B. Poole and D.N. Payne, *Electron. Lett.* **21** (1985) 738.
12. M.C. Brierley and P.W. France, *Electron. Lett.* **23** (1987) 815.
13. C.G. Young, *Proc. IEEE* **57** (1969) 1267.
14. R.J. Landry, E. Snitzer and R.H. Bartram, *J. Appl. Phys.* **42** (1971) 3827.
15. E.M. Vogel, *J. Amer. Ceramic Soc.* **72** (1989) 719.
16. F.W. Quelle, *SPIE Seminar on Laser Range Instrumentation*, El Paso, TX (1967).
17. J.P. Segre, *Proc. Electro-Optics Conf.*, Montreux, Switzerland (1974).
18. C.C. Robinson and J.T. Fournier, *J. Phys. Chem. Solids* **31** (1970) 895.
19. B.R. Judd, *Phys. Rev.* **127** (1962) 750.
20. G.S. Ofelt, *J. Chem. Phys.* **37** (1962) 511.
21. C.W. Nielson and G.F. Koster, *Spectroscopy Coefficients for p^n, d^n and f^n Configurations*, MIT Press, Cambridge, MA (1964).
22. W.F. Krupke, *Phys. Rev.* **145** (1966) 325.
23. S.E. Stokowski, B.A. Saroyan and M.J. Weber, *Nd-Doped Laser Glass Spectroscopic and Physical Properties*, Lawrence Livermore National Laboratory, University of California (1981).
24. R. Reisfeld and C.K. Jorgensen, 'Excited state phenomena in vitreous materials' in *Handbook*

on *The Physics and Chemistry of Rare Earths* (K.A. Gschneidner, Jr and L. Eyring, eds), Elsevier Science Publishers (1987).
25. C.B. Lane, W.H. Landermilk and M.J. Weber, *Phys. Rev. B* **16** (1977) 10.
26. J.Y. Allain, M. Monerie and H. Poignant. *Electron. Lett.* **26** (1990) 166.
27. J.Y. Allain, M. Monerie and H. Poignant, *Electron. Lett.* **26** (1990) 261.
28. W.J. Miniscalco, L.J. Andrews, B.A. Thompson, R.S. Quimby, L.J.B. Vacha and M.G. Drexhage, *Electron. Lett.* **24** (1988) 28.
29. F. Hakimi, H. Po, R. Tamminelli, B.C. McCollum, L. Zenteno, N.M. Cho and E. Snitzer, *Opt. Lett.* **14** (1989) 1060.
30. P.W. France, 'A review of fluoride glass fibre lasers', *Proc. 6th Internat. Symp. on Halide Glasses, Clausthal (FRG)*, (1989) p. 51.
31. S.K. Lyo, 'Dynamical theory of optical linewidths in glasses', in *Optical Spectroscopy of Glasses* (I. Zchokke, ed.) Reidel, Dordrecht (1986).
32. U. Osterberg and W. Margulis, *Opt. Lett.* **11** (1986) 516.
33. R.H. Stolen and A.W.K. Tom, *Digest of Conf. on Lasers and Electro-Optics, Baltimore*, Paper THL-2, (1987), p. 274.
34. H. Po, E. Snitzer, R. Tumminelli, L. Zenteno, F. Hakimi, N.M. Cho and T. Haw, *Optical Fiber Communications Conf., Houston, Texas* (1989), Postdeadline Paper PD7.
35. E. Snitzer, H. Po, F. Hakimi, R. Tumminelli and B.C. McCallum, *Optical Fiber Communications Conf. '88, New Orleans, Louisiana* (1988), Postdeadline Paper PD2.
36. V.P. Gapontsev, P.I. Sadovsky and I.E. Samartsev, *Conf. on Lasers and Electro-Optics, Anaheim, California* (1990), Postdeadline Paper CPDP38.
37. K.O. Hill, Y. Fujii, D.C. Johnson and B.S. Kawasaki, *Appl. Phys. Lett.* **32** (1978) 647.
38. L.F. Mollenaver and K. Smith, *Opt. Lett.* **13** (1988) 675.
39. M. Nakazawa, K. Suzuki and Y. Kimora, *Opt. Lett.* **14** (1989) 1065.

2 Introduction to glass fibre lasers and amplifiers

J.R. ARMITAGE

2.1 Introduction

Unlike all the following chapters in this book which review the current state-of-the-art of various areas of glass fibre laser technology, this chapter provides a basic introduction to the subject of rare-earth ion-doped glass fibre lasers and amplifiers. It aims to cover much of the background and introductory type material common to many of the remaining chapters in this book. The reader, it is hoped, should already possess a basic appreciation of laser physics and ought therefore to regard this chapter more as a résumé of useful information on rare-earth ion-doped laser media than an introductory text on solid-state lasers – many of which already exist (see, for example, [1–3]).

Even though the basic active media are identical, there are important differences between bulk glass lasers and glass fibre lasers. Section 2.2 discusses the advantages, and a few disadvantages, of taking a glass laser and constructing it in waveguide form. Section 2.3 then takes a look, from a theoretical viewpoint, at the spectroscopic properties of the rare-earth ions in glasses, starting with the electronic structure of isolated rare-earth ions and moving on via doped crystals to amorphous media. A simple overview of Judd–Ofelt theory is also included. This theory is important in that it can be used to seek out new laser transitions, to assess the suitability of new laser host materials, and in understanding the limitations imposed by the active medium on the device performance. Section 2.4 considers the various radiative and non-radiative processes, including ion–ion interactions, which affect the population dynamics in optically pumped rare-earth ion-doped media. In particular, a comparison is given of the merits of silica and of fluoride glasses as laser host media for the lanthanide ions. Numerical modelling of rare-earth ion-doped laser oscillators and amplifiers is important in understanding how the population dynamics affects device performance. However, rather than simply reproducing the work of other authors, Section 2.5 steps back from considering specific laser systems and gives an outline of the essential methods used in the modelling of active fibre devices, and comments upon many of the approximations that have been used in these analyses.

2.2 Comparison of bulk and fibre lasers

A glass fibre amplifier, rather than a bulk glass device, would seem to be the natural choice for the amplification medium to be used in fibre optic communications simply from the point of view of its physical compatibility with standard telecommunications fibre. There are nevertheless many potential lasing applications of rare-earth ion-doped fibres where the mere fact that the active medium is in fibre form is far less relevant. The advantages to be gained from constructing optically pumped lasers in waveguide form, rather than employing longitudinal pumping of the bulk material, will now be discussed. Although this book is solely concerned with glass fibre lasers and amplifiers, many of the arguments presented here apply equally well to crystal fibre lasers and to ion-implanted waveguides in crystalline media.

2.2.1 Threshold pump power requirements

Figure 2.1 shows a simplified diagram of a longitudinally pumped bulk laser. Light from the pump laser is brought to a focus (spot size ω_0) inside the active medium from which the light then diverges with a half-angle $\Delta\theta$. This divergence angle $\Delta\theta$ and the spot size are related. For the case of a TEM$_{00}$ pump beam

$$\Delta\theta = \frac{\lambda}{\pi\omega_0}.$$

If, however, the pump beam is F times diffraction limited in one plane, then in that plane the beam after the focus diverges at roughly F times the rate of a diffraction limited spot of the same size, thus

$$\Delta\theta \approx F \times \frac{\lambda}{\pi\omega_0}.$$

For a four-level laser system in which there is no ground-state depletion, the population in any of the excited levels, and hence the gain coefficient, scales linearly with pump intensity. In the more general case, however, where the

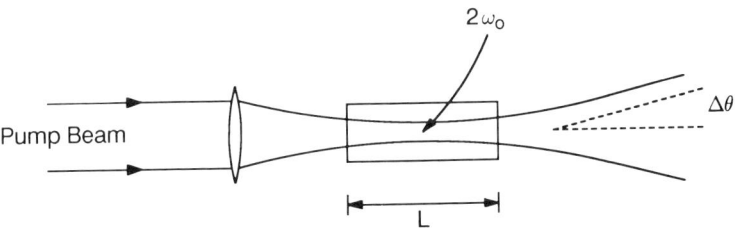

Figure 2.1 Schematic diagram of a longitudinally pumped bulk glass (or crystal) laser.

population in the ground state is depleted, all that can be said is that the gain coefficient will scale monotonically with pump intensity, so that the total single pass gain through the medium scales as $\langle I \rangle L$, where $\langle I \rangle$ is the pump intensity averaged along the length of the gain medium. Obviously by focusing more strongly into the gain medium, a smaller pump spot can be produced but only at the expense of a more rapidly diverging pump beam. It is not hard to see that for a given length of gain medium, there exists an optimum focusing condition given approximately by

$$L_{opt} \Delta \theta \approx 3 \omega_0$$

which maximises the product $\langle I \rangle L$ for a given pump power. For the case of a TEM_{00} pump laser, this yields the familiar result [4]

$$L_{opt} \approx \frac{3\pi\omega_0^2}{\lambda}.$$

The choice of L is dictated by the fact that most of the pump power needs to be absorbed by the active medium. If in a laser only 10% of the pump power is absorbed by the gain medium, then the quantum slope efficiency of that laser can never exceed 10%. In the case of rare-earth ion-doped materials, L can be reduced simply by increasing the doping density of the active ion. Nevertheless at some stage, the separation of neighbouring rare-earth ions becomes sufficiently small that ion–ion interactions start adversely affecting the population dynamics (see Section 2.4). Since the length of active medium needed is determined by pump absorption requirements, when designing a bulk device, the length and the pump spot size are *not* independent parameters.

The situation in a fibre laser is somewhat different (Figure 2.2). Whatever the dimensions of that fibre, once power has been launched into the core, then that power continues to propagate down the fibre until it is completely absorbed. Since glass fibre cores can be made only a few microns in diameter, the pump intensities in a fibre laser may well be around 100 times larger than those in a bulk device. Everything else being equal, the lasing threshold can be expected to decrease by this same factor. In the case of a four-level laser, this threshold reduction might not be overly important as the threshold pump power may only be a small fraction of the maximum launched pump power. However, for three-level systems where significantly higher pumping intensities are typically needed just to reach threshold, a fibre device may be

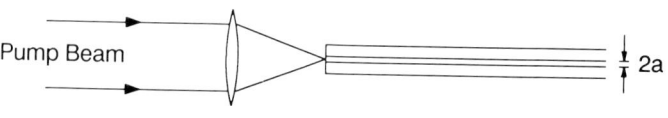

Figure 2.2 Schematic diagram of a fibre laser.

the only way to achieve laser action at sensible pump powers.* Another advantage of a fibre geometry is that the pump spot size and the device length are now *independent* parameters. If the rare-earth ion-doping density is reduced by a factor of 2, the lasing efficiency can be maintained simply by a doubling of the fibre length. This extra degree of freedom allows for instance the rare-earth ion doping density to be kept sufficiently low that there are no ion–ion interactions adversely affecting the device performance. Also having decoupled the pumping intensity from the device length allows weak absorption features to be used to excite the lasing transition. This can be of practical importance if that weak pump band were to coincide with the wavelength of commercially available diode lasers.

One very important problem to consider when designing fibre laser oscillators and amplifiers is the launch efficiency of the pump laser light into the fibre core. This launch efficiency depends basically on two quantities; the number of transverse modes supported by the fibre and the spatial quality of the pump beam. When launching a TEM_{00} beam into a single-mode fibre, one would typically expect to achieve around 50% launch efficiency given the correct choice of lens. This figure may well be even higher when launching that same pump beam into a multimode guide. If the beam from the pump laser is multi-transverse mode, as in the case of a diode laser array, then the launch efficiency into a single-mode guide is small. As a rule of thumb, for a beam that is Φ times diffraction limited, one would expect a launch efficiency around $50/\Phi$%. Even though in many cases the use of single-mode fibres is preferred, a multimode fibre laser may still have an advantage over a bulk device in terms of lasing threshold powers, but only if $L \gg 3\pi a^2/\lambda$. Although the arguments presented have been phrased in terms of threshold pump powers of laser oscillators, they are equally applicable to laser amplifiers.

2.2.2 *Other advantages of fibre lasers*

One advantage of making a fibre support only one transverse mode is that the output beam from such a guide is necessarily diffraction limited, and will remain so even at very high pump powers. Multimode fibre lasers, although they may well oscillate with a diffraction limited output just above threshold – the LP_{01} mode will almost always be the mode with the highest gain – tend to become multimoded with increasing pump powers, with the higher order modes feeding off the population inversion in the outer part of the core. Other reasons for considering a fibre geometry, either with single-mode or multimode fibres, are the thermal advantages [5]. Simply as a result of the fibre dimensions, heat dissipation from the fibre core is far more efficient than from

---

* It is for this reason that many three-level laser transitions have been demonstrated for the first time in fibre form.

the pumped regions of bulk glass lasers. The thermal lensing and stress problems, and decreases in fluorescence yield at the raised temperatures that can be encountered in bulk devices, are rarely a problem for fibre devices. Indeed, even if the raised temperature of the core were worrying, a trivial solution would be to reduce the rare-earth ion doping density keeping the doping density length product constant. The last point of note is that given the right transition and choice of pump wavelength, fibre laser oscillators can easily be assembled with efficiencies approaching 100% [6,7,8]. Such high efficiencies are a result of several features specific to fibre lasers. The pump and lasing modes overlap well throughout the whole of the gain region, and so the intracavity lasing flux can efficiently extract energy from the population inversion. Since rare-earth ion-doped glass fibres are high-gain devices, this then allows the possibility of designing a laser cavity with a low reflectivity output coupler [6, 8]. Although this low-reflectivity mirror will increase the lasing threshold slightly, since that threshold is typically so low to start with, the overall laser efficiency is barely affected. This use of a low Q cavity then permits one to incorporate slightly lossy elements (e.g. wavelength selective filters, phase modulators, etc.) into the cavity *without* any significant decrease in the overall lasing efficiency. Such design flexibility is not really possible with bulk glass lasers.

2.3 Energy levels of the rare-earth ions

Before discussing the population dynamics in rare-earth ion-doped glass lasers in some detail, a closer look needs to be taken at the energy levels of these triply ionised ions in glasses, and in particular at the radiative transitions between these levels.

2.3.1 *'Free-ion' energy levels*

Table 2.1 lists the ground-state electron configurations of the 15 lanthanide elements in their $+3$ ionisation state. Since the spectroscopy of the first and last ions in this table is uninteresting – their electron configuration consists entirely of closed shells and they have no absorption features until well into the ultraviolet region of the spectrum – these ions will not feature any further in this book.

Figure 2.3 shows the energy levels of the remaining lanthanide ions. Atomic and ionic energy levels are conventionally labelled according to the angular momentum properties of that atom or ion. An atom or ion existing in an energy level designated as $^{2S+1}L_J$ would have a spin quantum number S, a total angular momentum quantum number J and an orbital angular momentum quantum number L defined by the following letters $S, P, D, F, G, H, I,...$ which correspond to $L = 0, 1, 2, 3, 4, 5, 6,...$ respectively. Such a

Table 2.1 The electron configuration of the 15 lanthanide elements in the +3 ionisation state. (Xe) corresponds to the electron configuration of a neutral Xenon atom, i.e. $1s^2 2s^2 2p^6 3s^2 3p^6 3d^{10} 4p^6 5s^2 4d^{10} 5p^6$.

Element	Electron configuration	Element	Electron configuration
La	(Xe)	Tb	(Xe)4f^8
Ce	(Xe)4f^1	Dy	(Xe)4f^9
Pr	(Xe)4f^2	Ho	(Xe)4f^{10}
Nd	(Xe)4f^3	Er	(Xe)4f^{11}
Pm	(Xe)4f^4	Tm	(Xe)4f^{12}
Sm	(Xe)4f^5	Yb	(Xe)4f^{13}
Eu	(Xe)4f^6	Lu	(Xe)4f^{14}
Gd	(Xe)4f^7		

labelling scheme implies *LS*-coupling. In the case of the rare-earth ions even though such a scheme is widely used, L and S are not in fact good quantum numbers because the magnitude of the spin–orbit effect in these heavy ions is comparable to the non-central component of the electron–electron interaction. However, the most important feature to note regarding the energy levels in Figure 2.3 is that all the levels of a particular ion have the same electron configuration, and are consequently all of the same parity.* Since the electric dipole matrix element between two states of the same parity is identically zero, electric dipole transitions between any two levels of the ions shown in Figure 2.3 are totally forbidden. Magnetic dipole transitions, which are typically many orders of magnitude weaker than fully allowed electric dipole transitions, may nevertheless take place.

2.3.2 *Rare-earth ions in a crystal field*

When a lanthanide ion is embedded in a host glass, or indeed a crystalline medium, it is subjected to electric fields, known as the crystal field, due to the neighbouring atoms in the host lattice. One notable characteristic of the fluorescence and absorption spectra of rare-earth ion-doped media, particularly crystals, is the sharpness of the spectral features. This results from the fact that the optically active electrons (i.e. the partially filled 4f shell) are to be found inside the $5s^2$ and $5p^6$ shells, and are therefore to a certain extent shielded from the ion's immediate environment. By way of contrast, transition metal ion-doped media exhibit very broad absorption and fluorescence

---

* The parity of a single configuration wavefunction is either even or odd depending on whether the sum of the single electron orbital angular momentum quantum numbers, $\sum l_i$, is even or odd. Parity is a good quantum number when the atom or ion is in an environment which remains unaltered under an inversion of the coordinate system through the origin.

20 OPTICAL FIBRE LASERS AND AMPLIFIERS

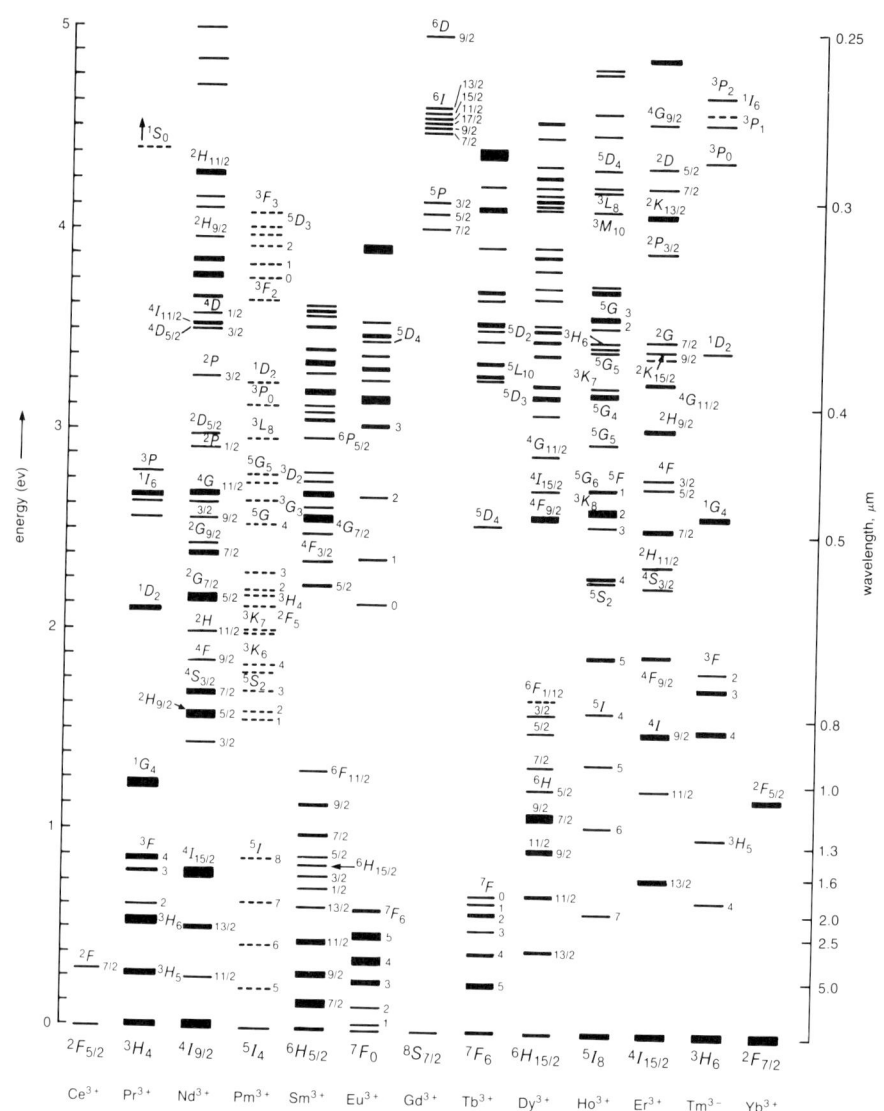

Figure 2.3 Energy level diagram of the triply ionised lanthanide ions.

features due to the fact that the optically active d-electrons interact very strongly with the crystal field.

There are two important effects that the crystal field has on the energy levels of the rare-earth ions. Firstly, the previously $(2J + 1)$-degenerate LSJ levels are split into a number of Stark levels – the precise number depending on the symmetry properties of the crystal field and on the value of J. As a result of the shielding of the 4f electrons, however, these Stark shifts are quite small, only around a few hundred inverse centimetres, and are usually smaller than the intermultiplet separations. Probably more importantly from the point of view of laser action, the crystal field breaks the inversion symmetry of the ion's environment and this now permits electric dipole transitions to occur between Stark levels in different *LSJ* multiplets. The oscillator strengths for these transitions are still nevertheless very small, typically of the order of 10^{-6}, which reflects the weakness of the interaction of the 4f electrons with the crystal field.

En route to considering the emission and absorption spectra of the lanthanide ions in glasses, it is worth while taking a look first at the simpler case of a perfect single-site crystal, where all the rare-earth ions see exactly the same crystal field and so have exactly the same set of energy levels. Figure 2.4 shows just two of the many Stark split *LSJ* manifolds. This Stark splitting of an *LSJ* level is typically within the lattice phonon spectrum of most solid-state media. At room temperature, as a result of interactions between the rare-earth ion and the vibrating lattice, rapid phonon-induced transitions can take place between different Stark levels within a given multiplet. If an ion is optically excited into one of the Stark levels of an excited *LSJ* manifold, then it will have time to make many transitions to the other Stark levels within that manifold, before decaying radiatively down to a lower lying *L'S'J'* multiplet. Since this intramultiplet transition rate is so fast, then, to a very good approximation, the population in any one manifold can be considered to be in local thermal equilibrium with the lattice. Consequently the probability that an ion in manifold *j* is to be found in Stark level α is given by

$$P_{j_\alpha} = \frac{e^{-E_{j_\alpha}/kT}}{Z_j}, \tag{2.1}$$

where the partition function $Z_j = \sum_\alpha e^{-E_{j_\alpha}/kT}$.

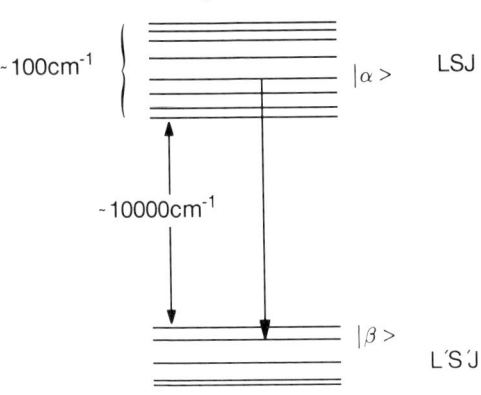

Figure 2.4 Partial energy level diagram for a rare-earth ion-doped crystal showing the splitting of two *LSJ* manifolds into many Stark levels.

When considering laser action in any medium, the question of homogeneity of the laser transition always crops up. Textbooks frequently categorise transitions as being either homogeneously or inhomogeneously broadened, although such a hard and fast distinction is somewhat idealistic. As we shall see with the rare-earth ions in glasses, the real situation lies somewhere between these two extremes. The question that really needs to be asked is if a narrow-linewidth high-intensity laser beam passes through an inverted medium, with what fraction of the upper laser level population can that laser beam interact? For instance, in the case of a doppler broadened gas laser – the 'classic' case of inhomogeneous broadening – a high-power narrow-linewidth laser only interacts with those atoms/ions doppler shifted into resonance with this laser. In the absence of velocity changing collisions, this laser can then only extract a fraction of the energy stored in the population inversion. In our case of the perfect crystal, suppose that some ions have been excited to one of the higher lying LSJ levels and that a narrow-linewidth laser beam is being amplified on the transition between the states α and β as shown in Figure 2.4. Although the laser only directly depopulates level α in the upper manifold, since the thermalisation time is always much smaller than the inverse of the stimulated emission rate, this laser beam effectively interacts with the population in *all* the Stark levels of that manifold. In some senses therefore this array of individual Stark transitions can be considered as a single homogeneous band.

In both the absorption and emission spectra of this perfect crystal, the band of radiative transitions between two different LSJ levels will consist of a large number of sharp spectral lines, each line corresponding to transitions between a pair of Stark levels in those two LSJ manifolds. The width of each of these Stark transitions will be determined by the lifetimes of the ion in the two Stark levels concerned. It can be shown from quantum mechanics that the probability per unit time that an ion in state α of the upper manifold j will spontaneously decay to state β in a lower manifold i is given by the following standard expression*

$$A_{j_\alpha i_\beta} = \frac{16\pi^3 v_{j_\alpha i_\beta}^3 n^2}{3 h \varepsilon_0 c^3} \left[\frac{(n^2+2)^2}{9n} \right] |\langle j_\alpha | \mathbf{D} | i_\beta \rangle|^2$$

where $\langle j_\alpha | \mathbf{D} | i_\beta \rangle$ is the electric dipole matrix element between states $|j_\alpha\rangle$ and $|i_\beta\rangle$, n is the refractive index of the medium and the other symbols have their usual meanings. The term in the square brackets is the local field correction; it aims to account for the fact that the electromagnetic fields acting on the

---

*This equation and some of the following ones in this section contain electric dipole matrix elements. It should be noted that magnetic dipole transitions can also occur between Stark levels in different manifolds, and that similar quantum mechanical expressions can be written for the strength of such transitions. The total radiative transition rate is then the sum of the electric and magnetic dipole contributions. However, the vast majority of rare-earth ion transitions are predominantly electric dipole. This section will concentrate only on such transitions.

lanthanide ion are modified by the polarisability of the lattice. The probability that an ion in any of the Stark levels in the upper manifold spontaneously decays to any of the Stark levels in the lower manifold is then simply given by

$$A_{ji} = \sum_{\alpha,\beta} A_{j\alpha i\beta} = \frac{16\pi^3 \bar{v}_{ji}^3 n^2}{3h\varepsilon_0 c^3 g_j}\left[\frac{(n^2+2)^2}{9n}\right]\sum_{\alpha,\beta} g_j P_{j_\alpha}|\langle j_\alpha|\mathbf{D}|i_\beta\rangle|^2 \quad (2.2)$$

where $\bar{v}_{ji}^3$ is some appropriately weighted average value of v^3 and where g_j is the degeneracy of manifold j.* If there are several lower manifolds to which that ion can decay, then the total spontaneous emission rate out of the manifold j is

$$\Gamma_j = \sum_i A_{ji}. \quad (2.3)$$

The radiative lifetime of this level, τ_j, and the branching ratio β_{ji}† to level i (i.e. the probability that a radiative decay from level j goes to level i) are then given by

$$\tau_j = \frac{1}{\Gamma_j} \quad (2.4)$$

$$\beta_{ji} = \frac{A_{ji}}{\Gamma_j} = A_{ji} \times \tau_j. \quad (2.5)$$

The summation over the dipole matrix elements on the right-hand side of Equation (2.2) is often defined as the line strength of the transition from j to i, S_{ji}‡

$$S_{ji} = \sum_{\alpha,\beta} g_j P_{j_\alpha}|\langle j_\alpha|\mathbf{D}|i_\beta\rangle|^2. \quad (2.6)$$

Let us now consider absorption of photons by rare-earth ions either in the ground-state manifold or in an excited state. The probability per unit time that an ion in Stark level β of the manifold i will absorb a photon of frequency v and be excited to Stark level α in a higher manifold j is

$$\sigma_{i_\beta j_\alpha}(v)I(v)/hv$$

where $\sigma_{i_\beta j_\alpha}(v)$ and $I(v)/hv$ are the absorption cross-section and the photon flux

---

* Even though the upper manifold consists of Stark levels each with different decay rates, because these levels are strongly coupled together via phonon transitions, the population in the whole manifold decays exponentially with a time constant being an appropriate average of all the individual decay times. Since the Stark level populations will change with temperature, A_{ji} and τ_j will in general be functions of temperature.
† This expression for the branching ratio will need to be expanded later to take account of non-radiative decays from level j.
‡ This definition of line strength differs from that used by some other authors in that it does not assume equal population in all the Stark levels.

respectively at frequency v. The probability therefore that an ion in any of the Stark levels of manifold i is excited to any of the Stark levels in the upper manifold j is then

$$\sum_{\alpha,\beta} P_{i_\beta}\sigma_{i_\beta j_\alpha}(v)I(v)/hv.$$

If N_i is the number density of ions in level i, then the total number of absorptions per unit time between these two manifolds can be written in the form

$$\frac{N_i \sum_{\alpha,\beta} P_{i_\beta}\sigma_{i_\beta j_\alpha}(v)I(v)}{hv} = \frac{k_{ij}(v)I(v)}{hv} = \frac{N_i\sigma_{ij}(v)I(v)}{hv} \tag{2.7}$$

where $k_{ij}(v)$ is the absorption coefficient at frequency v. Again, from basic quantum mechanics, this absorption coefficient can be related to sums of electric dipole matrix elements

$$\int k_{ij}(v)dv = \frac{2\pi^2 N_i \bar{v}_{ij}}{3\varepsilon_0 hc}\left[\frac{(n^2+2)^2}{9n}\right]\sum_{\alpha,\beta} P_{i_\beta}|\langle i_\beta|\mathbf{D}|j_\alpha\rangle|^2$$

$$= \frac{2\pi^2 N \bar{v}_{ij}}{3\varepsilon_0 hcg_i}\left[\frac{(n^2+2)^2}{9n}\right]S_{ij} \tag{2.8}$$

where $\bar{v}_{ij}$ is an appropriately averaged value of the absorption frequency. Similar expressions to Equations (2.7) and (2.8) can be written for stimulated emission from manifold j to manifold i. In Equation (2.7) an effective cross-section, $\sigma_{ji}(v)$, has been defined to take account not only of the strength of the individual Stark transitions, but also the probability of occupation of the individual Stark levels. Whether one decides to work in terms of cross-sections between individual Stark levels and the relative populations of those Stark levels, or in terms of an effective cross-section, depends on the transition under investigation. In the case where the number of individual Stark transitions within a band is small and they do not overlap to any significant extent, there is little benefit in defining an effective cross-section as it is just as easy to discuss populations in the individual Stark levels. The formulation of an effective cross-section is particularly beneficial when a band consists of a large number of overlapping transitions, since light at any one frequency will, more than likely, excite several Stark transitions simultaneously. One such example, which will be discussed again later, is the $^4I_{15/2} \to {}^4I_{13/2}$ band around 1.5 μm in Er^{3+}. In an Er^{3+} doped crystal with a low degree of symmetry, the $^4I_{13/2}$ and $^4I_{15/2}$ levels will be split into 7 and 8 Stark levels respectively. The 56 resulting Stark transitions of differing strengths are spread over a wavelength range of roughly 400 cm^{-1}. Assuming a thermal broadening for each Stark level of around 20 cm^{-1}, we see that a considerable degree of overlap is likely to occur.

Finally, since at room temperature there is a tendency for the population in a band to congregate in the lowest Stark levels, the absorption spectra shift towards shorter wavelengths and the fluorescence spectra towards longer wavelengths. This resulting asymmetry of the fluorescence and absorption bands would not occur if all the Stark levels in a manifold were equally populated, and it is this that enables a 1.48 μm pumped Er^{3+}-doped fibre to produce gain at 1.54 μm.

2.3.3 Rare-earth ions in glasses

Broadening of the rare-earth ion fluorescence and absorption bands in glasses is necessarily more complicated than that in crystals, since glasses have by definition a random structure, with each rare-earth ion seeing a different electric field and therefore having a different set of Stark levels. Each ion site, labelled by some parameter q, will have associated with it absorption and emission spectra, $k_{ij}^{(q)}(v)$ and $k_{ji}^{(q)}(v)$, and radiative decay rates $\Gamma_j^{(q)}$. The measurable absorption and emission spectra from a sample of the material, together with the radiative decay rates of all the excited levels, will now be averaged in some way over all the different ion sites, i.e.

$$\Gamma_j = \langle \Gamma_j^{(q)} \rangle \qquad k_{ij}(v) = \langle k_{ij}^{(q)}(v) \rangle \qquad k_{ji}(v) = \langle k_{ji}^{(q)}(v) \rangle.$$

This random structure severely limits the number of quantitative statements that can be made regarding rare-earth ion spectra in glasses.

The fact that different ions in the sample have different environments might suggest that the radiative transitions would be predominantly inhomogeneously broadened. Indeed, there are several easily observable consequences of these site-to-site variations in the lanthanide ion environments. When excited by a narrow-linewidth laser for instance, because different ion sites have different absorption cross-sections, certain subsets of the ions will be preferentially excited. This then leads to the shape of the fluorescence and gain spectra being dependent on the excitation wavelength [9, 10]. Also the decay of the fluorescence intensity following a short excitation pulse will not be purely exponential, as the fluorescence from each ion site will decay at different rates.

Nevertheless, the following two important experimental observations should be noted. Firstly, in 1.48 μm pumped Er^{3+} fibre amplifier experiments, the shapes of the gain spectra obtained are consistent with those expected from a purely homogeneously broadened line and show no evidence of the spectral hole burning that one would associate with an inhomogeneously broadened transition [11]. There have also been several demonstrations of narrow-linewidth fibre laser oscillators [6, 8] and of power amplifiers [12] with near quantum limited performance, i.e. once threshold has been exceeded, for every extra pump photon absorbed, one extra lasing photon is emitted. Such behaviour would seem to suggest that a high-intensity laser beam of narrow linewidth can, at least for the transitions mentioned, interact with all the ions in the upper laser manifold and can extract all the energy stored in the population inversion.

Such quasi-homogeneous behaviour ought not to be too surprising, for it was noted that if within a band between two *LSJ* levels there are a large number of Stark transitions, they tend to overlap forming a homogeneously broadened quasi-continuum. In such circumstances, $k_{ji}^{(q)}(v)$ and $k_{ij}^{(q)}(v)$ will probably be of significant value all across the band for *all* the ion sites and so to a first approximation all the ion sites are equally excited. Although some inhomogeneous effects have been seen, there does seem to be a growing array of evidence that suggests that at room temperature the broad absorption and emission bands of rare-earth ion-doped glasses can be considered, at least to a first approximation, as being homogeneously broadened [13, 14].

It is on the basis of this conclusion that the results to be derived in the next subsection on Judd–Ofelt theory, and also the modelling work to be considered in Section 2.5, are still all applicable. It should be remembered, though, that quantities such as cross-sections and lifetimes actually represent averaged quantities and that homogeneous saturation effects, characterised by $(1 + I/I_{sat})^{-1}$, where I_{sat} is some appropriately defined saturation intensity, may not hold exactly.

2.3.4 *Judd–Ofelt theory*

It will become apparent in the remainder of this chapter that a knowledge of the strength of the radiative transitions between different excited *LSJ* levels is of great importance when looking for new laser transitions, for assessing new host glasses and in understanding any limitations on device efficiency. Direct measurements of radiative transition rates between excited levels are unfortunately extremely difficult to perform. In the early 1960s, Judd [15] and Ofelt [16], working entirely independently, developed a quantum mechanical theory of rare-earth ion-doped solid-state media, which allowed estimates to be made of the electric dipole transition strengths between any pair of *LSJ* manifolds. A Judd–Ofelt analysis of a rare-earth ion-doped material essentially involves using the line strengths of all the observable transitions from the ground state to excited manifolds, which can be determined from white light absorption measurements, and using these line strengths to estimate line strengths of transitions between pairs of excited levels. The theory has been used in the past with a considerable degree of success [17, 18] which must justify, to a certain extent, the many approximations contained within the theory.

The full quantum-mechanical calculations behind Judd–Ofelt theory are extremely involved and consequently will not be presented here. The underlying philosophy behind the theory is quite simple however [19]. Judd–Ofelt theory takes as its starting point the eigenfunctions of an isolated rare-earth ion. Since *L* and *S* are not good quantum numbers, it is first necessary to expand these eigenfunctions, $|4f^n\alpha[L,S]JM_J\rangle$,* in terms of pure LS-coupled wavefunctions using the known energies of the states to empirically determine the mixing coefficients. The non-centrosymmetric terms in the expansion of the crystal field potential are only small perturbations to the hamiltonian of the 'free-ion', and so standard degenerate perturbation theory techniques can be used to calculate the first-order corrections to the ionic wavefunctions. Using these perturbed wavefunctions, which contain admixtures of eigenfunctions from excited electron configurations of opposite parity to the $4f^n$ configuration, the electric dipole matrix elements between any pair of Stark levels, and hence electric dipole transition strength, S_{ij}, between any two manifolds can be written down.†

At this stage in the calculation, the expressions derived contain summations over *all* the excited states of the ion. Both Judd and Ofelt realised that by assuming that all the excited electron configurations were effectively degenerate and that all the Stark levels within a manifold were equally populated, then various closure relationships could be invoked, and the line strength between two manifolds i and j could be simplified to an expression of the form

$$S_{ij} = \sum_{t=2,4,6} \Omega_t |(4f^n\alpha[L,S]J \| U^{(t)} \| 4f^n\alpha'[L',S']J')|^2. \tag{2.9}$$

The squared reduced matrix elements $|(4f^n\alpha[L,S]J \| U^{(t)} \| 4f^n\alpha'[L',S']J')|^2$ are essentially independent of the host material and their values, for most of the transitions of interest, can be found in the published literature (see, for example, reference [19]). All the information about the crystal field potential and its effect on the ionic wavefunctions is contained in the three Judd–Ofelt parameters Ω_t. It is these values that will change between different host materials. From the measured absorption spectrum of a rare-earth ion-doped material, Equation (2.8) can be used to determine the line strengths of all the observed transitions from the ground state. By assuming that all the Stark levels in the ground state manifold are equally populated, then a least-squares fit of these measured line strengths to the expression given in Equation (2.9) yields the values for the three Judd–Ofelt parameters Ω_t. The line strength for any transition between excited manifolds can then be estimated using these 'best fit' values for Ω_t and tabulated values of the reduced matrix elements. Branching ratios and radiative lifetimes can be estimated using

---

* Eigenfunctions of isolated ions can always be labelled by their values of J and M_J. The label '$4f^n$' implies that the state originates from a single electron configuration. The use of L and S in square brackets is meant to denote the conventional labelling of the manifolds by values of L and S, and not that L and S are to be taken as good quantum numbers. Finally, α denotes any other quantum numbers required to uniquely define the state.

†As in Section 2.3.2, Judd–Ofelt theory only considers electric dipole transitions; nevertheless, inclusion of magnetic dipole transitions into the line strengths is only a minor extension of the theory.

Equations (2.3) and (2.5). In the approximation that the line strengths S_{ji} and S_{ij} are equal, the radiative transition strength and the integrated absorption and stimulated emission coefficients can be related to each other as follows:

$$\frac{1}{N_i}\int k_{ij}(v)\,dv = \frac{c^2}{8\pi n^2 v_{ij}^2}\left(\frac{g_j}{g_i}\right)A_{ji} \qquad (2.10)$$

$$\int k_{ji}(v)\,dv = \left(\frac{g_j}{g_i}\right)\int k_{ij}(v)\,dv \quad \text{or} \quad \int \sigma_{ji}(v)\,dv = \left(\frac{g_j}{g_i}\right)\int \sigma_{ij}(v)\,dv. \qquad (2.11)$$

It can now be seen that the fact that the local field correction has not been included by some authors in their calculations is not too worrying in that it cancels out in the above expression relating integrated absorption coefficient to radiative line strength. The fact that A_{ji} can be related to $\int k_{ij}(v)$ has been used by several groups to determine absolute values for absorption and emission cross-sections. This will be considered further in Chapter 4.

2.4 Population dynamics in optically pumped rare-earth ion-doped media

To optimise the efficiency of any optically pumped laser device requires an understanding of the population dynamics of that laser system. Of particular interest in the case of rare-earth ion-doped devices is the fraction of those ions excited out of the ground state by the pump radiation that is channelled into the upper laser level. This section looks at the processes by which excited ions decay back down to the ground state, into which LSJ levels population can accumulate and hence on which transitions laser action might occur.

2.4.1 Radiative and non-radiative decay processes

In a weakly doped medium (i.e. one in which the rare-earth ions are sufficiently far apart that each ion decays independently of all the other ions in the sample) an ion in an excited LSJ manifold can decay to a different manifold via one of the following four processes:

(i) Spontaneous emission to any lower lying manifold.
(ii) Multiphonon decay to the next lower lying manifold.
(iii) Absorption/stimulated emission of a pump photon.
(iv) Absorption/stimulated emission of a lasing photon.

Radiative transitions were discussed in some detail in the previous section. In principle, an ion in any excited level can spontaneously decay to any one of the lower lying manifolds of that ion by emitting a photon of the relevant frequency. The absolute strength of each of these radiative decay paths, A_{ji}, depends on the host material. One particularly important point to note is that an ion in a level *above* the upper laser level can quite happily fluoresce to a level *below* the upper laser level. This bypassing of the upper laser level obviously results in inefficient transfer of absorbed pump energy to the upper laser level ions.

Phonon-assisted transitions were discussed in the previous section in connection with the thermalisation of the Stark level populations within a manifold. The lattice vibration spectrum of most solid-state materials only extends for a few hundred inverse centimetres, whereas the typical spacing between *LSJ* levels is often several thousand wavenumbers. There is nevertheless a probability that an excited ion sitting in the lattice will undergo multiple phonon emission and decay to the *next* lower *LSJ* manifold. This multiphonon decay rate varies between different glass hosts, and for a given glass the decay rate is a strong function of the proximity of the next lowest lying electronic level. It is, however, almost independent of the specific rare-earth ion or of the *LSJ* levels involved [17, 18]. In silica, for instance, as a result of the high multiphonon emission probabilities, fluorescence from certain levels is completely quenched (i.e. the non-radiative decay probability is so high that any ion arriving in that level will undergo multiphonon emission before having a chance to fluoresce). That same level in a fluoride glass host, where multiphonon emission rates are much lower, might well fluoresce quite happily.

In the energy levels of the lanthanide ions, there is always the possibility of accidental degeneracies existing, where the spacing of two excited energy levels exactly matches the energy of either the pump or the lasing photons. Absorption of pump photons from any excited level is generally bad in that it wastes some of the absorbed pump power. But there are situations where excited-state absorption of pump photons can be beneficial. The sequential absorption of two infrared photons may in some circumstances offer the possibility of an infrared pumped visible laser [20].

Absorption of lasing photons from any level, other than the lower laser level, is always a loss mechanism and must be avoided if at all possible. In the case of absorption of lasing photons by upper laser level ions, this absorption cross-section may be so large as to prevent the transition from lasing in the first place, as in the 1.3 μm transition in Nd^{3+}-doped silica. Even if there is sufficient gain on the laser transition to overcome such a loss, the laser slope efficiency will almost certainly be degraded as a result of this unwanted absorption.

Whereas the transition rate for excited-state absorptions depends not only on the precise pump and lasing wavelengths but also on their intensities, once a particular host material has been selected, the spontaneous emission and the multiphonon decay rates from all the excited levels are fixed. The relative sizes of A_{ji} and Γ_j^{nr} determine to which levels any population in level *j* decays, and need to be known in order to understand how the population dynamics limits device efficiency. Although the fluorescence lifetimes of excited levels can be measured with relative ease, they only yield the sum of the radiative and non-radiative decay rates from that level j

$$\tau_j^{fluor} = (\Gamma_j + \Gamma_j^{nr})^{-1} = (\sum_i A_{ji} + \Gamma_j^{nr})^{-1}. \qquad (2.12)$$

Since non-radiative decay processes are not directly observable, Γ_j^{nr} must be determined by indirect methods. This is an area where Judd–Ofelt theory finds probably its most important use. A Judd–Ofelt analysis can be used to make reasonable estimates of the values for A_{ji}, from which the total radiative lifetime for level j can be calculated. Any discrepancy between $(\sum A_{ji})^{-1}$ and the fluorescence lifetime can then be attributed to non-radiative decay processes. Using such a method, multiphonon decay rates have been estimated for different rare-earth ion energy levels in a wide variety of different glasses (see Figure 8.2) [17, 18].

2.4.2 Which transitions will lase?

The following expression needs to be satisfied if there is to be gain on the transition between level j and level i at a frequency v, and hence the possibility of laser action,

$$N_j \sigma_{ji}(v) - N_i \sigma_{ij}(v) > 0. \qquad (2.13)$$

It can be seen that to achieve gain at low pump powers requires a large population to accumulate in level j and the population in level i to be kept as small as possible. The requirement for N_j to be large implies two things. Firstly, ions excited out of the ground state need to be channelled efficiently into level j, and, secondly, the fluorescence lifetime of level j should be long so that the ions arriving in that level remain there for a reasonable length of time. To be a sensible candidate for an upper level of a laser transition, a level j ideally needs to be well separated from the next lowest lying energy level, in order that the non-radiative decay rate out of level j is small. These metastable levels of a rare-earth ion in a particular host material are easily identified experimentally, simply by seeing which of the excited levels of that ion fluoresce. A level which, when pumped efficiently, does not fluoresce or only fluoresces very weakly, cannot be seriously considered as an upper level for a laser transition.

The mere fact that a rare-earth ion level fluoresces, however, is still no indication that gain will exist on a laser transition originating from that level, even if the sample were to be pumped extremely hard. The population in the lower laser level needs to be considered. Even in the case where the excited state branching ratios are favourable so that ions are preferentially channelled into level j, there must always be some minimum population in level i simply as a result of spontaneous emission on the laser transition itself. In the 'best possible' case (see Figure 2.5), where the level i is pumped only by spontaneous

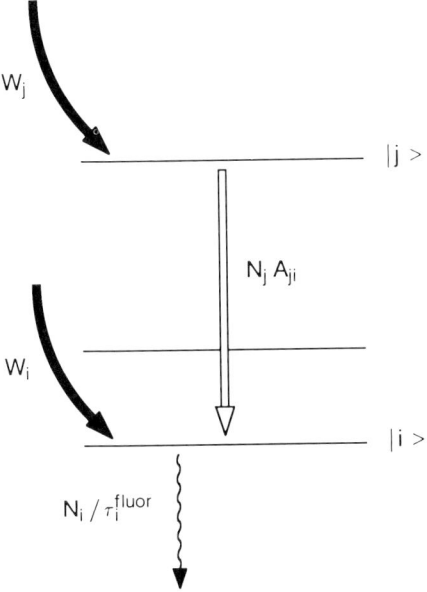

Figure 2.5 Partial energy level diagram showing a self-terminating laser transition. W_i and W_j are pumping rates into levels $|i\rangle$ and $|j\rangle$ respectively and τ_i^{fluor} is the fluorescence lifetime of level $|i\rangle$.

emission on the laser transition (i.e. $W_i = 0$), then under CW pumping conditions the rate at which ions are excited into level i is equal to the rate at which ions decay out of level i. This implies that

$$N_j A_{ji} = \frac{N_i}{\tau_i^{\text{fluor}}}.$$

Combining this expression with Equation (2.13), it can be shown that the following result must hold in order for gain to be sustainable on a CW basis

$$\frac{1}{\tau_i^{\text{fluor}} \times A_{ji}} > \frac{\sigma_{ij}}{\sigma_{ji}} \qquad (2.14)$$

i.e. the ratio of the total decay probability out of the lower laser level to the spontaneous decay probability on the laser transition must exceed the ratio of cross-sections [21]. This is, however, the minimum condition that needs to be satisfied. If level i is inadvertently pumped from any source other than spontaneous emission on the laser transition (i.e. $W_i \neq 0$), then the decay rate of the lower laser level population needs to be even larger than Equation (2.14) would suggest. Even if this inequality is not satisfied, it is still possible for pulsed lasing to occur on this transition. Population must be excited rapidly into level j, and then gain will exist for a short period of time before level i has filled up. However, as soon as laser action commences, the population in level i increases rapidly as a result of stimulated emission. This then reduces the population inversion to zero, and so the laser turns itself off. Such transitions are known, for obvious reasons, as self-terminating laser transitions.

Rare-earth ion laser transitions are often categorised as being three-level or four-level transitions depending on whether or not the lower laser level is part of the ground-state manifold.* An unpumped three-level transition is absorbing at the lasing wavelength, and thus there needs to be a significant depopulation of the ground-state manifold before there is any gain on the lasing transition. In order to reach this transparency position – where there is no nett gain or loss down the fibre – the pumping rate of ions out of the ground state needs to be roughly equal to the fluorescence rate on the laser transition. Typically, this implies pumping intensities of the order of $1\,\text{mW}/\mu\text{m}^2$, which are difficult to achieve in bulk glasses. In single-mode fibres, however, such intensities are readily achieved with laser diode pump sources. Three-level laser transitions are therefore readily accessible in fibre form. Because there is always a significant fraction of the ion population in the upper laser level of a three-level transition, such transitions are, unlike four-level transitions, always susceptible to pump excited state absorption problems. Lastly, as a result of there being a minimum pump power required for gain, there exists for three-level fibre lasers an optimum length of fibre for maximum gain from an amplifier or for maximum output power from a laser oscillator. At this optimum length the pump power leaving the fibre is approximately equal to the transparency power.

2.4.3 Multi-ion effects in rare-earth ion-doped glasses

In addition to the single-ion processes described in Section 2.4.1, as the rare-earth ion doping density is increased, there comes a point where the mean ion separation is sufficiently small that spatial migration of excitation from one rare-earth ion to a neighbouring ion becomes possible. Such a transfer of excitation can be used to enhance the pumping of a laser transition and so increase the lasing efficiency. However, there are also situations where the presence of multi-ion phenomena can impair laser performance. These multi-ion effects can be divided into two categories depending on whether energy migration takes place between similar or dissimilar ions.

2.4.3.1 Co-doping.
The idea of deliberately incorporating two, or more, types of rare-earth ion simultaneously into solid-state lasing media has been in

---

* The term 'quasi four-level transitions' has also been used by some authors. These are laser transitions which terminate on one of the high lying Stark levels of the ground-state manifold. Because such levels are only weakly populated at room temperature, the upper laser level population required to reach transparency is still small, unlike a true three-level laser where fractional upper level populations around 50% are required for lasing.

INTRODUCTION TO GLASS FIBRE LASERS AND AMPLIFIERS 33

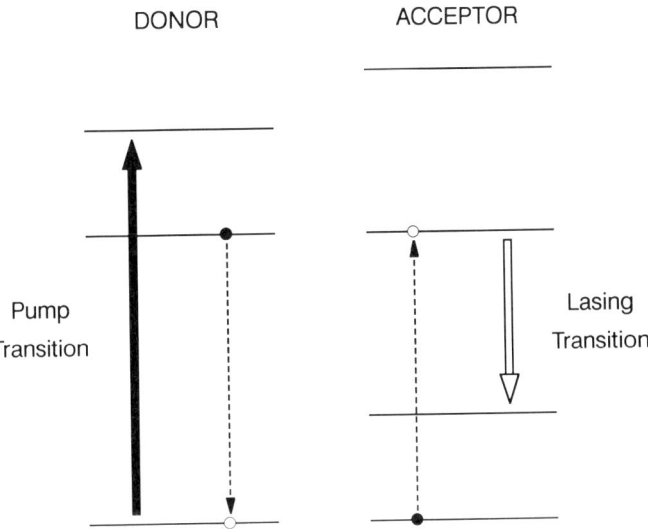

Figure 2.6 Schematic energy level diagram showing how energy transfer between donor and acceptor ions (represented by the dashed arrows) can lead to a population inversion in the acceptor subsystem.

existence for a long time. Although it is used in some flashlamp pumped solid-state lasers, co-doping has yet to be fully exploited in the context of glass fibre lasers. Probably the simplest, and most common use of co-doping is shown schematically in Figure 2.6. The laser transition of interest takes place between two levels of the so-called 'acceptor' ion. The other rare-earth ions present, known as the donor ions, are there to absorb the pump radiation. Following the absorption of a pump photon, a donor ion relaxes from the pumped level down to a metastable level. If the fluorescence from that metastable level overlaps with the absorption spectrum of the acceptor ion, and if the acceptor ion density is sufficiently high, then there is a distinct probability that a donor–acceptor pair will undergo a radiationless transition* leaving the donor ion back in the ground state and 'pumping' the acceptor ion to an excited level.

The idea of introducing the donor ions into the gain medium of a flashlamp-pumped laser is to increase the absorption of the active medium in those spectral regions not absorbed by the acceptor ions. By absorbing a greater proportion of the broad spectral emissions from the flashlamp, the laser slope efficiency ought to be increased, assuming that the donor–acceptor transfer is reasonably efficient. The reasons for co-doping in the case of glass fibre lasers

---

*Although the efficiency of this ion–ion transfer is related to the degree of overlap between the donor fluorescence band and the acceptor absorption band, the process itself does *not* involve the emission of a photon by the donor ion, and subsequent reabsorption by the acceptor. In a fibre geometry, such a process would be extremely unlikely and could never be efficient.

are somewhat different. The pump source for a fibre laser is invariably a laser, and a laser emits only over a very limited spectral range. One obvious situation where co-doping a fibre laser may be worth while considering is when the lasing ion does not absorb at a convenient pump wavelength. Being able to pump a fibre laser, for instance, with a cheap high-power GaAlAs diode laser has obvious practical advantages. Yet not all the rare-earth ions have absorption bands coinciding with GaAlAs emission wavelengths.

A more subtle version of this basic scheme is where, although the acceptor ion *does* absorb at the pumping wavelength, there are pump excited-state absorption problems in the acceptor ion which need to be avoided. By utilising a co-doped pumping scheme and by carefully tailoring the concentration of the donor and acceptor ions, it should be possible to ensure that most of the pump photon absorptions occur in the donor subsystem, and not from an excited state of the acceptor ions. Such a scheme was considered in an attempt to produce efficient laser action at 1.5 μm from a GaAlAs diode pumped Yb^{3+}–Er^{3+} co-doped silica fibre [22]. Similarly, in the case of a very weak pump absorption, where the pump absorption coefficient is comparable to the background fibre losses, a co-doped pumping scheme may improve the laser performance for much the same reasons.

The other situation where co-doping may well be beneficial is when one of the metastable levels of the lasing ion requires to be depopulated. A specific example of this is the case of a self-terminating laser transition (Section 2.4.1) where the fact that the lower laser level is long-lived prevents

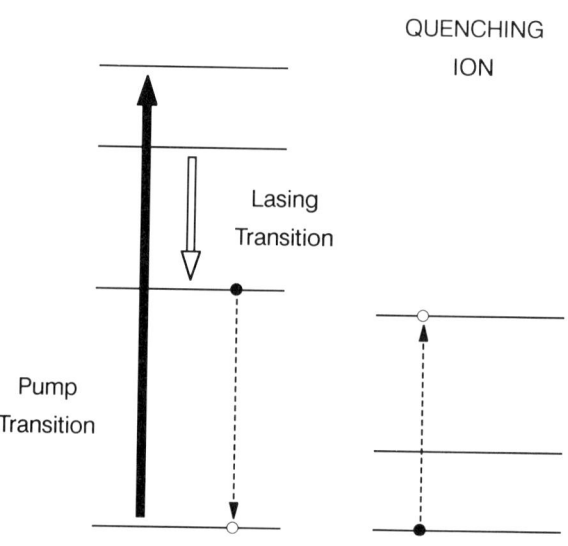

Figure 2.7 Schematic energy level diagram showing how energy transfer to a quenching ion can lead to a depopulation of the lower laser level.

CW gain from being established. By incorporating a second rare-earth ion, or quencher, into the medium and by arranging for ion–ion transfer to depopulate the lower laser level (Figure 2.7), it should be possible to convert a self-terminating transition into a CW one. Depopulation of the lower laser level of the 2.7 μm transition in Er^{3+} using Tb^{3+} ions as the quenching species is one situation where the above scheme has been considered [23]. An argument similar to this could be applied to the quenching of a metastable level that was the source of excited-state absorptions.

2.4.3.2 *Like-ion effects.* In a single ion doped medium, ion–ion interactions can also affect the population dynamics if the rare-earth ion density is sufficiently high. There are really two cases to consider here.* Figure 2.8(a) shows two neighbouring ions both initially existing in the same level $|b\rangle$. The energy levels of this particular ion are such that $E_{ba} \approx E_{db}$. There exists the possibility of a radiationless transition whereby one of the ions is excited to a higher state $|d\rangle$ while the other ion returns to the ground state $|a\rangle$. Such a situation occurs for the $^4I_{13/2}$ level of Er^{3+} – the upper level of the 1.5 μm laser transition. Cooperative up-conversion can, however, be put to good use if the energy difference between state $|d\rangle$ and $|a\rangle$ is greater than the pump photon energy, as there now exists the possibility of an up-conversion laser, i.e. one which lases at a shorter wavelength than the pumping wavelength.

Figure 2.8(b) shows the process which is essentially just the reverse of that shown in Figure 2.8(a).† Here two neighbouring ions, one in the ground state and the other in an excited state, undergo a radiationless transition, both ending up in the same state $|j\rangle$. This is essentially a quenching process in that it basically involves the de-excitation of a high lying *LSJ* level. Again this process may or may not benefit laser action depending on the circumstances. In Nd^{3+}, for example, where state $|k\rangle$ corresponds to the $^4F_{3/2}$ upper laser level of the 1 μm transition, such a mechanism depopulates the upper laser level and so increases laser thresholds. It is this mechanism which is believed to be responsible for the dramatic shortening of the fluorescence lifetimes in so-called 'clustered' Nd^{3+} fibres (see Chapter 3). However, in the case of Tm^{3+}, ions can be pumped to the $^3F_4$ level ($|k\rangle$) by GaAlAs lasers and then for every pump photon absorbed, there are *two* ions in the $^3H_4$ level ($|j\rangle$) – effectively a 200% quantum yield [24].

Lastly in this section, it is worth while mentioning, for the sake of completeness, an ion–ion interaction that does not really affect the population

---

*In both the schematic examples given here, either before or after the non-radiative transfer, both ions are in the same energy level. There is, however, no need for the two transitions to share a common level. All that is required is that the ion has two different radiative transitions which are roughly degenerate in frequency and that the two initial states have long lifetimes,

†The difference between case (a) and case (b) is really just one of classification. In the first case, one of the ions ends up in a higher state than either of the ions prior to the non-radiative transfer taking place.

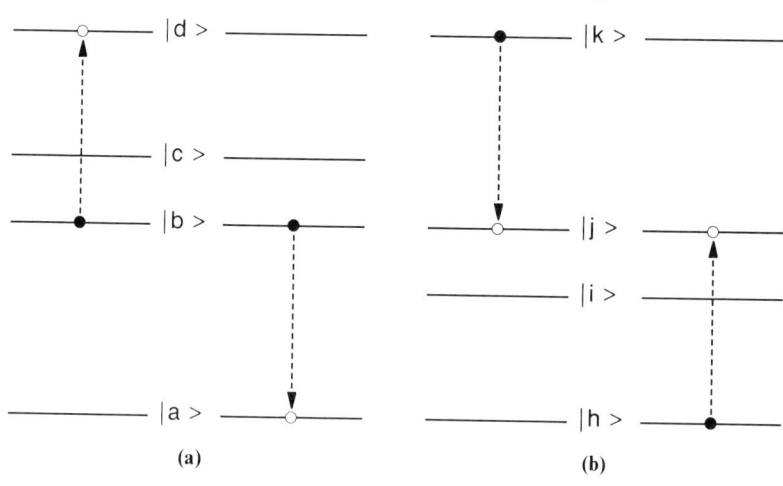

Figure 2.8 Schematic diagram showing up-conversion (a) and quenching (b) energy transfer processes in highly doped media.

dynamics. Consider the situation of two neighbouring ions, one of which is initially in the ground state and the other in an excited state, changing places as the result of a radiationless transition. Although such a process leads to the spatial migration of excitation away from the position where the pump photon was originally absorbed, and may well improve the homogeneity of the lasing transition, it does not affect the population dynamics since the population distribution of ions among the excited states is exactly the same before and after the non-radiative decay takes place.

2.4.4 Comparison of rare-earth ion-doped silica and fluorozirconate glasses

The overwhelming majority of glass fibre laser research reported to date has involved either silica or fluorozirconate ('fluoride') glasses as the host medium for the rare-earth ions. Although many other laser glasses are available, this somewhat restricted choice of host glass is simply due to the fact that silica and, to a lesser extent, fluorozirconate glass fibre fabrication are relatively mature technologies that have been developed over the last two decades, not for fibre laser applications but for use in optical communications. Notable differences exist between these two glasses as laser host materials due to their vastly differing multiphonon emission rates. Because of this tendency to work with just these two host materials, some general comparisons between silica and fluorozirconate glasses will now be given, from the point of view of the rare-earth ion population dynamics.

The fundamental vibrational modes of fluoride glasses lie further into the infrared than those of silica. Two important properties follow directly from this fact. Firstly, fluoride glasses remain transparent right out to around 3.5 μm whereas silica becomes absorbing beyond about 2 μm. Secondly, the multiphonon emission probability for an ion in any excited *LSJ* level is much higher in silica than in fluorozirconate glasses. Only when the energy gap to the next lower level is larger than around 4600 cm^{-1} does a level in silica fluoresce to any significant degree [25], as opposed to 3100 cm^{-1} in fluoride glasses [26]. This severely limits the number of possible laser transitions in rare-earth ion-doped silica.

At the same time, however, most of the rare-earth ion population dynamics in silica are determined by a level-by-level cascade of non-radiative transitions. There are few level bypassing radiative decays, and consequently there is frequently a 100% efficient feeding of the population from the upper pumped level into the upper laser level. As an example of this, consider the case of a 532 nm pumped Er^{3+}-doped silica fibre (Figure 2.9). Even though the ions are pumped to a manifold four levels above the upper laser level, the population cascades down through each of the intervening levels, since none of those excited levels fluoresces, so that every pump photon absorbed on the $^4I_{15/2} \rightarrow {}^4S_{3/2}$ transition gives rise to an extra upper laser level ion.

The situation in a fluoride glass is radically different in that the non-radiative decay probabilities are very much smaller. The population dynamics

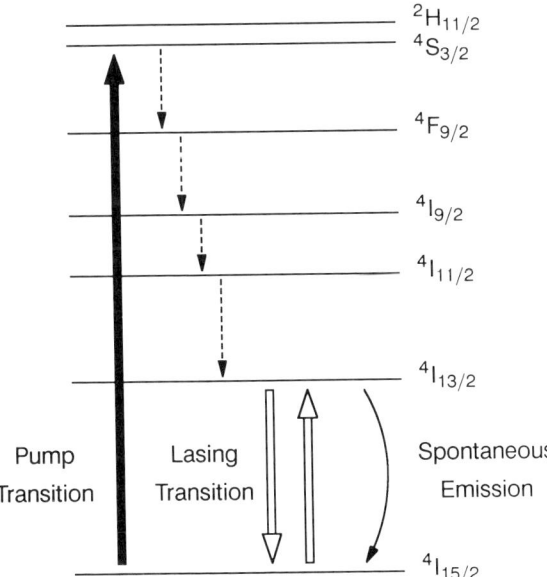

Figure 2.9 Er^{3+} energy level diagram showing the radiative and non-radiative transitions in silica which affect the population dynamics when pumped at 532 nm.

are now more strongly determined by the radiative transitions rates, A_{ji}. Taking the same example as for silica, all the Er^{3+} levels up to the $^4S_{3/2}$ level fluoresce to a certain degree in fluoride glasses and there is now ample scope for population in any of these levels to decay directly to the ground state so bypassing the $^4I_{13/2}$ level. In order, therefore, to ensure efficient pumping of the upper laser level it is necessary in fluoride glass systems to carefully choose the pump transition either to pump directly into the upper laser level or at least into a level which transfers non-radiatively, and hence efficiently, into that upper laser level. The recently reported 800 nm pumped lasing transitions at 1.05 μm and 1.3 μm in Nd^{3+}-doped fluorozirconate glasses are efficient [7] for this latter reason.

The fact that there are in principle several metastable levels in rare-earth ion-doped fluoride glasses means there are potential 'bottlenecks' in the radiative decay chain where large fractions of the total ion population may accumulate. Self-terminating laser transitions are just one example of this problem, where the lower laser level is also a metastable level. The fact that population can accumulate in a level not associated with either the pump or the laser transition gives rise to a nonlinear laser characteristic [23]. Such a behaviour is in itself not a problem, as an efficient laser can still be produced, at any given pump power, simply by choosing the correct fibre length. The accumulation of population in a metastable level only degrades the laser performance if an excited-state absorption originates from this level or if the fibre exhibits background losses due to impurities, scattering, etc.

2.5 Modelling of the fibre laser oscillators and amplifiers

2.5.1 Introduction

A fair number of theoretical papers have been published to date, which analyse the performance of active fibre devices [4, 23, 27–37]. Of these analyses, a significant proportion have, not surprisingly, dealt with various aspects of Er^{3+} doped silica fibre amplifiers, although modelling of fibre laser oscillators and of superfluorescent sources have not been totally forgotten. Rather than simply duplicate the results of other workers in this area, the aim in this last section is to outline the essential method used in *all* these models, and to comment upon many of the approximations that have been invoked.

The analysis breaks down naturally into two parts: the first part looks at the population dynamics of the rare-earth ions while the second is concerned with the amplification or attenuation of the pump and signal, and maybe noise, beams as they propagate up and down the fibre. If the pump and signal intensities at a particular point in the fibre are known, then a set of rate equations can be set up describing the temporal evolution of the populations in all the excited levels of that ion at that point. A steady-state solution of these equations then allows the populations in all the different *LSJ* levels to be determined.* The absorption and gain coefficients for each wavelength propagating down the fibre can then be worked out, thus enabling one to determine how the pump and signal beam intensities change down the fibre. In a laser oscillator, there are two signal beams to consider – namely, the forward and backward travelling waves within the laser cavity. For an amplifier, there might be many signal beams all at slightly different wavelengths to consider, all of which will have their own propagation equations. Furthermore, there are circumstances where it may also be necessary to consider the noise generated by this laser device. A certain amount of fluorescence at all wavelengths across the laser transition will propagate in both directions down the fibre and be amplified as it goes. So in addition to propagation equations for the signal beams deliberately launched into the fibre, there also needs to be a set of propagation equations describing the evolution down the fibre of the amplified spontaneous emission (ASE). All that remains to be done is a simultaneous solution of these equations subject to the boundary conditions at the fibre ends. Except for one or two very special cases, analytic solutions to the problem do not exist.

2.5.2 Rate equation analysis of the population dynamics

Although multi-ion effects were described in Section 2.4, they will not be considered any further in this section, not because they are unimportant but because they just complicate the situation without adding anything to the understanding of the principles involved. The most general expression describing the rate of increase of the population density in level j is

$$\dot{N}_j = \begin{bmatrix} \text{Rate at which ions are} \\ \text{excited into level } j \end{bmatrix} - \begin{bmatrix} \text{Rate at which ions are} \\ \text{de-excited from level } j \end{bmatrix}$$

$$= \left[\sum_{i<j} \frac{N_i \sigma_{ij} I_p}{h\nu_p} + \sum_{l>j} \frac{N_l \sigma_{lj} I_p}{h\nu_p} + \sum_{i<j} \frac{N_i \sigma_{ij} I_0}{h\nu_0} \right.$$

$$\left. + \sum_{l>j} \frac{N_l \sigma_{lj} I_0}{h\nu_0} + \sum_{l>j} N_l A_{lj} + N_k \Gamma_k^{nr} \right]$$

$$- \left[\sum_{j<l} \frac{N_j \sigma_{jl} I_p}{h\nu_p} + \sum_{j>i} \frac{N_j \sigma_{ji} I_p}{h\nu_p} + \sum_{j<l} \frac{N_j \sigma_{jl} I_0}{h\nu_0} \right.$$

$$\left. + \sum_{j>i} \frac{N_j \sigma_{ji} I_0}{h\nu_0} + \sum_{j>i} N_j A_{ji} + N_j \Gamma_j^{nr} \right].$$

---

*To see some of the extra approximations that need to be invoked in a time-dependent analysis, see reference [29].

40 OPTICAL FIBRE LASERS AND AMPLIFIERS

The first pair of terms in both square brackets refer to absorption and stimulated emission of pump photons, whereas the second pair of terms correspond to absorption and stimulated emission of lasing photons. The fifth term in both cases represents decays due to spontaneous emission and the last term non-radiative decays to the next lower lying level. This equation is the most general that can be written and significant simplifications will be possible for most of the energy levels of interest.

In a silica host where the decay rate from the vast majority of excited levels is dominated by non-radiative processes, the population dynamics can be vastly simplified. To demonstrate this, consider the example shown in Figure 2.10. The laser transition is between level $|2\rangle$ and level $|j\rangle$. Pump radiation not only excites ions from the ground state to state $|n\rangle$ but as a result of an accidental degeneracy, there is the possibility that ions might be excited out of level $|m\rangle$ into level $|p\rangle$. The upper laser level $|2\rangle$ is well separated from the next lower lying level, so that Γ_2^{nr} can be taken as being zero. For the excited levels $|n\rangle$, $|m\rangle$, $|k\rangle$ and $|j\rangle$, the non-radiative decay rate is very much larger than the spontaneous decay rate, and so fluorescence from these levels can be ignored. In addition to non-radiative decay to level $|m\rangle$, an ion in level $|n\rangle$ can also return to the ground state via stimulated emission at the pump wavelength. However, for typical pump intensities of around $10\,\mathrm{mW}/\mu\mathrm{m}^2$ and with cross-

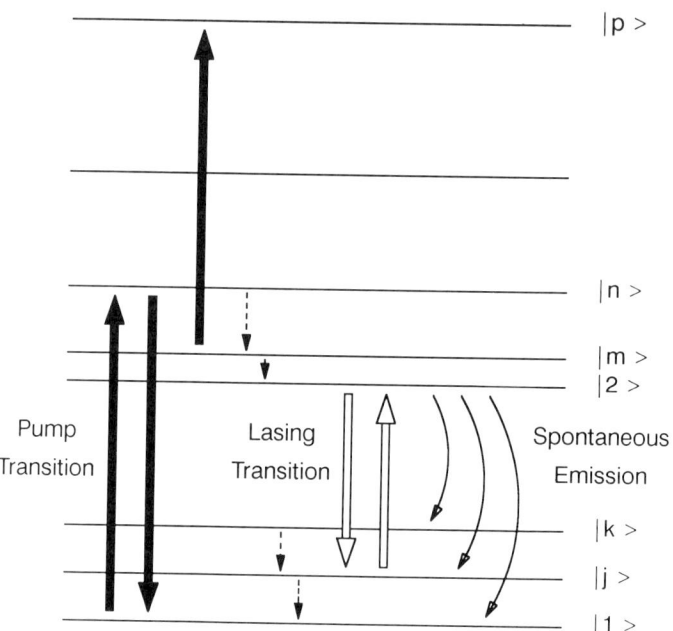

Figure 2.10 Schematic rare-earth ion energy level scheme showing various radiative and non-radiative decay processes.

sections of the order of 10^{-20} cm^{-2}, pumping rates work out to be around 10^3 s^{-1}. This should be compared with Γ^{nr} which will be somewhere around 10^6 s^{-1} or larger. So at sensible pump powers, the non-radiative decay probability out of level $|n\rangle$ is very much greater than the stimulated emission probability. We can therefore ignore pump stimulated emission from the populations dynamics and write

$$\dot{N}_n = \frac{N_1 \sigma_{1n} I_p}{h\nu_p} - N_n \Gamma_n^{nr}.$$

It is also possible to ignore excited-state absorption of pump photons from the state $|m\rangle$ for the same reason, and then write

$$\dot{N}_m = N_n \Gamma_n^{nr} - N_m \Gamma_m^{nr}.$$

The equations for the remaining levels are then simply

$$\dot{N}_2 = N_m \Gamma_m^{nr} + \frac{N_j \sigma_{j2} I_0}{h\nu_0} - \frac{N_2 \sigma_{2j} I_0}{h\nu_0} - N_2(A_{2k} + A_{2j} + A_{21})$$

$$\dot{N}_k = N_2 A_{2k} - N_k \Gamma_k^{nr}$$

$$\dot{N}_j = N_2 A_{2j} + N_k \Gamma_k^{nr} + \frac{N_2 \sigma_{2j} I_0}{h\nu_0} - \frac{N_j \sigma_{j2} I_0}{h\nu_0} - N_j \Gamma_j^{nr}$$

and

$$N_{\text{total}} = N_1 + N_j + N_k + N_2 + N_m + N_n.$$

In the steady state, all the time derivatives can be set to zero and so the following results can be derived:

$$\frac{N_1 \sigma_{1n} I_p}{h\nu_p} = N_n \Gamma_n^{nr} \tag{2.15a}$$

$$N_n \Gamma_n^{nr} = N_m \Gamma_m^{nr} \tag{2.15b}$$

$$N_m \Gamma_m^{nr} = \frac{N_2 \sigma_{2j} I_0}{h\nu_0} - \frac{N_j \sigma_{j2} I_0}{h\nu_0} + N_2(A_{2k} + A_{2j} + A_{21}) \tag{2.15c}$$

$$N_2 A_{2k} = N_k \Gamma_k^{nr} \tag{2.15d}$$

$$N_j \Gamma_j^{nr} = \frac{N_2 \sigma_{2j} I_0}{h\nu_0} - \frac{N_j \sigma_{j2} I_0}{h\nu_0} + N_2 A_{2j} + N_k \Gamma_k^{nr}. \tag{2.15e}$$

By eliminating various pairs of terms we can arrive at the equation

$$\frac{N_1 \sigma_{1n} I_p}{h\nu_p} = \frac{N_2 \sigma_{2j} I_0}{h\nu_0} - \frac{N_j \sigma_{j2} I_0}{h\nu_0} + N_2 \Gamma_2.$$

But from Equations (2.15d) and (2.15e), $N_j \ll N_2$ and so

$$\frac{N_1 \sigma_{1n} I_p}{h\nu_p} = \frac{N_2 \sigma_{2j} I_0}{h\nu_0} + N_2 \Gamma_2.$$

The original multilevel problem has now been reduced to a very much simpler two-level problem involving only those levels from which the non-radiative decay rates are small. As a general rule when modelling rare-earth ion-doped silica lasers, stimulated emission of the pump radiation only needs to be considered when pumping directly into a metastable level (e.g. as in the 1.48 μm pumped Er^{3+} fibre amplifier) and similarly excited state absorption is only a problem from long-lived levels. If the lower level of a four-level system empties very quickly then we need not consider the population there, at least for CW situations. Because the vast majority of silica fibre laser situations simplify to two-level problems, this makes modelling of these systems particularly simple. There are relatively few parameters which need to be known; basically only the pump absorption and maybe emission cross-sections, the lasing emission and absorption cross-sections and the fluorescence lifetime of the laser transition. The same model, but with different values for the cross-sections, can be applied immediately to many different laser transitions, even though they may have vastly differing energy level schemes.

The population dynamics in rare-earth ion-doped fluoride fibres can be significantly more complicated. In principle all the levels shown in Figure 2.10 will fluoresce, and so a set of six equations will need to be set up. Fluorescence branching ratios to all lower levels will now need to be considered. Since sizeable fractions of the total ion population may accumulate in any one of these excited levels, it should be obvious that absorption of either pump or signal photons may well pose a problem. Even a relatively simple situation with no excited-state absorptions is complicated [23]. Also, it should not be overlooked that far more experimental input is needed in modelling a fluoride system in the form of branching ratios, non-radiative decay rates (probably derived from Judd–Ofelt theory), etc., in order to attempt a realistic model of the excited level populations.

2.5.3 *Propagation equations with gain and loss*

In a doped fibre, not only do the pump and signal beams vary in intensity transverse to their direction of propagation, but also the rare-earth ion density need not be uniform across the fibre core. As a consequence, the local gain and absorption coefficients for the various beams in the fibre will vary radially. When considering the propagation of the pump, signal and noise beams down the fibre, there are two approaches that can be taken. The simpler of these is to take the infinite plane wave approximation where one effectively ignores any transverse variations of the pump, signal and rare-earth ion profiles. Propagation equations are set up for the case of infinite plane waves propagating through a uniformly doped medium, and the core size is introduced basically as a normalising factor between the intensity of the infinite plane wave and the power in the fibre. The alternative to this is to calculate the pump and lasing mode profiles in the fibre and to include these and the rare-earth ion density profile into the calculations.

2.5.3.1 *Infinite plane wave approximation.* Consider an infinite plane wave propagating through a uniformly doped medium. The difference in pump (or signal) intensity between planes z and $z + \Delta z$ is given by

$$I(z + \Delta z) - I(z) = \begin{pmatrix} \text{Number of} \\ \text{stimulated emissions} \\ \text{per unit volume} \end{pmatrix} - \begin{pmatrix} \text{Number of} \\ \text{absorptions} \\ \text{per unit volume} \end{pmatrix} \times \begin{pmatrix} \text{Photon} \\ \text{energy} \end{pmatrix} \times \Delta z$$

$$- \alpha I(z) \Delta z \qquad (2.16)$$

where α is the background loss of the medium. The number of absorptions and the number of stimulated emissions per unit volume are given by expressions of the form $N_i \sum \sigma_{ij}$ summed over all the levels in which there is significant population. For the specific example of the laser transition in the rare-earth ion-doped fluoride glass considered in Section 2.5.2, in which pump stimulated emission and pump excited-state absorption are present, the pump propagation equation would be

$$I_p(z + \Delta z) - I_p(z) = -(N_1 \sigma_{1n} - N_n \sigma_{n1} - N_m \sigma_{mp}) I_p(z) \Delta z - \alpha_p I_p(z) \Delta z.$$

This equation is in terms of intensity, i.e. the number of photons per unit area per unit time $\times h\nu$, but can readily be converted to pump powers by multiplying by an effective core area. This plane wave approximation amounts to assuming square pump and signal intensity mode shapes and a square rare-earth ion profile, and should apply well to highly multimoded fibres [4]. A similar equation exists to describe the signal beam propagation through this medium

$$I_0(z + \Delta z) - I_0(z) = \pm (N_2 \sigma_{2j} - N_j \sigma_{j2}) I_0(z) \Delta z - \alpha_0 I_0(z) \Delta z$$

where the choice of sign depends on whether the signal and pump beams are co-propagating or counter-propagating.

When considering the noise beams propagating within the fibre, each frequency across the laser transition needs to be considered separately. For each noise frequency component, two propagation equations (for the forward and backward directions down the fibre) need to be set up, similar to Equation (2.16) but with an extra term added to the right-hand side corresponding to the fluorescence added to the noise beam in the frequency range ν to $\nu + \Delta \nu$ between the planes z and $z + \Delta z$.

$$I_n^\nu(z + \Delta z) - I_n^\nu(z) = \text{Gain} \times I_n^\nu \Delta z$$
$$+ \begin{pmatrix} \text{Fluorescence generated in the length} \\ \Delta z \text{ that is guided by the fibre} \end{pmatrix} \Delta z.$$

To generate the full ASE spectrum of a small signal amplifier, or the output spectrum of a superfluorescent laser source, such equations have to be set up and solved for each of the frequency components across the laser transition (see reference [27]).

2.5.3.2 *Guided modes in lossless fibres.* A full theoretical analysis of the guided modes in optical fibres is beyond the scope of this book [38]; instead, the basic properties of fibre modes will now be summarised here. Consider the case of a perfect cylindrically symmetric optical fibre consisting of a core of diameter $2a$ and with a refractive index n_2 surrounded by a cladding of infinite extent of a lower refractive index n_1. A solution of Maxwell's equations in this structure will produce a finite number of guided solutions, or modes of the fibre. Although exact mathematical solutions exist for the perfect fibre described above, it was realised that in the limit $n_2 - n_1 \ll 1$ [39], these exact solutions simplify to a far more amenable set of solutions, the so-called linearly polarised, or LP-modes. One key fibre parameter, its V value, is defined by the expression

$$V = \frac{2\pi a}{\lambda}(n_2^2 - n_1^2)^{1/2} \approx \frac{2\pi a}{\lambda}(2n\Delta n)^{1/2}. \quad (2.17)$$

All the properties of the fibre modes are determined by the value of this parameter. In particular, the number of guided modes supported by the fibre depends on its V value. For V values less than 2.405, only a single transverse mode, the LP_{01} mode, is supported by the fibre.

For fibre lasers and amplifiers, our interest is in the radial and azimuthal intensity profiles of these guided modes. The normalised intensity profiles, $f(r, \theta)$ for the $LP_{\nu\mu}$ modes are

$$f(r,\theta) = \begin{cases} AJ_\nu^2\left(\frac{ur}{a}\right)\cos^2\nu\theta & r < a \\ A\left[\frac{J_\nu(u)}{K_\nu(w)}\right]^2 K_\nu^2\left(\frac{wr}{a}\right)\cos^2\nu\theta & r > a \end{cases} \quad (2.18)$$

where A is chosen so that

$$\int_{r=0}^{\infty}\int_{\theta=0}^{2\pi} f(r,\theta)r\,dr\,d\theta = 1.$$

The normalised transverse propagation constants, u and w, are the solutions of the following characteristic equation*

$$\frac{uJ_{\nu+1}(u)}{J_\nu(u)} = \frac{wK_{\nu+1}(w)}{K_\nu(w)} \quad (2.19)$$

---
*The LP subscript μ indicates the particular root of Equation (2.18) to which that mode corresponds.

where

$$u^2 + w^2 = V^2.$$

In the majority of cases, the aim is to work with single-mode fibres. The obvious resemblance of the LP_{01} mode profile to a Gaussian beam has prompted many authors to use a Gaussian mode representation rather than the Bessel function solutions [40, 41]. Although this approximation does not really simplify the mathematics, it often reduces the amount of computing time required.

2.5.3.3 Overlap of the pump and signal modes with the rare-earth ion profile. The populations in all the excited rare-earth ion levels at a position (r, θ, z) depend on the pump and signal intensities at that same point, hence the local gain and absorption coefficients vary across the pump and signal beams. For instance, since the pump intensity is generally largest on the fibre axis, it might be imagined that the central part of the signal mode would experience greater gain than the outer part of that mode so distorting the fibre mode as it propagated down the fibre. Furthermore, there is no reason why the doping density should be uniform across the fibre core. There is indeed a significant interest in the use of fibres where the dopant ions are confined to the central region of the fibre core. In this case there is most certainly a greater local gain coefficient on the fibre axis. The question then arises: How does this non-uniform absorption or amplification across the pump and signal beams distort the respective mode shapes from those expected for a lossless fibre? Because the oscillator strengths of the rare-earth ion transitions are so small, the $1/e$ gain and absorption lengths are consequently quite long – typically 10 cm to 10 m. Such effects are therefore only a minute perturbation to Maxwell's equations describing the propagation of the modes within the lossless guide. It is because the gain and absorption effects are so small in comparison to the lengths associated with the guiding properties of the fibre, which are determined by the core-cladding refractive index difference Δn and the wavelength, that the fibre modes are essentially unperturbed by the presence of this non-uniform gain or absorption. The modes can still be perfectly well represented by the Bessel function solutions or Gaussian beams as previously described. The nett power absorbed or added to a beam between planes z and $z + \Delta z$ is then simply the gain or absorption averaged across the beam profile, i.e.

$$P(z + \Delta z) - P(z) = \int_{r=0}^{\infty} \int_{\theta=0}^{2\pi} \left[\sum_i N_i(r)\sigma - \alpha \right] I(r, \theta) r \, dr \, d\theta \quad (2.20)$$

where $I(r, \theta)$ is the intensity profile of the mode as given by Equation (2.18). The effective gain and absorption coefficients at any plane z down the fibre are therefore simply the local gain and absorption coefficients averaged across the respective mode profile.

Obviously inclusion of these transverse variations of the pump, signal and dopant ion profiles into a model dramatically increases its complexity, and it is only natural to ask whether this increased degree of sophistication of the model is worth the extra effort. One should not lose track of the fact that the modelling is being performed only as an aid to understanding trends in device performance. There are situations where it has been found that the trend under investigation does not depend at all, or maybe only very weakly, on the precise details of the overlaps between the pump and the signal modes and the rare-earth ions – the slope efficiency of a fibre laser oscillator or the extraction efficiency of a power amplifier are examples of this. In such cases, the use of a far simpler model would achieve the same level of understanding of the problem. There are indeed some situations when the plane wave approximation can actually yield analytic results [31] showing explicitly how various parameters affect device performance. A small amount of thought will show that, in the vast majority of cases, there is probably no need to include transverse profiles. It is really only when questions need to be asked regarding the pumping efficiency of different pump modes, and when looking at the benefits of confining the rare-earth ions to the centre of the fibre core [28], that a full calculation needs to be undertaken.

2.5.3.4 *Solution of the propagation equations.* The typical pump intensities at which the ground-state ion populations start to be depleted are of the order of a few milliwatts per square micron. In longitudinally pumped bulk lasers such pump intensities are rarely encountered and so, to a good approximation, the pump power can be taken as decaying exponentially along the length of the gain medium. In single-mode fibre devices, however, only a few milliwatts of pump power are needed to bleach the ground-state absorption and therefore it is imperative that the pump propagation equation be solved explicitly. There are two exceptions to this which occur when the upper laser level is the only metastable level of the ion. Firstly, in a four-level laser with a high-Q cavity, lasing can be achieved with only a very small fraction of the ion population in the upper laser level and hence most of the rare-earth ions are still in the ground state. Secondly, if excited-state absorption of pump photons is present, and if the excited-state absorption cross-section is equal to the ground-state absorption cross-section, then the fibre absorption coefficient will be the same independent of which energy levels the rare-earth ions are occupying.

The simplest active fibre device to model is the small-signal amplifier. The definition of 'small signal' implies that the populations in the excited levels are unperturbed by the signal fluxes through the amplifier. As a direct result of this, the pump propagation equation is decoupled from the signal equation and can be integrated relatively simply along the length of the device. Inclusion of noise into such a model is also simple but only if the total noise power is small and does not start to perturb the population inversion in the amplifier. For an amplifier in this small-signal regime, even if there is background loss in the fibre, the signal propagation equations and the overall amplifier gain are the same whether the pump and signal beams are co-propagating or counter-propagating down the fibre. The noise spectra emanating from the two ends of a three-level fibre amplifier are not the same due to the different inversion levels at the two ends of the device. Hence the relative direction of pump and signal propagation in a three-level amplifier affects the signal-to-noise ratio [34].

As the signal intensity is increased, at some point the signal at the output of the amplifier will be sufficiently intense to affect the population dynamics. The pump and signal propagation equations are now genuinely coupled and must be solved simultaneously. Even so, a solution of both the co-propagating and counter-propagating cases is still relatively easy because the two propagation equations can be integrated in the *same* direction down the fibre. It is when the noise power from the amplifier starts to perturb the populations that the problem becomes far more complex in that the noise from one end of the amplifier now affects the pump (and signal) propagation at the other end of the amplifier and vice versa. To solve this problem, some sort of iterative procedure is needed to give a self-consistent set of solutions which satisfy the amplifier input conditions.*

The input (or boundary) conditions for the amplifier problem are simply the pump and signal intensities launched into the fibre. For an oscillator with reflectors at both ends of the cavity, there are two boundary conditions to be satisfied by the intracavity lasing fluxes (Figure 2.11), namely

$$I_0^{(+)}(0) = R_1 I_0^{(-)}(0) \quad \text{and} \quad I_0^{(-)}(L) = R_2 I_0^{(+)}(L) \tag{2.21}$$

in addition to the pump power launched into the fibre. To determine the output power from one of the mirrors for a given input pump power requires a certain amount of iteration to find a self-consistent solution. However, it should be noted that the gain in the oscillator is isotropic, i.e.

$$\frac{1}{I_0^{(+)}} \frac{dI_0^{(+)}}{dz} = -\frac{1}{I_0^{(-)}} \frac{dI_0^{(-)}}{dz}$$

and this equation when integrated gives

$$I_0^{(+)} I_0^{(-)} = \text{constant}. \tag{2.22}$$

In addition, by combining Equations (2.21) and (2.22), the following general result for any oscillator is obtained

$$\text{Fraction of the total output power from the laser emitted through mirror 1} = \left[1 + \frac{(1-R_2)}{(1-R_1)}\left(\frac{R_1}{R_2}\right)^{1/2}\right]^{-1}. \tag{2.23}$$

* A superfluorescent laser is really just a high gain amplifier with no input signal where the noise power has become so large that it strongly saturates the amplifier gain and is, usually, a sizeable fraction of the pump power launched into the device.

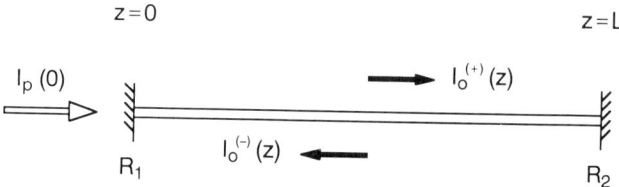

Figure 2.11 Schematic diagram of a laser oscillator.

These results can often be used to simplify the problem at hand, and reduce the amount of computing time required.

2.5.4 *Conclusions*

Modelling of active fibre devices is often a complex task, even though the basic principles behind a model are fairly simple. Although a 'complete' model of a fibre device could be envisaged, incorporating many of the ideas described in this section, it is important to emphasise that modelling devices simply to obtain a 'good agreement' with a set of experimental data is in itself of little benefit. Indeed, many of the relevant device parameters (e.g. cross-sections, branching ratios, rare-earth ion-doping densities, dopant distributions across the fibre core, etc.) are not known very accurately, and the homogeneity of the rare-earth ion transitions in glasses must always be questioned. It should always be remembered, therefore, that the real motivation behind device modelling is to understand trends in the device performance when different parameters are varied, which should then allow optimisation of the required properties of an oscillator or amplifier.

References

1. A. Yariv, *Optical Electronics*, Holt-Saunders (1985).
2. O. Svelto, *Principles of Lasers*, Plenum Press (1982).
3. K. Shimado, *Introduction to Laser Physics*, Springer Series in Optical Sciences, Springer-Verlag (1984).
4. M.J.F. Digonnet and C.J. Gaeta, 'Theoretical analysis of optical fiber laser amplifiers and oscillators', *Appl. Opt.* **24** (1985) 333.
5. D.C. Hanna, M.J. McCarthy and P.J. Suni, 'Thermal considerations in longitudinally pumped fiber and miniature bulk lasers', *Proc. Soc. Photo-Opt. Instrum. Eng.* **1171** (1989) 160.
6. R. Wyatt, 'High power broadly tunable erbium doped silica fibre laser', *Electron. Lett.* **25** (1989) 1498.
7. M.C. Brierley and M.H. Hunt, 'Efficient semiconductor pumped fluoride fiber lasers', *Proc. Soc. Photo-Opt. Instrum. Eng.* **1171** (1989) 157.
8. J.R. Armitage, R. Wyatt, B.J. Ainslie and S.P. Craig-Ryan, 'Highly efficient 980 nm operation of an Yb^{3+} doped silica fibre laser', *Electron. Lett.* **25** (1989) 298.
9. K. Liu, M.J.F. Digonnet, K. Fesler, B. Y. Kim and H.J. Shaw, 'Broadband diode-pumped fibre laser', *Electron. Lett.* **24** (1988) 838.

10. D.C. Hanna, I.R. Perry, R.G. Smart, P.J. Suni and A.C. Tropper, 'Efficient superfluorescent emission at 974 nm and 1040 nm from an Yb-doped fibre', *Opt. Commun.* **72** (1989) 230.
11. C.G. Atkins, J.F. Massicott, J.R. Armitage, R. Wyatt, B.J. Ainslie and S.P. Craig-Ryan, 'High gain, broad spectral bandwidth erbium doped fibre amplifier pumped near 1.5 μm', *Electron. Lett.* **25** (1989) 910.
12. J.F. Massicott, R. Wyatt, B.J. Ainslie and S.P. Craig-Ryan, 'Efficient, high power, high gain Er^{3+} doped silica fibre amplifier', *Electron Lett.* **26** (1990)
13. S. Zemon, G. Lambert, W.J. Miniscalco, L.J. Andrews and B.T. Hall, 'Characterisation of Er^{3+} doped glasses by fluorescence line narrowing', *Proc. Soc. Photo-Opt. Instrum. Eng.* **1171** (1989) 219.
14. E. Desurvire and J.R. Simpson, 'Evaluation of $^4I_{15/2}$ and $^4I_{13/2}$ Stark level energies in erbium-doped aluminosilicate glass fibers', *Opt. Lett.* **15** (1990) 547.
15. B.R. Judd, 'Optical absorption intensities of rare-earth ions', *Phys. Rev.* **127** (1962) 750.
16. G.S. Ofelt, 'Intensities of crystal spectra of rare-earth ions', *J. Chem. Phys.* **37** (1962) 511.
17. M.J. Weber, 'Probabilities for radiative and non-radiative decay of Er^{3+} in LaF_3', *Phys. Rev.* **15** (1967) 262.
18. R. Reisfeld and Y. Eckstein, 'Dependence of spontaneous emission and non-radiative relaxations of Tm^{3+} and Er^{3+} on glass host and temperature', *J. Chem. Phys.* **63** (1975) 4001.
19. A.A. Kaminskii, *Laser Crystals*, Springer Series in Optical Sciences, Springer-Verlag (1981).
20. F. Tong, W.P. Risk, R.M. MacFarlane and W. Lenth, '551 nm diode laser pumped upconversion laser', *Electron Lett.* **25** (1989) 1389.
21. R.S. Quimby and W.J. Miniscalco, 'Continuous-wave lasing on a self-terminating transition', *Appl. Opt.* **28** (1989) 14.
22. W.L. Barnes, S.B. Poole, J.E. Townsend, L. Reekie, D.J. Taylor and D.N. Payne, 'Er^{3+}–Yb^{3+} and Er^{3+} doped fiber lasers', *J. Light. Technol.* **7** (1989) 1461.
23. R.S. Quimby, 'Output saturation in fiber lasers', *Appl. Opt.* **29** (1990) 1268.
24. E.W. Duczynski, G. Huber, V.G. Ostroumou and I.A. Shcherbakov, 'CW double cross pumping of the $^5I_7 \rightarrow {}^5I_8$ laser transition in Ho^{3+}-doped garnets', *Appl. Phys. Lett.* **48** (1986) 1562.
25. B.J. Ainslie, S.P. Craig and S.T. Davey, 'The absorption and fluorescence spectra of rare-earth ions in silica based monomode fibre', *J. Light. Technol.* **6** (1988) 287.
26. S.T. Davey and P.W. France, 'Rare-earth doped fluorozirconate glasses for fibre devices', *Brit. Telecom Technol. J.* **7** (1989) 58.
27. E. Desurvire and J.R. Simpson, 'Amplification of spontaneous emission in erbium doped single mode fibers', *J. Light. Technol.* **7** (1989) 835.
28. J.R. Armitage, 'Three-level fiber laser amplifier: a theoretical model', *Appl. Opt.* **27** (1988) 4831.
29. E. Desurvire, 'Analysis of transient gain saturation and recovery in erbium doped fiber amplifiers, *IEEE Phot. Tech. Lett.* **1** (1989) 196.
30. E. Desurvire, 'Analysis of Er^{3+} doped fiber amplifiers pumped in the $^4I_{15/2} - {}^4I_{13/2}$ band', *IEEE Phot. Tech. Lett.* **1** (1989) 293.
31. J.R. Armitage, 'Spectral dependence of the small-signal gain around 1.5 μm in erbium doped silica fiber amplifiers', *IEEE J. Quant. Electron.* **26** (1990) 423.
32. M.J.F. Digonnet and K. Liu, 'Analysis of a 1060 nm $Nd:SiO_2$ superfluorescent fiber laser', *J. Light. Technol.* **7** (1989) 1009.
33. M.J.F. Digonnet, 'Theory of superfluorescent fiber lasers', *J. Light. Technol.* **4** (1986) 1631.
34. R. Olshansky, 'Noise figure for erbium doped optical fibre amplifiers', *Electron. Lett.* **24** (1988) 1363.
35. E. Desurvire, J.R. Simpson and P.C. Becker, 'High gain erbium doped travelling wave fiber amplifier', *Opt. Lett.* **12** (1987) 888.
36. P.R. Morkel and R.I. Laming, 'Theoretical modelling of erbium doped fibre amplifiers with excited state absorption', *Opt. Lett.* **14** (1989) 1062.
37. A. Bjarklev, S.L. Hansen and J.H. Povlsen, 'Large signal modelling of an erbium doped fibre amplifier', *Proc. Soc. Photo-Opt. Instrum. Eng.* **1171** (1989) 118.
38. D. Marcuse, *Theory of Dielectric Optical Waveguides*, Academic Press, New York (1974).
39. D. Gloge, 'Weakly guiding fibers', *Appl. Opt.* **10** (1971) 2252.
40. W.T. Anderson, 'Consistency of measurement methods for the mode field radius in a single mode fiber', *J. Light. Technol.* **2** (1984) 191.
41. D. Marcuse, 'Loss analysis of single mode fiber splices', *Bell Syst. Technol.* **56** (1977) 703.

3 Glass structure and fabrication techniques

S.P. CRAIG-RYAN and B.J. AINSLIE

3.1 Introduction

Before examining fibre fabrication methods it is important to review the basics of glass structure and to explore the most efficient materials from which to make doped fibres. Fibres can be made out of plastics, liquids and glasses, all of which can be doped, but the most suitable is glass since it offers a very low loss transmission medium which can be made in long lengths to give a continuous guiding medium. This chapter is limited to glasses and concentrates on oxides primarily based on silicon. Elements that can be incorporated into the glass structure to give the important fluorescing properties necessary for some applications are also considered. The rare-earth and transition metals, with their unique property of partially filled inner electron levels, are discussed here. Firstly, each of these groups of materials is considered separately and then doped glass properties are examined. Finally, the different technologies available for producing doped fibres are assessed and some of the results of this work discussed.

3.2 Doped glass structures and properties

3.2.1 Glass properties

What is a glass? A glass is an amorphous material that has been formed from materials cooled so rapidly from the liquid phase that crystallisation has been prevented. Glasses have short-range order and form a three-dimensional matrix, but lack the uniformity, symmetry and structure of the crystalline material, having no long-range periodicity. Glass is therefore in a state somewhere between that of a crystalline solid and a liquid.

Looking at the materials that can form glasses (Figure 3.1) it is found that by far the most important glass-forming elements are non-metals in groups IIIA, IVA and VA of the periodic table combining with oxygen to form oxide glasses. These elements are important in forming glasses, because they are all capable of forming a three-dimensional network with oxygen, giving very strong covalent bonds that give glasses their characteristic properties. For

GLASS STRUCTURE AND FABRICATION TECHNIQUES 51

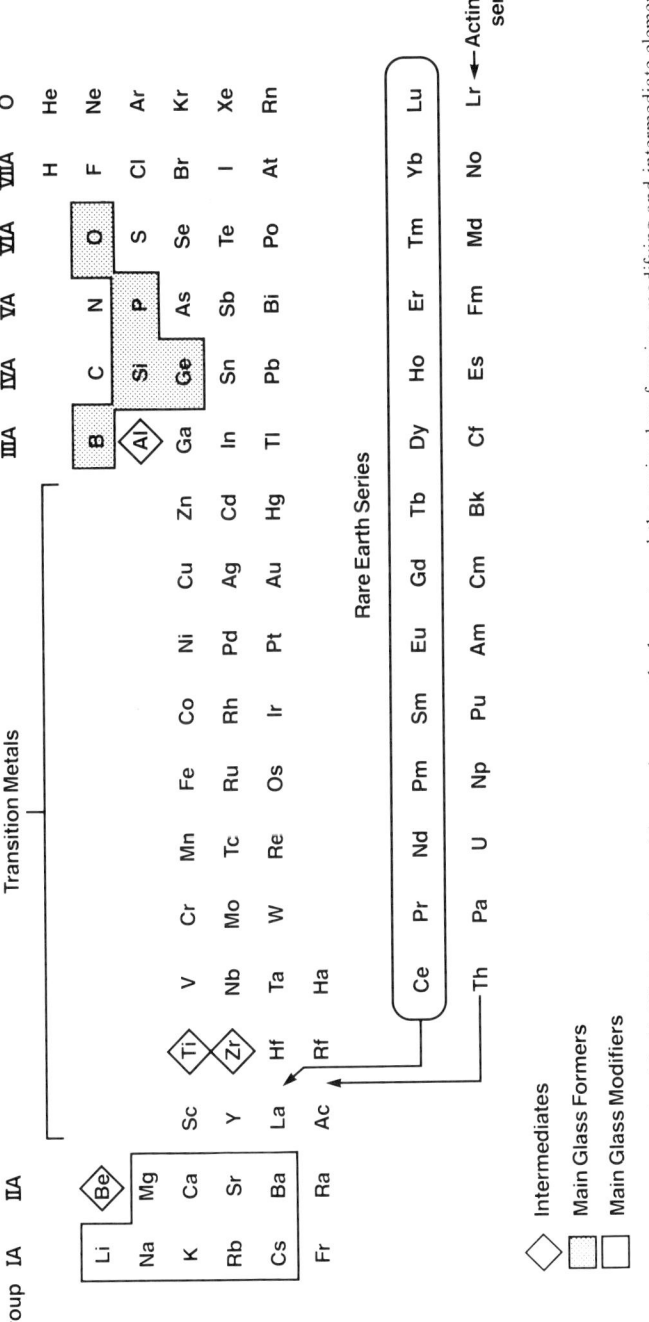

Figure 3.1 Periodic table highlighting the transition and rare-earth elements and the main glass-forming, modifying and intermediate elements.

present purposes only this group will be considered, although other materials will form glasses – for example, chalcogenides, halides, nitrates and some organic compounds [1]. Of these the fluorides are another important group of glasses that will be considered later (see Chapter 8).

When considering the properties of glasses one thing that is immediately apparent is that glasses have no precise melting point; instead, over a large temperature range, they soften and their viscosity changes. It is this key property that allows the formation of a long pathlength waveguide, i.e. a long length of fibre. One way of comparing different glasses is to define certain temperatures with set viscosity, η [2,3] and to see how these temperatures change with different host glasses. These temperatures are:

Strain temperature	$\eta = 10^{13.5}$ N s m^{-2}
Anneal temperature	$\eta = 10^{12.5}$ N s m^{-2}
Softening temperature	$\eta = 10^{6.5}$ N s m^{-2}
Working temperature	$\eta = 10^{3}$ N s m^{-2}
Fibre drawing temperature	$\eta = 10^{2} - 10^{4}$ N s m^{-2}.

The viscosity is a very important property since if glasses are to form in preference to the lower energy, more stable, crystalline state then the rate of crystallisation and the rate of nucleation must be very slow. Glasses therefore need to be cooled very rapidly, especially through the temperature range of nucleation and crystal growth. Doremus [1] gives an upper limit for the crystallisation rate of 10^{-4} cm s^{-1}, explaining that at this rate a glass would need to be cooled through the maximum crystallisation range in less than 10^{-2} s if crystals of > 100 Å are to be prevented from forming. Materials with a high viscosity near their melting point generally have low crystallisation rates and are therefore more likely to form glasses. One of the lowest crystallisation rates is for silica glass, which has a very high viscosity compared to other glass-forming materials for the same temperature.

If the oxide glasses are now considered in more detail, the simplest oxides are made up from just two elements: a glass-forming atom (Figure 3.1) and oxygen. Multicomponent structures can be made by modifying the simple structure and either introducing a modifier atom to alter the basic structure or adding an element to directly replace some of the glass-forming atoms in the matrix. The question of oxide structure was considered by Zachariasen [4], who suggested a number of rules regarding the spatial arrangement of glass forming oxides, thus allowing a prediction of the oxides that tend to form glasses:

1. An oxygen atom cannot be linked to more than two glass-forming atoms.
2. The coordination number of glass-forming ions must be small.
3. The polyhedral formed by the oxygen must only share corners and not edges or faces.
4. The polyhedra must form a three-dimensional network.

Using this traditional view of glass formation we see that the simplest glass-forming materials satisfying these conditions are SiO_2, GeO_2, P_2O_5 and B_2O_3, which are the main glass-forming elements shown in Figure 3.1. More contemporary approaches to glass formation have been postulated [5,6] but the above explanation is adequate for the present purpose.

3.2.2 Bonding in glasses

If we look at one of the simplest and well-known glass structures, which is vitreous silica (SiO_2), we see from Figure 3.2(a) that the silicon atom in the centre is covalently bonded to four oxygen atoms. The oxygen atoms are arranged tetrahedrally around the silicon atom, and each tetrahedron is bonded to four others via silicon–oxygen bonds, thus obeying Zachariasen's rule and only sharing corners. This structure is very similar to the crystalline form of silica, crystobalite (Figure 3.2(b)), but the structure, being slightly more random, leads to the lack of long-range periodicity or symmetry. The oxygen atoms in the lattice are predominantly bridging (i.e. linking one tetrahedron to another), however, non-bridging groups may be formed which allow ionic bonding with other species.

The silicon–oxygen glass structure is disrupted by the introduction of network modifiers; these ions tend to be from group IA or IIA of the periodic table. When these elements are added to the glass, the silicon–oxygen structure is opened up (Figure 3.3) lowering the density of the glass, weakening the bond strength and lowering the fusion temperature and the

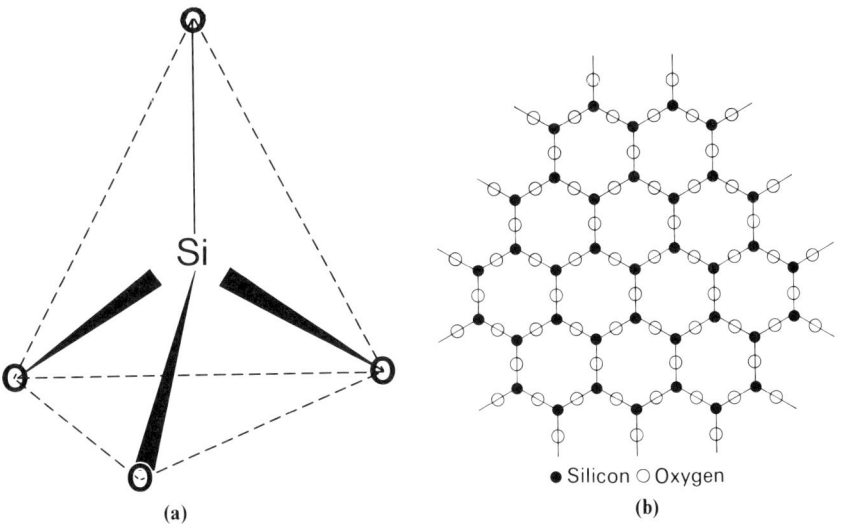

Figure 3.2 (a) A schematic of three-dimensional tetrahedral arrangement in fused silica glass; (b) a diagrammatic two-dimensional matrix of silicon and oxygen atoms, as found in crystobalite.

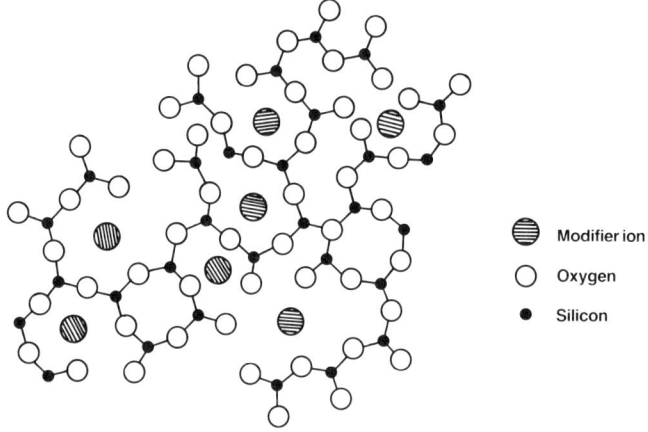

Figure 3.3 A schematic showing the silica glass matrix disrupted by the addition of modifier ions.

viscosity of the glass. In general, as long as the ratio of $SiO_2:RO$ or R_2O (where R is the metal ion) is < 1:1 then the basic glass structure is maintained. Since each tetrahedron can still be linked to at least three others, the random network will still be present [1]. Some of the silicon–oxygen bridging bonds will, however, have been broken thus increasing the number of non-bridging oxygens present, and the spare negative charges on the oxygen ions are used to compensate for the positive charge of the metal ion. The addition of network modifiers is important to allow processing of the glass at workable temperatures; as pure silica requires very high temperatures for working [7].

Other elements, such as germanium, aluminium, phosphorus or boron, may substitute a number of silicon sites in fused silica and we shall consider these substituted glasses as they have been the most important to date in rare-earth doped fibre work.

If silicon is replaced by germanium, the tetrahedral structure is retained and its properties are similar to fused silica glass [1]. In telecommunications fibre, germanium is commonly used to dope the silica glass during the core deposition as it gives an increase in the refractive index over that of the cladding glass and enables a wide range of guiding structures to be made [8]. Aluminium oxide cannot easily form a glass on its own, as it is an intermediate type of element [9]. It can be incorporated into silica glass, replacing some silica ions, but as it is a trivalent ion some charge compensation is necessary to restore the charge equilibrium. In fibre for telecommunications, aluminium has been used as an alternative to germanium for increasing the refractive index of the core [10]. As will be seen in the next section, the rare-earth ions are positively charged, and on introduction into an Al_2O_3–SiO_2 glass system they tend to preferentially congregate near aluminium sites [11]. This can be viewed as aluminium charge deficiency being equilibrated by the rare-earth ion.

The phosphate glass system is also based on tetrahedra, with four oxygen atoms being bonded to each phosphorus atom [1]; however, since phosphorus has a valency of 5, one oxygen has a double bond to the phosphorus, as shown in Figure 3.4. This reduces to three the number of tetrahedra that can bond to any one tetrahedron. Phosphate glasses are well known, however, and are less dense than fused silica. Phosphorus is very important in fibre making since it is commonly used in conjunction with fluorine to dope the silica, giving a cladding material with the same refractive index as fused silica but of lower viscosity, allowing a lowering of the process temperature [12]. The viscosity of silica is very effectively lowered by small additions of phosphorus to the network.

With borate glasses, the basic building block is B_2O_3. Since boron is trivalent it forms only three bonds to neighbouring oxygen atoms and X-ray work on borate glasses shows that planar rings are formed of alternate boron

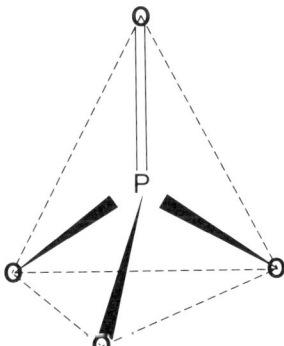

Figure 3.4 A diagrammatic form of a three-dimensional phosphorus and oxygen tetrahedron showing the double bond between one oxygen and the phosphorus.

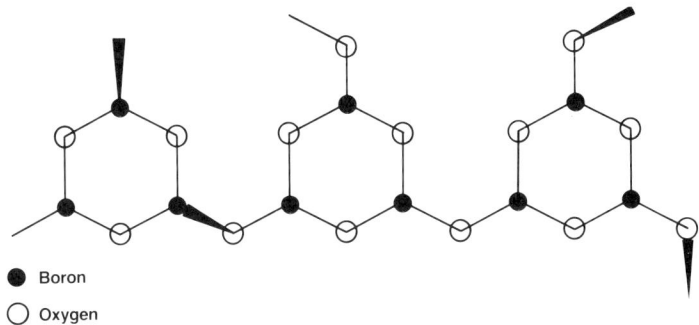

Figure 3.5 A two-dimensional arrangement of the boron and oxygen atoms found in borate glass.

and oxygen atoms which are linked together via bridging oxygen atoms to form the three-dimensional lattice (Figure 3.5) [1, 3]. The addition of B_2O_3 to fused silica decreases the refractive index and lowers the processing temperature. It has been used as a cladding glass, but as it has a very short phonon edge at about 7 μm it is not now commonly used for long wavelength telecoms applications [7, 13].

We shall now turn our attention to the rare-earth dopants and have a brief look at their chemistry.

3.2.3 Rare-earth compounds

As we have seen in Chapter 2, the rare-earth elements are a series of fifteen transition metals, beginning with lanthanum, atomic number 57, and ending with lutetium, atomic number 71 (see Figure 3.1). The chemistry of all the elements in the series is very similar and is based on the gradual filling of the 4f subshell along the series [14]. All the elements have the configuration [1], [2], [3], $4s^2$, $4p^6$, $4d^{10}$, $4f^x$, $5s^2$, $5p^6$, $5d^y$, $6s^2$, where [1], [2], [3] represent closed shells for these levels, $x = 0$ to 14 and $y = 0$ or 1, but since the 4f level is an inner orbital the effect of filling it has very little effect on the chemistry of the elements. All the rare-earth elements form trivalent ions with the electronic structure [1], [2], [3], $4s^2$, $4p^6$, $4d^{10}$, $4f^x$, $5s^2$, $5p^6$ (see Table 3.1). Some of the elements (Ce, Sm, Eu, Dy, Tm) can form divalent ions and Ce, Tb, Pr and Dy can also form tetrapositive ions in certain compounds, although in general these latter two valence states are less stable than the trivalent ions.

Table 3.1 shows that the ionic radii of the trivalent ions reduce very slightly down the series, except for ytterbium. The ionic radii are large and the field

Table 3.1 Structure and some properties of the rare-earth elements.

Element	Symbol	M	M^{3+}	Colour	Ionic radii Å
Lanthanum	La	$5d^14f^0$	f^0	colourless	1.060
Cerium	Ce	$5d^14f^1$	f^1	pale yellow	1.034
Praseodymium	Pr	$5d^04f^3$	f^2	green	1.013
Neodymium	Nd	$5d^04f^4$	f^3	violet-pink	0.995
Promethium	Pm	$5d^04f^4$	f^4	—	—[a]
Samarium	Sm	$5d^04f^6$	f^5	colourless	0.964
Europium	Eu	$5d^04f^7$	f^6	colourless	0.950
Gadolinium	Gd	$5d^14f^7$	f^7	colourless	0.938
Terbium	Tb	$5d^04f^9$	f^8	colourless	0.923
Dysprosium	Dy	$5d^04f^{10}$	f^9	colourless	0.908
Holmium	Ho	$5d^04f^{11}$	f^{10}	yellow	0.894
Erbium	Er	$5d^04f^{12}$	f^{11}	pale pink	0.881
Thulium	Tm	$5d^04f^{13}$	f^{12}	pale green	0.870
Ytterbium	Yb	$5d^04f^{14}$	f^{13}	colourless	0.930
Lutetium	Lu	$5d^14f^{14}$	f^{14}	colourless	0.850

[a] Radioactive, most useful isotope is ^{147}Pm.

strength is low for rare-earth ions when compared with the first row transition elements, where the ionic radii are much smaller and more variable throughout the series since the valency of the most stable ion changes from M^{2+} through to M^{4+}. The transition metal ions are also highly coloured when they are complexed to other ligands and are greatly influenced by their environment because the valence electrons are not as well shielded by the outer orbitals from environmental effects. However, in the rare-earth series the compounds are not so highly coloured, being very pale shades of pink, yellow, green or even colourless in many cases [15]. The ion is not significantly affected by the ligand because the valency electrons are well shielded from outside influences and so the colour remains fairly constant irrespective of the complexing group. The rare earths, being such large ions, tend to have coordination numbers of between 6 and 9. Even 6 is unusual, with less than 6 being extremely rare due to the largely electrostatic bonding and the small amount of crystal field splitting which occurs in these compounds [14]. Again this is very different from the smaller transition metal ions which mainly have coordination numbers of 4 or 6 and have large crystal field splitting. The spectra of the rare-earth compound is due to transitions of electrons between energy levels. The metastable level in the rare earth is always well separated from the lower levels, making non-radiative transitions difficult and so fluorescence is observed between these levels. By examining the absorption and fluorescence properties of the rare-earth ions in glasses it is possible to match an ion to a specific application [16].

The rare-earth ions that have been most studied to date in fibres are Er^{3+}, which has a fluorescence band around $1.55\,\mu m$ corresponding to the third telecoms window, and Nd^{3+}, which has a very efficient emission at $1.06\,\mu m$ and a fluorescence band at $1.3\,\mu m$, corresponding to the second telecoms window. Both ions have convenient absorption bands into which to pump and thus excite the species.

3.2.4 Rare earths in glass

When the rare earths are incorporated into glasses they do so as network modifiers and, as we have seen before, break up the covalently bonded glass structure [15]. Since all the rare-earth ions are incorporated in the trivalent state, there must be charge compensation within the network from the oxygen atoms which are now non-bridging. If the concentration of rare-earth is low, then the ion will be in an isolated site and there will be little or no effect from other rare-earth ions. As the concentration is increased, then the distance between ions will become smaller and concentration quenching (see Chapters 2 and 4) could result with certain rare-earth ions. The stability of rare-earth ions in glasses depends very much on the host glass [17]. It is possible to achieve very high solubility in certain glass types where network modifiers are being replaced. In glasses made solely of network formers, as is the case with the doped silica glasses, the stability can be limited, with phase separation and consequential devitrification occurring at relatively low rare-earth concentrations.

The absorption and fluorescence spectra of rare-earth ions in glasses is very different to that of the crystalline hosts. This is due to the amorphous matrix and sites occupied by the individual ions being very slightly different, and so the spectrum tends to be very much smoother and broader than that for crystals. The absorption and fluorescence wavelengths are only slightly dependent on the host glass composition, but the line shapes, cross-sections and lifetimes are more markedly affected by the host glass [18]. A more detailed discussion on this subject with Nd^{3+} and Er^{3+} in various silica-based host glasses follows in the results section.

3.2.5 *Glass compositions*

For many years laser glasses have been manufactured for use in high-powered, Q-switched lasers. These multicomponent glass structures are mainly based on silicate, phosphate or fluorophosphate glasses. A typical silicate host glass composition might be: SiO_2 60 mol%, Li_2O 27.5 mol%, CaO 10 mol%, Al_2O_3 2.5 mol%, with the rare-earth concentration up to 8 mol% [18-20]. The geometry of the glass used takes the form of a short rod suitable for flashlamp pumping. To achieve sufficient absorption for the required high output powers, the rare-earth concentrations need to be very high and so most efforts have been directed in developing stable, high-concentration glasses. This work is not necessarily of direct relevance in fibre technology, where long continuous lengths of fibre with small cores are available for efficient longitudinal pumping. Fibres allow much lower rare-earth ion concentrations to be employed, while still maintaining efficient pump power absorption. In addition, the problem of heat dissipation seen in laser glass rods or the problem of ion-ion interactions can be overcome. The major direction in recent years, therefore, has been to develop the doped silica-based glass system capitalising on knowledge acquired in telecoms fibre technology.

3.3 Fabrication methods

The first record of a rare-earth doped fibre dates back to 1961 [21, 22] when Snitzer published results of laser action in a Nd^{3+} fibre laser. A core rod of the laser glass (a Nd^{3+} doped barium crown glass) was clad with a soda-lime silicate glass to give a refractive index difference between the core and cladding (Δn) of 0.02 and 3 in. long fibres were drawn to make fibre lasers. Experimental details on the fibre production were not given; however, fibres based on phosphate laser glasses have recently been reported [23]. The process used here was very simple (called 'rod in tube') and consists of a core laser glass rod

GLASS STRUCTURE AND FABRICATION TECHNIQUES

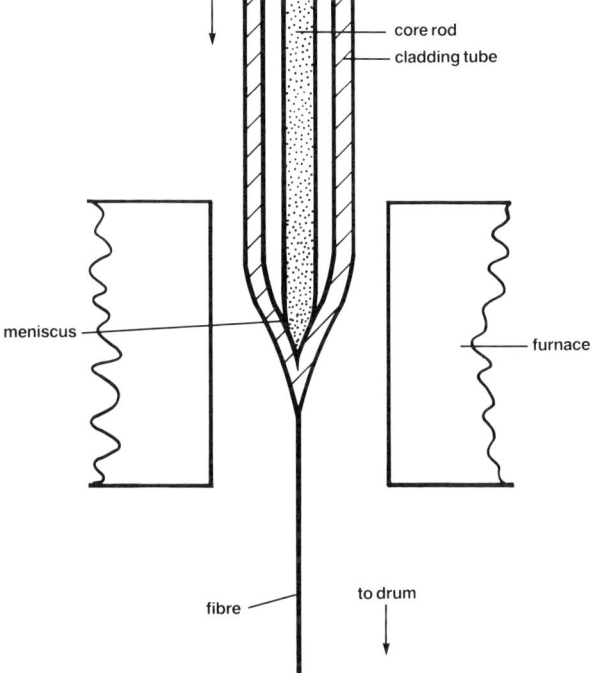

Figure 3.6 Fibre fabrication by the 'rod in tube' method.

surrounded by a cladding glass tube and drawing the whole assembly into fibre with the core-clad interface forming a meniscus in the hot zone of the furnace, as shown in Figure 3.6. Fibre drawing is usually carried out at a glass viscosity of $10^3 - 10^4$ N s m^{-2}, which can be achieved for most multicomponent glasses at fairly modest temperatures, thus allowing conventional furnaces to be used. This method is probably the most applicable for fabricating fibres from standard multicomponent laser glasses (these types of glass are also sometimes referred to as 'soft glasses'). However the rod-in-tube method has also been extended very recently by Snitzer and Tumminelli [24] to enable soft core glass fibre to be drawn with pure silica glass claddings.

This very simple method for fibre making was not pursued to make low-loss undoped telecoms fibre for a number of reasons including problems with interface scatter between the core and cladding glass and —OH impurity. Almost all fabrication methods for mainstream telecoms fibre today are based on vapour phase techniques, which produce high purity silica fibre [7]. The starting materials consist of volatile halides, e.g. $SiCl_4$, $GeCl_4$, $POCl_3$, CCl_2F_2, which are carried in a stream of O_2 at about room temperature, to be subsequently oxidised to make pure glass. There are a number of reasons for this development, and some of the primary ones include:

(i) the availability of ultra pure starting materials (already developed for the semiconductor industry);
(ii) in-built purification step; the volatile precursors are several orders of magnitude higher in vapour pressure than harmful transition metal impurities;
(iii) the glass can be made extremely dry with Cl_2 drying processes;
(iv) the intrinsic loss of silica is among the lowest for oxide glasses;
(v) silica is a most durable glass and hence is desirable as a cladding.

Thus, in more recent years, it has been a primary aim for fabricators of rare-earth doped fibres to produce optically efficient fibres but to keep compatibility with the developed telecoms fibres. Figure 3.7 shows the vapour pressure of Nd^{3+} halides [25] (which is typical of these lanthanide compounds) as a function of temperature. These vapour pressures are several orders of magnitude lower than conventional glass-making halides, hence novel ways of incorporating controlled amounts of involatile rare-earth ions into core glass have had to be devised. Nearly all the methods attempted have been additions to the major fibre-making technologies. We shall consider each in turn.

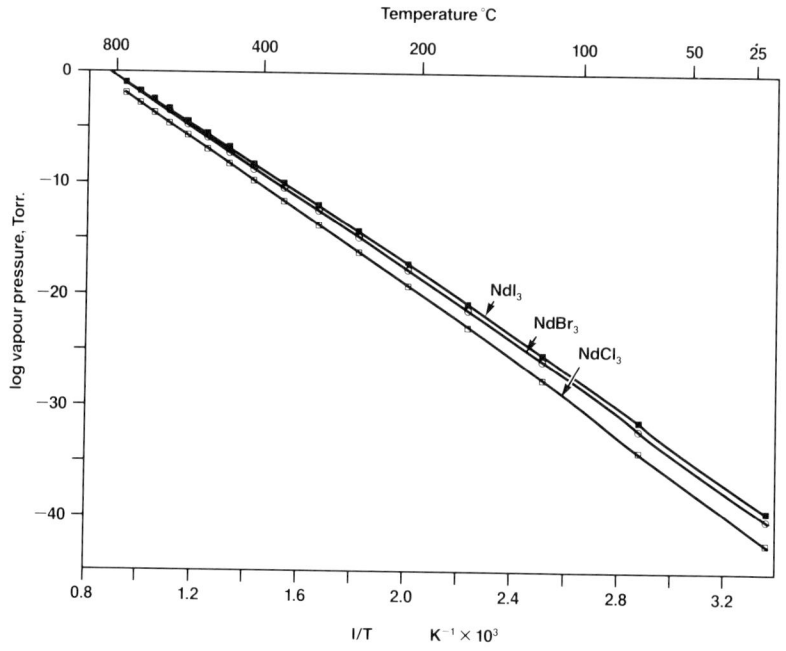

Figure 3.7 Vapour pressure of $NdCl_3$, $NdBr_3$ and NdI_3, as a function of temperature.

3.3.1 *Modified chemical vapour deposition (MCVD)*

In the basic MCVD process a silica deposition tube is mounted in a special glass working lathe and the reactant halides are passed down the inside of the tube while it is rotated. A burner is traversed along the tube, which produces a local hot zone and the halides are oxidised and fused on the inside walls of the tube. By changing the reactant vapours and concentrations, glasses of varying refractive indices can be deposited. Thus the core glass typically consists of GeO_2-doped silica while the cladding glass usually consists of P_2O_5 and F-doped silica (the P_2O_5 enables the layers to be fused at moderate temperatures but increases the refractive index, hence F is added to offset this increase; F-doped silica being a very low index material). The tube is subsequently collapsed to form a solid rod preform by ceasing reactant flow down the tube and traversing the burner at high temperatures very slowly along the length of the tube. Surface tension forces cause the collapse, which is usually carried out under a Cl_2-O_2 atmosphere to keep the innermost layers ultra dry. The resulting rod, termed a preform, is drawn into fibre at $\sim 2000°C$ by feeding into a very high temperature furnace. This is usually a carbon resistance furnace or an RF coupled zirconia-based furnace.

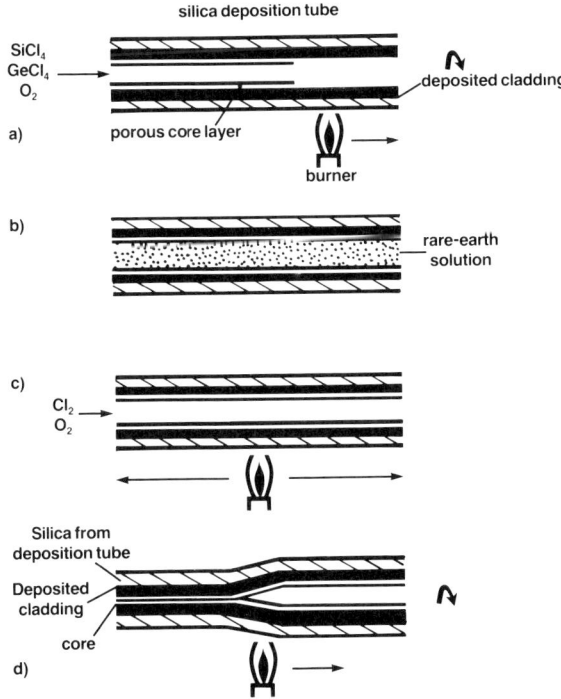

Figure 3.8 Rare-earth doped fibre fabrication by the 'solution doping' method.

3.3.1.1 *Solution-doping technique.* This method was first reported by Stone and Burrus in 1973 [26] and was not reportedly used again until 1987 [27] but is now a very popular technique [11, 28, 29]. The first stage of the process is the same as that for producing a standard MCVD preform except that the core material is deposited at a lower temperature than normal to prevent complete fusing of the glass (see Figure 3.8). Obviously the oxidation temperature of the halides must be exceeded for the reaction to take place, and thus with low-viscosity core glass compositions, e.g. heavily P_2O_5-doped silica, the prevention of complete fusing is difficult. This potential problem can be overcome and a porous core layer achieved by reversing the traverse direction of the core and afterwards lightly fusing the layer in the absence of reactants. This maintains the versatility inherent with the MCVD technique. A solution of the rare-earth to be incorporated is introduced into the tube and allowed to soak into the pores for approximately one hour. Various salts in either aqueous or alcoholic solutions have been used including halides, nitrates and sulphates. The liquid is subsequently expelled from the tube and the porous layer blown dry. This is followed by carefully fusing the layer under a Cl_2–O_2 atmosphere to completely dry the material and collapsing the tube in the usual way.

Slight modifications to the process include the deposition of a pure silica core layer instead of GeO_2–P_2O_5-doped silica and soaking the layer in an Al^{3+} rare-earth ion solution to make an Al_2O_3–SiO_2 host glass. This method has been used used to fabricate up to 15 wt% rare-earth in the core glass [30].

3.3.1.2 *Volatile halide methods.* The current interest in rare-earth doped fibres in silica host glasses was a result of the work of Poole *et al.* [31], who demonstrated that heating rare-earth halide crystals adjacent to the deposition tube was an efficient and convenient way to generate and transport a rare-earth vapour. Figure 3.9 shows the deposition process; in the first step a rare-earth chloride was fused to the walls of a dopant carrier chamber, upstream from the deposition tube, by gently warming. The cladding layers were then deposited in the usual way and during core deposition the dopant chamber was heated to $\approx 1000°C$ to increase the vapour pressure of the halide. The vapour was incorporated with the main reactants and included with the deposited core layers. As above, the tube was collapsed to form a preform and fibre was drawn in the usual way.

A modification to this procedure is to use a silica sponge which has been impregnated with a rare earth in place of the dopant carrier chamber [32]. This provides a more reproducible rare-earth vapour stream.

The addition of Al_2O_3 to the core host glass can also be effected by MCVD [33] by transporting Al_2Cl_6 in heated lines to the reaction zone, as shown in Figure 3.9(c). Note that co-doping with rare-earth ions is also possible in this case.

A transportation and oxidation method [34] has been reported where the

GLASS STRUCTURE AND FABRICATION TECHNIQUES

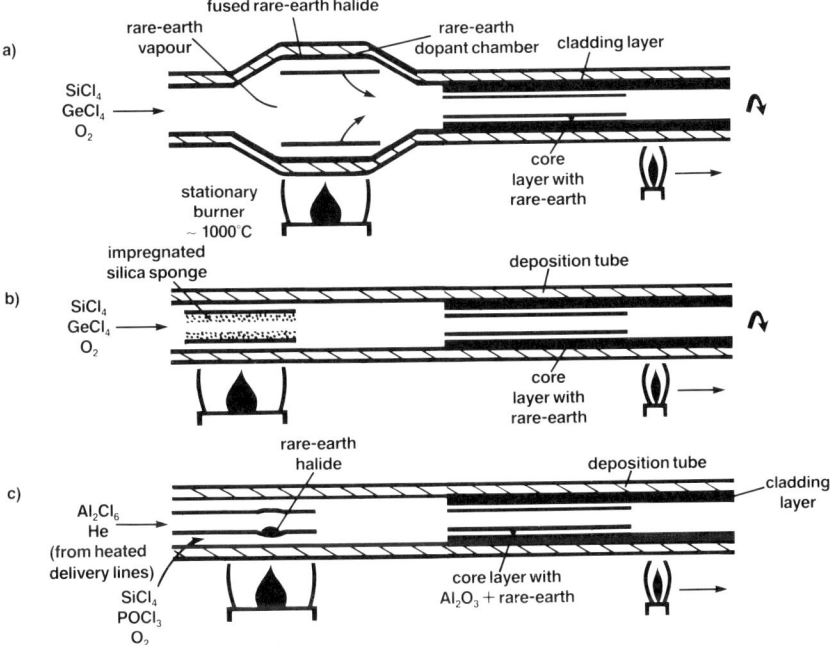

Figure 3.9 Rare-earth doped fibre fabrication by various forms of the 'volatile halide' method.

inner surface of a deposition tube, which has previously been coated with P_2O_5–SiO_2, is subsequently coated with a thin layer of $YbCl_3$ by driving the vapour from one end of the tube to the other. The rare-earth halide is then oxidised by heating in oxygen and the tube is collapsed conventionally.

3.3.1.3 *Aerosol doping.* Little work has been reported using aerosols, which is surprising as this method would seem to be useful for incorporating a wide range of involatile species into the core glass. However, work has been reported which discusses the incorporation of Nd^{3+} into silica-based core glass by generating an aerosol from an aqueous solution of $NdCl_3$ [35]. The generator consisted of a 1.5 MHz transducer which produced particles of several microns in diameter, which could be swept into the reaction zone without condensation. Further work on this interesting technique is likely to be reported in the future.

3.3.2 Outside vapour deposition (OVD)

OVD is similar to MCVD in that volatile halides are used as source materials for the preparation of a preform, but the halides are fed into a burner which introduces the vapours into a CH_4–O_2 flame. Here reaction takes place and very wet doped silica particles are formed. The solid reaction products are directed onto a mandrel, which is rotated and traversed in front of the burner. A porous layer is built up, and the different core and cladding compositions are produced by varying the reactants fed into the burner as in the MCVD process. The central mandrel is removed and the porous preform is dried and fused in a He–Cl_2–O_2 gas mixture at between 1400 and 1600°C to form a clear glass preform which can be drawn into fibre.

This technology was used to fabricate rare-earth doped fibres before the current interest in fibre lasers and amplifiers emerged. Ce^{3+}-doped fibre was prepared to produce fibre which was potentially more resistant to radiation effects than standard fibre. The resistance was based on the Ce^{3+} acting as a trap for electrons and holes which were generated from the radiation. The approach adopted here was to produce a reasonably volatile organometallic precursor and transportation of the vapour into the OVD burner for oxidation and deposition [36] and thus is the closest method to standard fibre fabrication discussed so far. The experimental arrangement is shown in Figure 3.10. The organometallic source prepared for this work was a β-diketonate, specifically tetrakis 6,6,7,7,8,8,8-heptafluoro-2,3-dimethyl-3,5-octanedione cerium(IV) (usually abbreviated to Ce(fod)$_4$). The vapour pressure of this material is still relatively low (see Figure 3.11) when compared to the host glass halides, $SiCl_4$ (236T 25°C) and $GeCl_4$ (84T 25°C), which necessitates temperature control of the material around 150–200°C and heating the delivery lines, to prevent condensation, directly to the burner. An inert carrier gas is also important due to the reactivity of the material. Over 1 wt% Ce^{3+} was successfully incorporated into the core glass, so despite the apparent limitation in source volatility, high rare-earth concentrations are possible by this method. It should also be mentioned that the control possible on source temperature and carrier gas flow rate make this method seem very attractive for accurate and reproducible rare-earth incorporation.

This work was extended to the rare-earth ion Er^{3+} [37] and various core host glasses were studied, including pure silica (with fluorosilicate cladding),

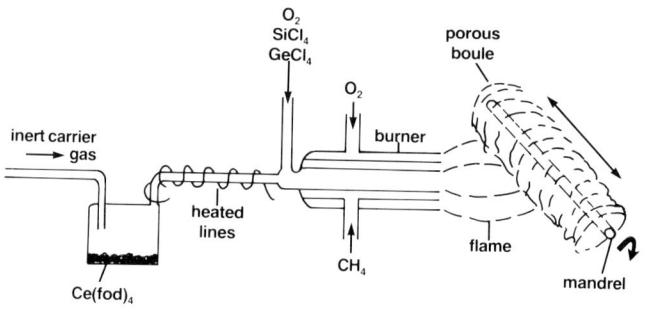

Figure 3.10 Fabrication of a rare-earth doped boule by the OVD technique, incorporating a volatile organometallic source material.

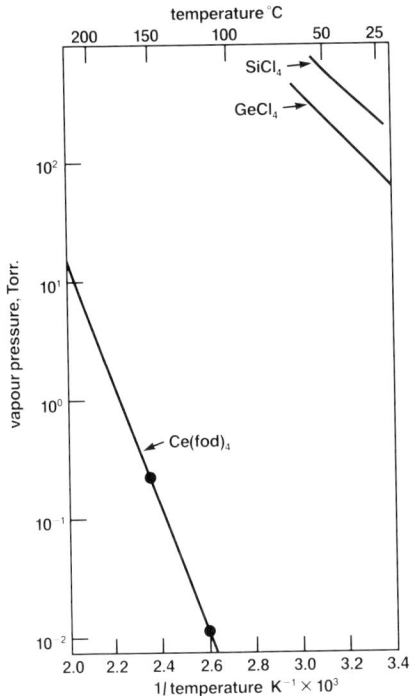

Figure 3.11 Vapour pressure of Ce(fod)$_4$ compared to SiCl$_4$ and GeCl$_4$ as a function of temperature. (Redrawn and adapted from [36].)

GeO$_2$–SiO$_2$ and Al$_2$O$_3$–SiO$_2$. Note that with the OVD process, the porous boules are fused and dried by flushing with Cl$_2$–He gas mixtures at high temperatures for up to several hours. This can lead to devitrification, especially in the case of Al$_2$O$_3$ containing glasses and when higher concentrations of rare earths are attempted.

3.3.3 Vapour axial deposition (VAD)

This is a similar technology to OVD and has been used for many years to produce standard telecoms fibre. The main difference with this technology is that the particles which grow in the torch flame are directed onto the end of a rotating boule and the core and cladding glasses are formed using separate burners. The sintering and drying of the porous boules is very similar to the OVD process. Two methods for incorporating rare-earth ions have been reported. The first, termed 'molecular stuffing' [38], is very similar to the MCVD solution doping technique (Figure 3.12 outlines the various steps). A GeO$_2$–SiO$_2$ core boule is prepared and soaked in an alcoholic solution of rare-earth chloride. The solvent is allowed to evaporate and the preform is

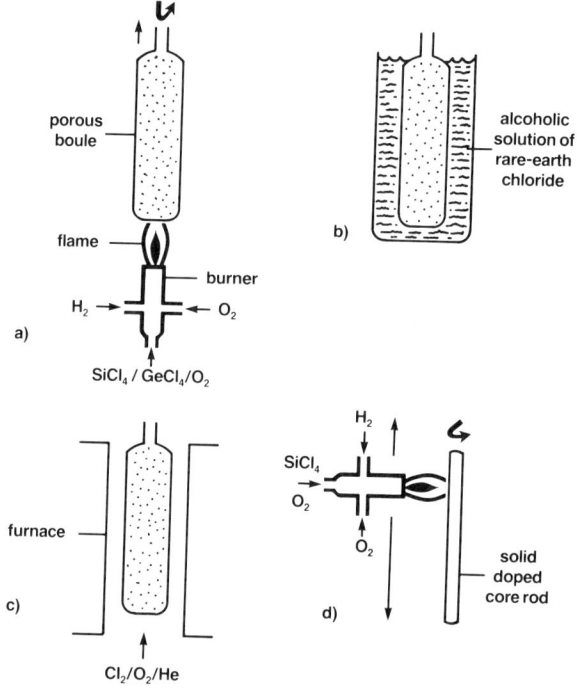

Figure 3.12 The fabrication of rare-earth doped VAD boules by 'molecular stuffing'.

sintered and dried in an electric furnace. This produces a core rod which has to be coated with silica to produce a preform. Up to 2 wt% Nd^{3+} incorporation was reported with good uniformity. As with the solution doping method, the concentration of the rare-earth ion was controlled by adjusting the solution strength.

The second method [39], by gas phase doping the porous core boule (Figure 3.13), has been demonstrated with Nd^{3+}. The core boule is placed in a consolidating furnace with $NdCl_3$ crystals, which are melted by heating above 784°C. The vapour generated diffuses into the pores of the boule and oxidises on the surface of the glass. By varying the $NdCl_3$ temperature, the dopant concentration could be controlled over a wide range up to 0.3 wt%. The atmosphere during consolidation affects dopant incorporation; using the commonly employed Cl_2 dehydrating agent inhibits surface oxidation of $NdCl_3$ and an O_2 atmosphere readily forms Nd_2O_3 on the outer surface of the boule, thus preventing uniform incorporation. The preferred sintering gas is therefore He.

Another VAD method, which has not been used for rare-earth incorporation but was used to incorporate involatile ions such as Pb^{2+}, uses a nebuliser to generate a nebulised aqueous solution of lead nitrate [40]. (This is

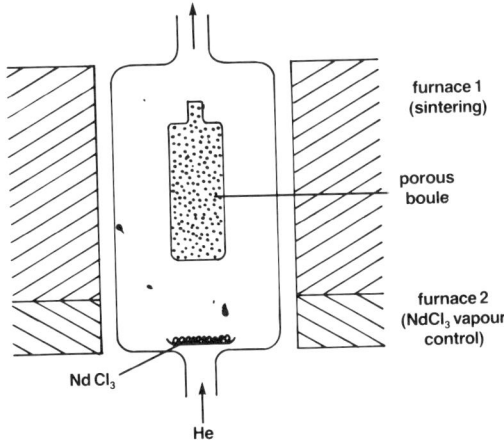

Figure 3.13 Doping a VAD boule with Nd^{3+} by a vapour phase process during the consolidation stage.

similar to the aerosol method discussed in the MCVD section.) It has been suggested that this method would also be suitable for rare-earth ion incorporation [29].

3.4 Results

3.4.1 *Rare-earth ions in silica-based host glasses*

Much work has been reported on the various properties of the rare earths in the high silica glasses. Of initial concern was the limiting solubility of the ion in silica, but it was shown [41] that high solubilities were possible by adding Al_2O_3 to the silica matrix. The tendency for the Al_2O_3–SiO_2 host to devitrify, when the Al_2O_3 concentration exceeded a few wt%, was prevented by adding P_2O_5 to the glass [10]. The effect of Nd^{3+} concentration on the optical properties of the doped SiO_2–Al_2O_3–P_2O_5 host glass has recently been reported [30] and serves to illustrate the general behaviour of the lanthanides in this host glass. At low concentrations the ions disperse in the glass matrix homogeneously and are widely separated. Time-resolved fluorescence measurements show that up to 0.5 wt% Nd^{3+} can be incorporated in this way. Figure 3.14(a) gives the fluorescence decay for a very low concentration of Nd^{3+}, which is clearly a single exponential indicating absence of both non-radiative transitions and ion–ion interactions. Figure 3.14(b) shows that as the Nd^{3+} concentration is increased, the decay rate ceases to be a single exponential and an increase in the early stages of the decay is evident. This is due to ion–ion coupling and is related to the nearness of adjacent Nd^{3+} ions,

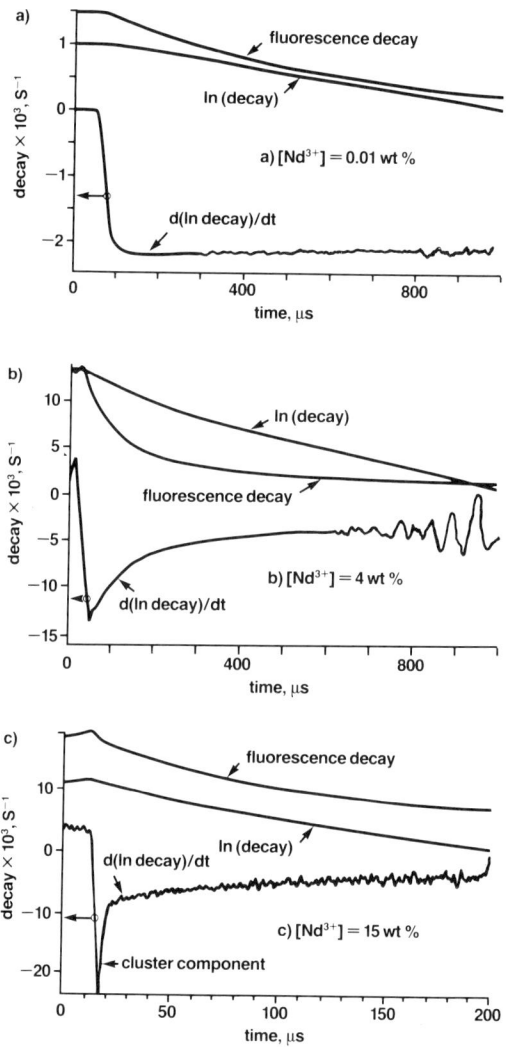

Figure 3.14 The fluorescence decay, ln(decay) and d(ln decay) for the $^4F_{3/2} \rightarrow {}^4I_{11/2}$ emission for various concentrations of Nd^{3+}. (From [30].)

but it does not mean that the Nd^{3+} ions are no longer homogeneously dispersed. Increasing the Nd^{3+} concentration further to ≈ 15 wt% leads to the onset of an additional very fast decay component immediately after excitation, as shown in Figure 3.14(c). This very fast component is thought to be due to decay from regions highly concentrated in Nd^{3+}, termed 'clusters'. Examination of the core glass by transmission electron microscopy (TEM) has confirmed the existence of these highly concentrated regions. Figure 3.15(c) shows a TEM micrograph of this highly doped material and the globular dark

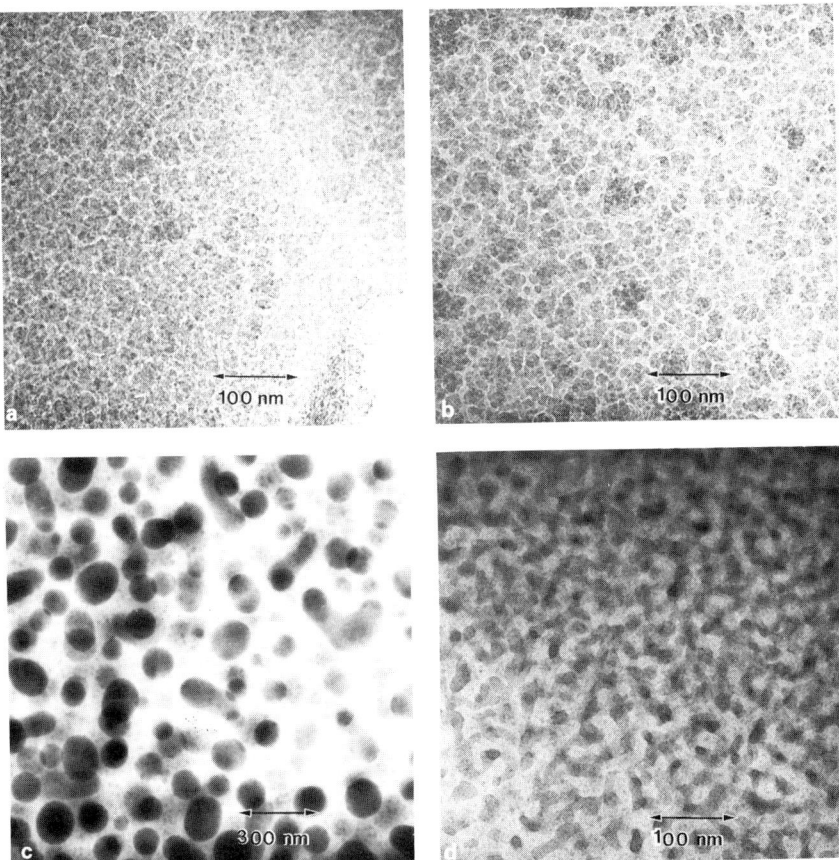

Figure 3.15 TEM of (a) 7 wt% Nd^{3+} preform core glass, (b) pure silica standard, (c) 15 wt% Nd^{3+} preform glass and (d) 15 wt% Nd^{3+} fibre core glass. (From [30].)

phases, which were analysed by the energy dispersive X-ray technique, were found to have a Nd^{3+} concentration of 40 wt%. The background glass contained a much lower concentration; around 6 wt% Nd^{3+}, which accounts for the non-exponentiality after the fast component. The 7 wt% Nd^{3+} core glass is shown in Figure 3.15(b) and this compares favourably with the pure silica cladding glass (Figure 3.15(a)), thus confirming uniform incorporation of the Nd^{3+} up to 7 wt%. It is interesting to note that a TEM of the 15 wt% core glass, when drawn into fibre, shows phase separation but on a much smaller scale (Figure 3.15(d)). This would account for the high transparency reported for the fibre, because the opacity of the core glass visible in the preform is largely due to Rayleigh scatter, which is dependent on the diameter of the higher index phase separated regions to the sixth power.

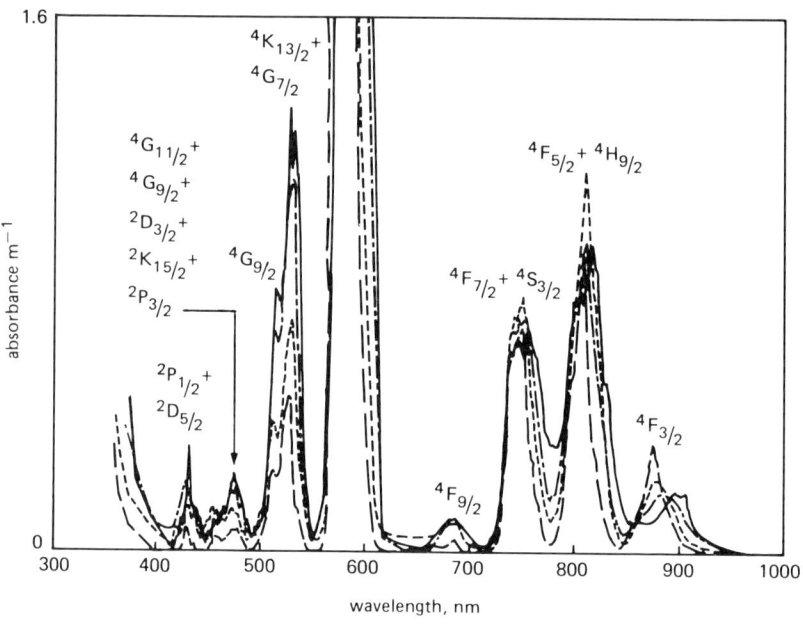
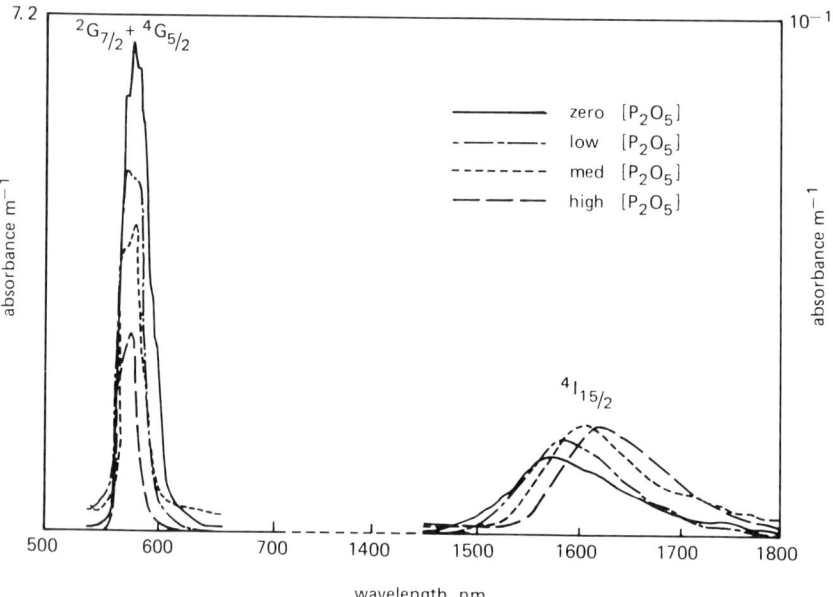

Figure 3.16 The absorption spectra for Nd^{3+} in SiO_2–GeO_2–P_2O_5 for various P_2O_5 concentrations. (From [32].)

GLASS STRUCTURE AND FABRICATION TECHNIQUES

Increases in solubility of Nd^{3+} in silica-based glasses have also been reported by adding P_2O_5 to the host glass [32]. Similar work has been reported, but for Er^{3+} in place of the Nd^{3+} [11], where it was shown that the solubility limits for the two ions are very close. Detailed examination of the fluorescence decay rates [42] on the Er^{3+} fibres has shown that for amplifier applications very low concentrations of Er^{3+} are desirable, which necessitates long fibre lengths being used to compensate for the low absorption. This in turn makes the background loss of the fibre a more important feature.

Another much researched area is the variation of absorption and fluorescence spectra as a function of silica-based host glass. The two ions, Nd^{3+} and Er^{3+}, have received most attention [11, 32]. Figure 3.16 shows the different absorption spectra for essentially 'telecommunications' GeO_2-SiO_2 core host glass with small successive additions of P_2O_5. The first point to make is that the absorption spectrum of the GeO_2-SiO_2 host in the absence of P_2O_5 strongly resembles the spectra from pure silica host glass [26]. This result is not unreasonable, as the substitution of Si for Ge will not significantly change the glass structure, as discussed in the early part of this chapter. However, for an addition of just 2 wt% P_2O_5, the spectrum changes significantly and for the higher P_2O_5 content fibres, greater resemblance to the multicomponent ED-2 and LG-650 laser glasses is apparent [43]. The fluorescence spectra show a similar trend (see Figure 3.17). In the case of the GeO_2-SiO_2 only host glass, a spectrum with broad bands at relatively long wavelengths, typical of pure silica, are evident. The addition of P_2O_5 narrows the emission bands and there is a shift to slightly shorter wavelengths. The main $^4I_{11/2} \rightarrow {}^4I_{9/2}$ transition is situated at 1060 nm, with the 2 wt% P_2O_5 co-doped host, which coincides with silicate laser glasses. An explanation for the changes in spectra as a function of

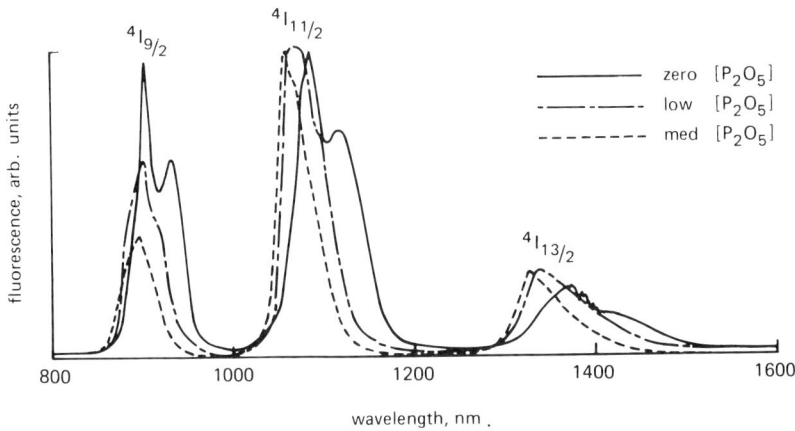

Figure 3.17 The fluorescence spectra of Nd^{3+} in SiO_2-GeO_2-P_2O_5 for various concentrations of P_2O_5. (From [32].)

P_2O_5 concentration is the existence of the P=O and Si—O⁻ groups in these glasses. The formation of the —O⁻ groups will clearly be important for coordination of the Nd^{3+}.

The addition of Al_2O_3 to the host glass is important not only from a concentration aspect, but also the important changes offered in fluorescence properties. This is illustrated with Er^{3+} in Figure 3.18 where the $^4I_{13/2} \rightarrow {}^4I_{15/2}$ (ground state) transition is shown compared to GeO_2–P_2O_5–SiO_2 host glass. Clearly shown is the broad emission from the Al_2O_3 containing host glass, which is nearly twice that of the other glass shown. This result ties in with the Er^{3+} being coordinated at least partially by Al—O⁻ groups to account for the change in fluorescence spectrum. Further evidence for the role of the Al—O⁻ is presented when we examine the fluorescence spectra of Er^{3+} in a series of Al_2O_3–GeO_2–SiO_2 host glasses in which the Al:Ge ratio is varied. Figure 3.19 shows the narrowing of the spectra as the Al is substituted by the Ge. However the important point is that the spectrum for the Ge free fibre is very similar for that where the Al:Ge ratio is 1:3, i.e. despite the relatively high concentration of Ge, it would seem that the Er^{3+} is still preferentially coordinating with the Al—O⁻. The broad fluorescence possible with appropriate choice of host glass, combined with fractional population of the metastable $^4I_{13/2}$ level (see Chapter 4) provides an important choice on potential amplifier bandwidth. In applications where many wavelengths require amplifying, some Al_2O_3 in the core glass is clearly essential; however,

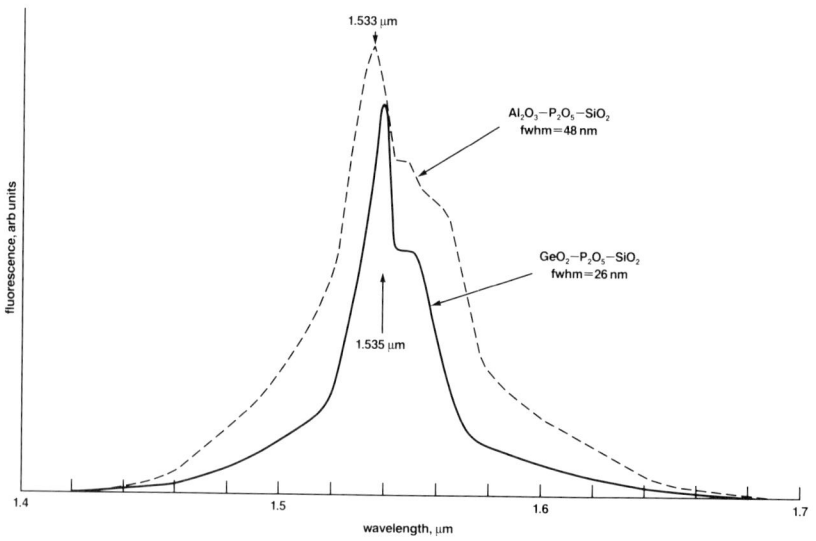

Figure 3.18 The fluorescence spectrum from the $^4I_{13/2} \rightarrow {}^4I_{15/2}$ transition in Er^{3+} in Al_2O_3–P_2O_5–SiO_2 and GeO_2–P_2O_5–SiO_2 host glasses. (From [11].)

GLASS STRUCTURE AND FABRICATION TECHNIQUES 73

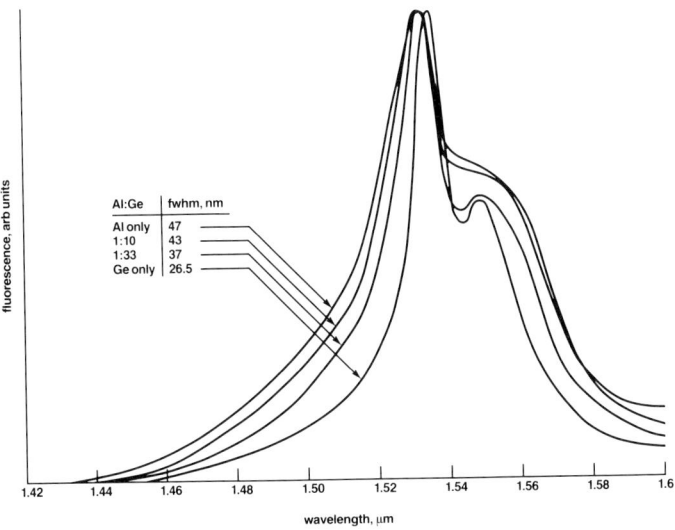

Figure 3.19 The fluorescence spectrum from the $^4I_{13/2} \rightarrow {^4I_{15/2}}$ transition in Er^{3+} in Al_2O_3–GeO_2–SiO_2 host glasses for various Al:Ge ratios.

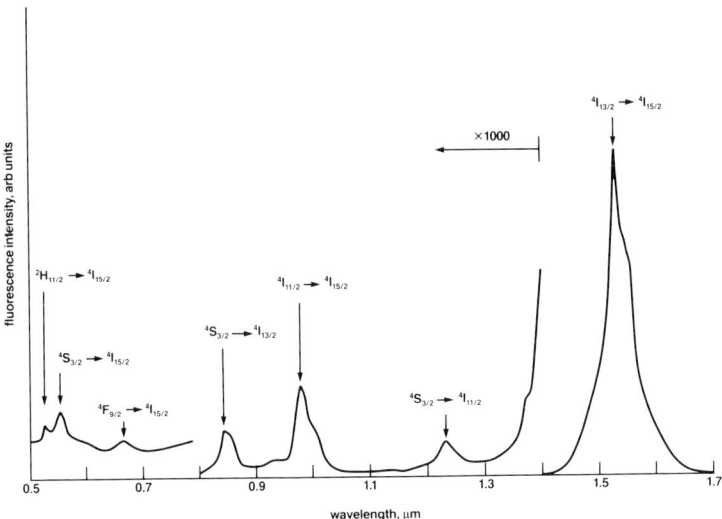

Fig 3.20 The fluorescence spectrum of Er^{3+} in a silica-based host glass showing high emission from the $^4I_{13/2} \rightarrow {^4I_{15/2}}$ transition, even when pumped at 514.5 nm.

for narrow band applications where, for example, spontaneous emission over a wide wavelength range increases amplifier noise, then the GeO_2–SiO_2 host would be more appropriate.

A further point to make is that the fluorescence observed from Er^{3+} in silica-based host glasses is very largely from the $^4F_{3/2}$ level, even when pumped at 514 nm. Emission from higher energy levels is very weak, due to the relatively close spacing of the levels and the position of the multiphonon edge. This is clearly shown in Figure 3.20 and will be an important feature for efficient amplifiers and lasers designed to function in the third telecommunications window. In glasses where this edge is at much longer wavelengths, e.g. fluorides, significant emission can be observed from many more levels. For more detailed discussion on this topic see Chapter 8.

3.4.2 Fibre properties

The incorporation of rare-earth ions in terms of concentration has been discussed above; however, the radial distribution can be important in three-level amplifier systems [44] where concentrating the ions in the centre of the core is seen as desirable. Various profile data have been reported: for VAD fibres, using the vapour-phase method, the rare earth is incorporated uniformly across the boule, thus a step profile results [39]; with MCVD fibres the choice of host glass is important. Figure 3.21 shows that in the GeO_2–SiO_2

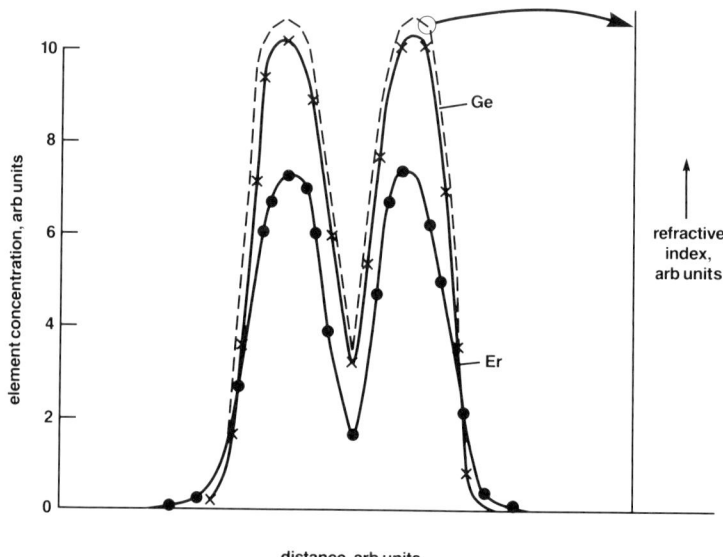

Figure 3.21 The elemental and refractive index profiles of a GeO_2–SiO_2 based Er^{3+} doped preform fabricated by the MCVD technique. (From [11].)

host fibres, a central refractive index depression results, which is due to the formation of the volatile suboxide of GeO_2 (this effect is sometimes known as burn-off). The reduction of GeO_2 concentration mirrors the rare-earth concentration profile [11]. This suggests that the rare-earth ions close to the inner surface of the tube are swept out during the collapse stage of preform fabrication along with some GeO_2. The rare-earth profile is therefore the opposite of that which is required for greatest efficiency, and although some compensation can be made by etching the innermost core regions prior to the final collapse stage [45], concentrating the rare earth in the centre would seem to be difficult by this method. A different result is evident using Al_2O_3 in the core glass, as shown in the case of a multimode fibre elemental scan in Figure 3.22. Here a step profile with the Al (which will give a step refractive index profile) is observed and an absence of a central depression. The reason for this is the involatility of Al_2O_3 and the lack of formation of a volatile suboxide. Furthermore, the Er^{3+} profile mirrors that of the Al_2O_3, which ties in with the Er^{3+} coordinating with the Al_2O_3. Also shown in this figure is the depletion due to P from the central core region and that in the cladding layers.

These features have been capitalised upon to confine the Er^{3+} to the central region of the core [46] and fabricate so-called 'confined fibre', as shown in Figure 3.23. The outer region of the core was deposited in the usual way and consists of GeO_2–SiO_2 glass. The inner core region was made up of Al_2O_3–

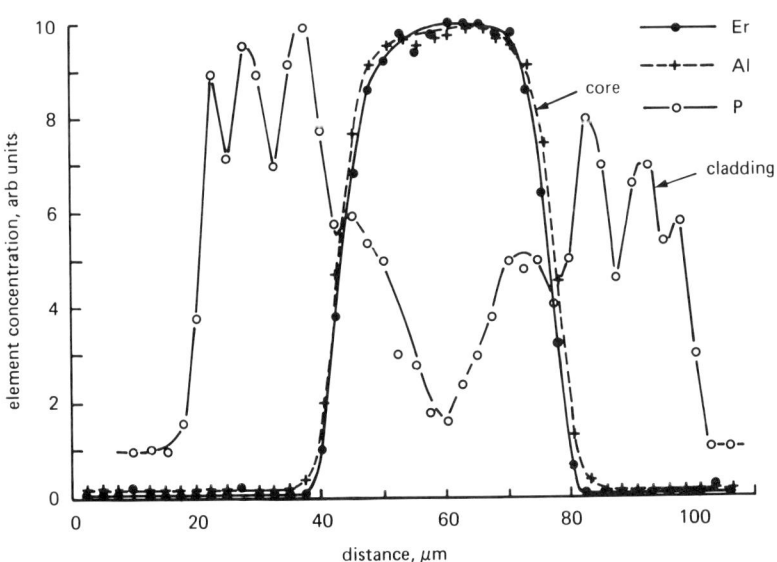

Figure 3.22 The elemental profiles of an Al_2O_3–P_2O_5–SiO_2 based Er^{3+} doped preform fabricated by the MCVD technique. (From [11].)

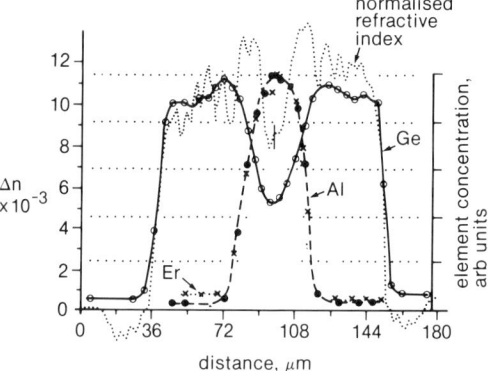

Figure 3.23 The elemental and refractive index profiles of a 'confined' preform fabricated by the MCVD technique (From [46].)

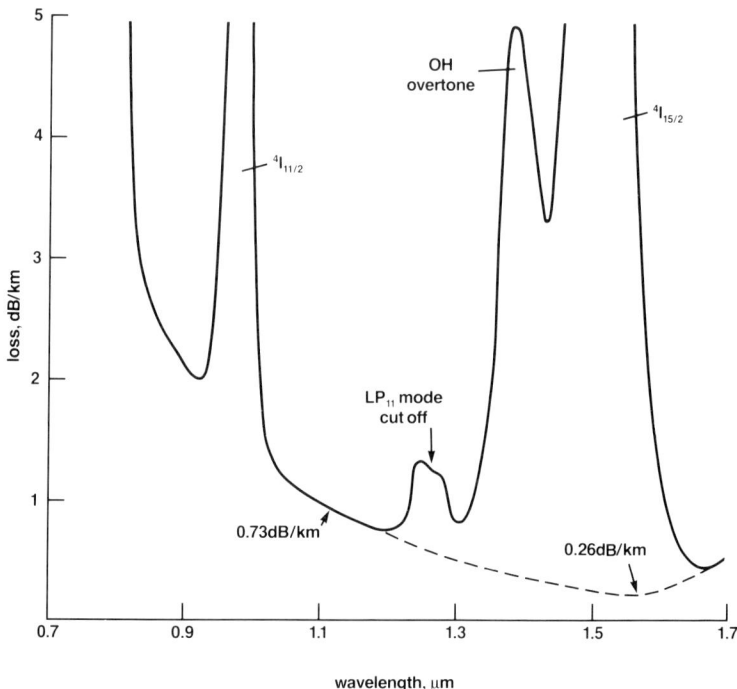

Figure 3.24 The total attenuation spectrum of a fibre doped with a few ppm Er^{3+} showing exceptionally low background loss. (From [47].)

P_2O_5–SiO_2 glass doped with the Er^{3+} and was made by the solution method. A high degree of confinement is evident and although the figure shows the results for the preform, the same degree of confinement was observed on resulting multimode fibre and can be inferred for single-mode fibre. (The dimensions of the core in the drawn single-mode fibre were such as to present resolution uncertainties with the elemental measurements.)

Recent proposals on distributed amplifiers for non-linear systems [47] and lossless links have focused attention on producing fibres that have very low background loss and very low concentrations of Er^{3+}. Very low background loss fibre has been achieved using the OVD technique, with a reported total loss of 0.68 dB/km at 1.2 μm [37]. Fibre prepared by MCVD has also been reported with an extrapolated loss (i.e. subtracting the loss due to the Er^{3+} absorption) of 0.26 dB/km at 1.55 μm.

Thus doped fibre, virtually free from excess loss, can be fabricated by these two techniques and in all probability VAD technology will be reporting similar results in the near future. Figure 3.24 shows the loss spectrum for 6 km of fibre with an Er^{3+} concentration of $\approx$ 1 ppm. The only observable impurity is —OH absorption at 1.38 μm, which will hardly contribute to the loss in the long wavelength region of interest. The Er^{3+} incorporation was uniform to $\pm 12\%$ and thus shows great promise as a key component for exciting new network and system possibilities.

3.5 Conclusion

Rare-earth doped fibres were first fabricated nearly three decades ago, but it is only very recently that their importance and potential have been recognised. The great progress made over the last year or so has centred on fibre that is based on standard telecoms technology and essentially telecoms 'high silica' compositions. It is of great interest to observe that such progress within the telecoms arena would not have been predicted by materials scientists a few years ago, and shows the considerable resourcefulness of the workers in this field. We wait with anticipation for further developments in the future.

References

1. R.H. Doremus, *Glass Science*, Wiley (1973).
2. J.E. Midwinter, *Optical Fibres for Transmission*, Wiley (1979).
3. H. Rawson, *Properties and Applications of Glasses*, vol. 3, Glass Science and Technology, Elsevier (1980).
4. W.H. Zachariasen, *J. Amer. Ceramic Soc.* **54** (1932) 3841.
5. D.R. Uhlmann, *J. Amer. Ceramic Soc.* **66** (1983) 95.
6. C.H.L. Goodman, *Glass Technol.* **28**, No. 1 (1987) 19.
7. Tingye Li, *Optical Fibre Communications*, vol. 1, *Fibre Fabrication*, Academic Press (1985).
8. S.R. Nagel, J.B. MacChesney and K.L. Walker, *IEEE J. Quant. Electron.* **QE18**, No. 4 (1982), 459.

9. D.R. Uhlmann, N.J. Kreidl, *Glass Forming Systems*, vol. 1, *Glass Science and Technology*, Academic Press (1983).
10. J.R. Simpson and J.B. MacChesney, *Electron. Lett.* **19** (1983) 261.
11. B.J. Ainslie, S.P. Craig, S.T. Davey and B. Wakefield, *Mater. Lett.* **6**, No. 5 (1988) 139.
12. B.J. Ainslie, K.J. Beales, C.R. Day and J.D. Rush, *IEEE J. Quant. Electron.* **QE18**, No. 4 (1982), 514.
13. G.W. Tasker, W.G. French, J.R. Simpson, P. Kaiser and H.M. Presby, *Appl. Opt.* **17**, No. 11 (1976) 1836.
14. S.A. Cotton and F.A. Hart, *The Heavy Transition Elements*, Macmillan Press (1984).
15. K. Patek, *Glass Lasers*, ILIFFE Books, Butterworth (1970).
16. B.J. Ainslie, S.P. Craig and S.T. Davey, *J. Light. Technol.* **6**, No. 2 (1988) 287.
17. E. Snitzer, *IEEE NEREM Record* (1968) 42.
18. M. Weber, *Handbook of Laser Science and Technology*, vol. 1, *Lasers and Masers*, CRC Press (1985).
19. Hoya Glass Data Sheet LHG-8 High Nd doped 8–3.
20. N. Neuroth, *Opt. Engng* **26**, No. 2 (1987) 097.
21. E. Snitzer, *J. Appl. Phys.* **72**, No. 1 (1961) 36.
22. E. Snitzer, *Phys. Rev. Lett.* **7**, No. 12 (1961) 444.
23. T. Yamashita, in *Proc. SPIE Conf. on Fibre Lasers Sources and Amplifiers* **1171** (1989), 291.
24. E. Snitzer and R. Tumminelli, *Opt. Lett.* **14**, No. 14 (1989) 757.
25. O. Kubaschewski and C.B. Alcock, *Metallurgical Thermochemistry*, Pergamon Press (1979).
26. J. Stone and C.A. Burrus, *Appl. Phys. Lett.* **23**, No. 7 (1973) 388.
27. J.E. Townsend, S.B. Poole and D.N. Payne, *Electron. Lett.* **23**, No. 7 (1987) 329.
28. L. Cognolato, B. Sordo, E. Modone, A. Gnazzo, and G. Cocito, in *Proc. SPIE Conf. on Fibre Lasers, Sources and Amplifiers* **1171** (1987) 202.
29. J.R. Simpson, ibid., 2.
30. B.J. Ainslie, S.P. Craig and R. Wyatt, *Mater. Lett.* **8**, No. 6 (1989) 204.
31. S.B. Poole, D.N. Payne and M.E. Fermann, *Electron. Lett.* **21**, No. 17 (1985) 737.
32. B.J. Ainslie, S.P. Craig and S.T. Davey, *Mater. Lett.* **5**, No. 4 (1987) 143.
33. J.B. MacChesney and J.R. Simpson, in *Proc. of Optical Fibre for Communications Conference, San Diego, CA, USA*, Paper WH5 (1985), p. 100.
34. M. Watanabe, H. Yokota and M. Hoshikawa, in *Proc. of European Conf. on Optical Communications, Venice, Italy* (1985), p. 15.
35. T.F. Morse, in *Proc. SPIE Conf. on Fibre Lasers, Sources and Amplifiers* **1171** (1989), p. 72.
36. D.A. Thompson, P.L. Bocko and J.R. Ganon, in *Proceedings of Fibre Optics in Adverse Environments* (II), SPIE, vol. 506 (1984), p. 170.
37. P.L. Bocko, in *Proc. Optical Fibre for Communications, Houston, USA*, Paper TUG2 (1989).
38. T. Gozen, Y. Kikukawa, M. Yoshida, H. Tanaka, and T. Shintani, in *Proc. Optical Fibre for Communication, New Orleans, USA*, Paper WQ1 (1988).
39. M. Shimizu, F. Hanawa, H. Suda and M. Horiguchi, *Japanese J. Appl. Phys.* **28**, No. 3 (1989) L476.
40. R. Sanada, T. Shioda, T. Moriyama, R. Inada, S. Takahashi and M. Kawachi, in *Proc. 6th European Conf. on Optical Communications York, UK* (1980), P. 14.
41. R. Arai, H. Namikawa, K. Kumata and T. Handa, *J. Appl. Phys.* **59** (1986) 3430.
42. B.J. Ainslie, S.P. Craig-Ryan, S.T. Davey, J.R. Armitage, C.G. Atkins and R. Wyatt, in *Proc. 7th Int. Conf. on Integrated Optics and Communication, Kobe, Japan*, Session 20A3 (1989), p. 22.
43. G.J. Linford, R.A. Saroyan, J.B. Trenholme and M.J. Weber, *IEEE J. Quant. Electron.* **15** (1979) 510.
44. J.R. Armitage, *Appl. Opt.* **27**, No. 23 (1988) 4831.
45. B.J. Ainslie, K.J. Beales, D.M. Cooper and C.R. Day, *Electron. Lett.* **18**, No. 19 (1982) 809.
46. B.J. Ainslie, J.R. Armitage, S.P. Craig and B. Wakefield, in *Proc. 14th European Conf. on Optical Communication, Brighton, UK* (1988), p. 62.
47. S.P. Craig-Ryan, B.J. Ainslie and C.A. Millar, *Electron. Lett.* **26**, No. 3 (1990) 185.

4 Spectroscopy of rare-earth doped fibres

R. WYATT

4.1 Introduction

This chapter describes techniques for measuring selected properties of rare-earth doped fibres that may affect their use and application in oscillators and amplifiers. (Note that fluoride fibres are covered in a separate chapter.) We shall concentrate on those dopants that are receiving most attention from a practical viewpoint: erbium (1.55 μm) and neodymium (0.9, 1.06, 1.3 μm). Topics covered in this chapter are absorption and emission cross-sections, and fibre gain characteristics, the measurement and effects of parasitic absorption from the upper laser level (known as excited-state absorption, or ESA) at both pump and signal wavelengths, and detailed measurements of fluorescence decay rates from the excited level, which can yield information about quenching, up-conversion and energy transfer processes.

4.2 Absorption and emission cross-sections

The measurement of absorption and emission cross-sections for rare earths in silica fibre will be illustrated for the erbium system, although the same basic techniques are applicable to any laser system.

Fibre absorption is usually measured using a standard white light cut-back method; light from a white light source is launched into one end of the test fibre, while light exiting the fibre is collected and analysed using a monochromator and detector. The fibre is then cut back to a shorter length, and the measurement repeated. The difference between the results for these two measurements can be attributed to the length of fibre removed, and hence the absorption per metre as a function of wavelength can be calculated. More accurate results are obtained from this method if large-core multimode fibres are used, because a much larger signal is obtained on the detector due to the increase in core area. Measurement using a highly multimode fibre will also require no correction for mode overlap factors (see Chapter 2) due to the essentially uniform excitation across the core. Because the dopant distribution in the fibre is rarely known to high accuracy, this removes one potential source of error.

The fluorescence spectrum can be measured using a laser source to excite the upper level of the test fibre and analysing the fluorescence exiting the fibre, as before. Figure 4.1 shows a partial energy level diagram for erbium; potential pump bands for convenient sources occur around 800 nm, and also for the blue-green argon ion laser lines. Multimode fibre would again be used and the length of fibre will have to be maintained sufficiently short such that reabsorption (in the case of a three-level system such as erbium) or amplification effects are minimised. The fluorescence decay time is also required for evaluation of the cross-section, and this may most conveniently be measured with a pulsed excitation source such as a modulated laser diode, and measuring the decay with a filtered detector placed transversely to the fibre. Decay measurements will be covered in Section 4.4.

Having now obtained the absorption and fluorescence spectra for the tested fibre, we can use Equations (2.10) and (2.11) to convert these into absolute cross-sections. These equations simply relate the integrated emission and absorption cross-sections by a ratio of level degeneracies, and use the radiative transition rate to scale the emission cross-section. To do this, we need to make the assumption that the spontaneous emission rate is equal to the inverse fluorescence lifetime; that is, there is no non-radiative contribution to the decay rate. This assumption holds well for the 1.5 μm transition in erbium, but may not hold so well for other systems, and in this case a measurement of the fluorescence quantum yield is also necessary. The other major assumption behind the use of this method is that all Stark levels are equally populated. This is clearly not strictly true for the broad erbium levels at room temperature, and the effect of this will be discussed later in the chapter.

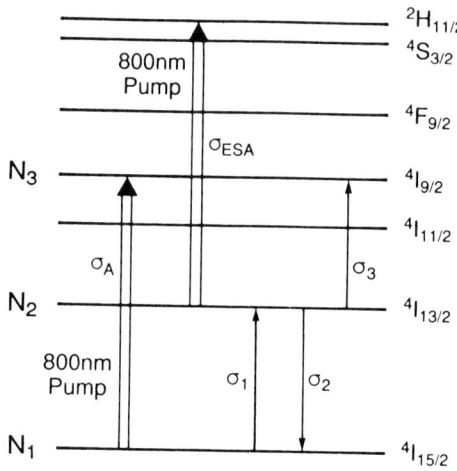

Figure 4.1 Partial energy level diagram for the erbium ion.

SPECTROSCOPY OF RARE-EARTH DOPED FIBRES

Figures 4.2 and 4.3 show calculated cross-sections in the 1.5 μm band for a range of lightly doped SiO_2–Al_2O_3–GeO_2–P_2O_5 fibres, with varying alumina to germania ratios, ranging from all alumina, to all germania, as discussed in Chapter 3. The addition of alumina broadens the transition and reduces the peak cross-sections for absorption and emission. The reduction in peak cross-section due to the broadening of the transition is partially compensated by the increase in integrated line strength; the measured fluorescence decay time reduces from 11.9 ms for the germania fibre to 10.3 ms for the alumina fibre. These calculated cross-sections, and the trends in spectral shape with glass composition changes, are in reasonable agreement with those presented by other workers [1–3]. Figure 4.4 shows measured absorption cross-sections in the range 0.6 to 1.1 μm for one of the above fibres. Peak cross-sections and transition widths for the three main absorption bands are given in Table 4.1. Again, an increasing amount of alumina in the fibre is found to broaden the transitions; the band around 980 nm is particularly sensitive to this. This is of relevance when considering the effectiveness of various pump sources, as present experimental high-power semiconductor lasers in the 980 nm region have broad spectral widths and unpredictable centre wavelengths.

An alternative method to measure cross-sections would be to measure the erbium distribution and concentration accurately in the fibre, and then, using this known concentration, calculate the absorption cross-section from the

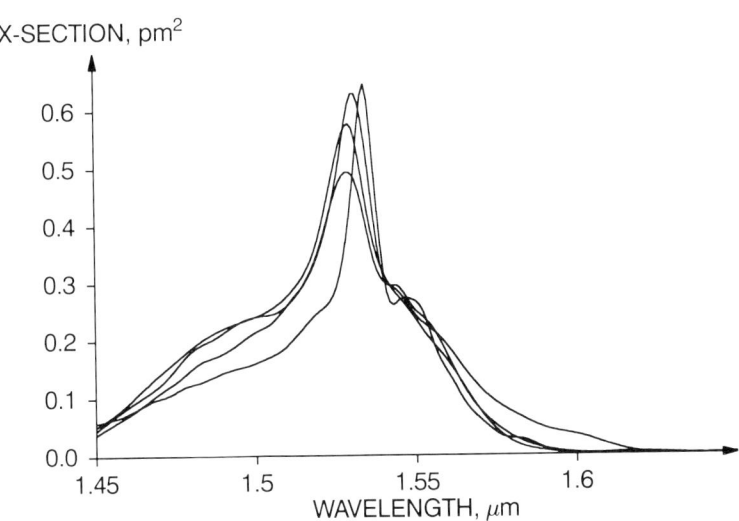

Figure 4.2 Absorption cross-sections around 1.5 μm for erbium in several different host glasses. The highest cross-section corresponds to a pure germania host, the lowest pure alumina, and those in between a mixture of the two, with Al/Ge ratios 1:10 and 1:33. Increasing germania results in an increase in peak cross-section.

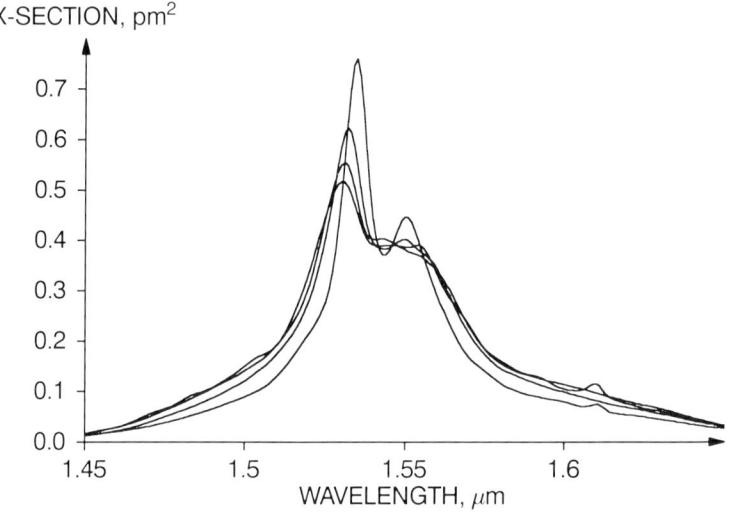

Figure 4.3 Emission cross-sections for fibres in Figure 4.2.

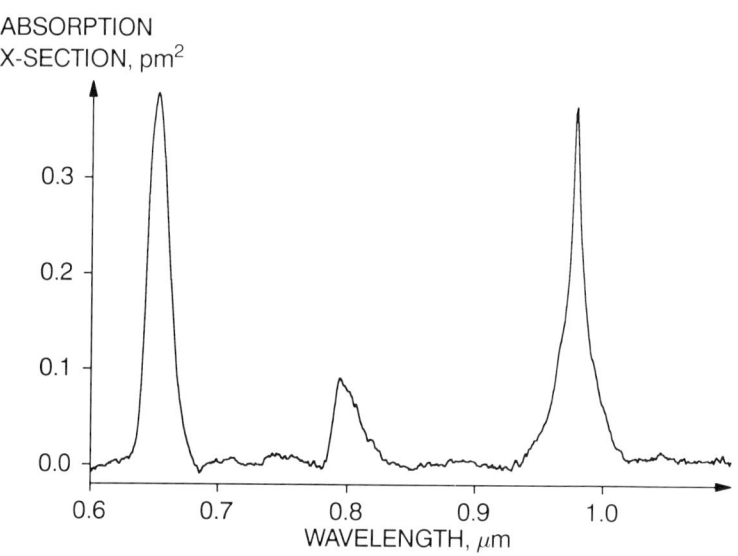

Figure 4.4 Absorption spectrum from 0.6 to 1.1 μm for fibre of Figure 4.2 with Al/Ge ratio of 1:10.

Table 4.1 Peak absorption cross-sections (X), pm², centre wavelengths (P), nm, and transition widths (W), nm, from 0.6 to 1.1 μm for erbium-doped fibres.

Ratio Al:Ge	650			800			980		
	P	W	X	P	W	X	P	W	X
Al only	651	17	0.44	795	20	0.067	978	19	0.24
1:10	652	16	0.49	794	21	0.086	978	17	0.29
1:33	653	18.5	0.4	795	20	0.089	979	12	0.38
Ge only	653	20	0.27	802	20	0.073	980	10	0.36

measured fibre absorption spectrum. The difficulty with this method is that accurate measurements of rare-earth concentrations are extremely difficult in fibre form, and so errors in the cross-sections are quite large. However, it is possible to obtain reasonable data for rare-earth distribution in the fibre preform, and this has been used [3] to check the validity of Equations (2.10) and (2.11), and hence obtain cross-sections applicable to rare earths in preforms, by the following procedure. Having measured the erbium profile, and generated an effective doping density, absorption and fluorescence data are obtained for the preform. (It should be noted here that the authors of reference [3] found considerable differences between preform and fibre spectra, and fluorescence decay properties. This is probably attributable to the different heat treatments involved in preform making and fibre drawing.) From the absorption coefficient, the absorption cross-section may be calculated and the emission cross-section obtained using Equation (2.11) to normalise the data. Equation (2.10) now allows an estimate to be made of the radiative transition rate, and in every case considered this was $\sim 20\%$ larger than that obtained by directly measuring the fluorescence lifetime; that is, the calculated lifetimes are shorter than those measured experimentally. Non-radiative transitions can only shorten the measured lifetime, so the presence of a significant amount of non-radiative decay could only increase this discrepancy. This indicates the breakdown of Equations (2.10) and (2.11), almost certainly because of the unequal populations in the Stark levels, as mentioned earlier. It seems then that cross-sections calculated in the above fashion may only be accurate to $\sim 20\%$, although considerable further work is necessary to clarify this point.

Having calculated the emission and absorption cross-sections, we can now produce additional information useful for the design of erbium amplifiers. In a three-level amplifier system, the net gain at any wavelength is a strong function of the fractional inversion (defined as the population density in the upper laser level relative to the total ion density) integrated along the length of the amplifier (Equation 2.16). Calculation of absolute gain values from the emission and absorption cross-sections would require a full solution of the propagation equations, and accurate knowledge of the ion profile; however, in most cases the shape of the gain spectrum is of most interest, rather than its absolute magnitude. This is a far simpler quantity to derive, as illustrated in Figure 4.5, where the evolution of the gain spectrum with inversion has been calculated for the fibre of Figures 4.2 and 4.3 with an Al:Ge ratio of 1:10. At full inversion, the relative gain spectrum follows the emission cross-section, while for lesser inversions, the gain spectrum becomes modified and substantially flatter. This interplay between the absorption and emission processes in a three-level amplifier allows a certain degree of spectral tailoring to be achieved. It is apparent from Figure 4.5 that the emission cross-section alone is not a good indication of the spectral response of a practical amplifier, and that substantially broader bandwidths may be produced by suitably tailoring the inversion factor [4]. This is clearly of practical importance when optimising fibre composition for a particular spectral response. The validity of this method is confirmed by the experimental results shown in Figure 4.6 [5], where measured small-signal gain data for a 10 m length of Er^{3+}-doped SiO_2–Al_2O_3–P_2O_5 fibre are compared with the calculated gain spectra for several pump powers. The good agreement between the measured and calculated spectra, if appropriate inversion factors are taken, shows that the technique is capable of accurately predicting spectral shapes for three-level amplifiers.

Another important factor for erbium amplifiers pumped directly into the $^4I_{13/2}$ metastable level between 1.45 µm and 1.5 µm is the maximum degree of

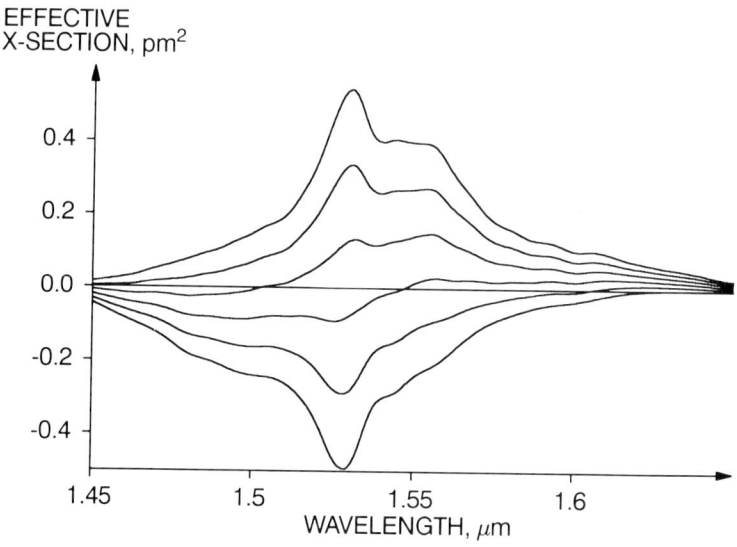

Figure 4.5 Evolution of effective cross-section with inversion parameter for erbium. Inversion parameter varies in steps of 0.2 from 0.0 for bottom spectrum, up to 1.0 for top spectrum.

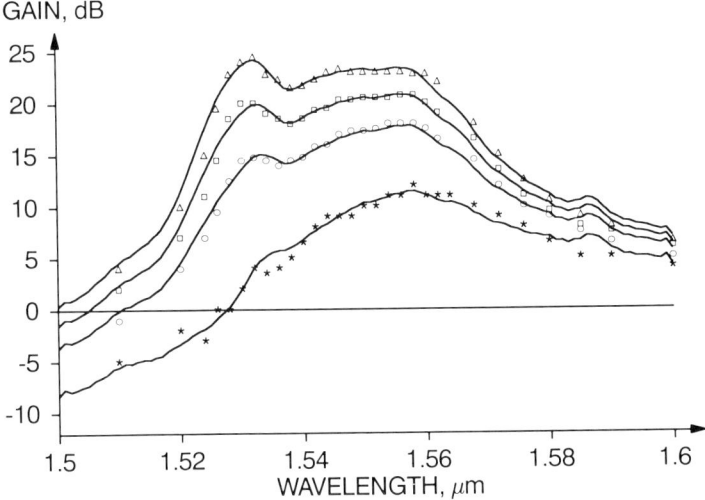

Figure 4.6 Measured and calculated small-signal gain spectra for an erbium amplifier pumped at 1.49 μm. Experimentally measured gains are depicted for pump powers of 16 mW, 25 mW, 33 mW and 50 mW (stars, circles, squares and triangles), while the corresponding calculated curves (solid lines) are for inversion factors of 0.49, 0.56, 0.59 and 0.62 respectively.

inversion achievable. The presence of a finite emission probability at the pump wavelength, as well as the wanted absorption, results in a maximum achievable inversion of less than unity, no matter how hard the amplifier is pumped. In the limit of infinite pump power, the inversion parameter (μ) reaches a maximum value, which can be shown to be

$$\mu_{max} = \frac{\sigma_1}{\sigma_1 + \sigma_2} \tag{4.1}$$

where σ_1 and σ_2 are the absorption and emission cross-sections respectively. This value is of importance as it limits the ultimate noise figure for an amplifier to greater than the 3 dB minimum for an ideal amplifier [6] and restricts the range of integrated inversions available. This implies that only a subset of spectral profiles is possible for $\mu_{max} < 1$. The maximum inversion level for the fibres of Figure 4.2 is shown in Figure 4.7, in the range from 1.45 to 1.55 μm. It can be seen that the maximum inversion in the short wavelength absorption wing is obtained using the germania fibre, while the lowest value is obtained for the alumina fibre. This is consistent with the change in width of the absorption and emission spectra for these fibres.

In this section we have presented methods and measurements for erbium in silica. Corresponding results for erbium in fluoride glasses are given in Chapter 8.

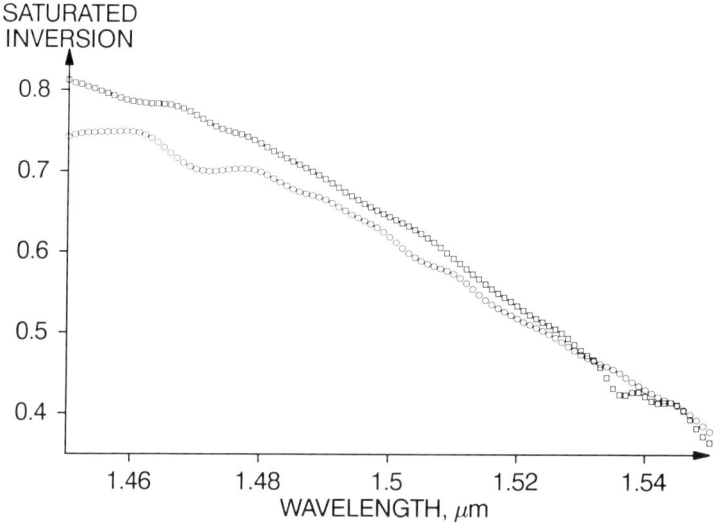

Figure 4.7 Maximum obtainable inversion as a function of pump wavelength for erbium fibres as in Figure 4.2. Pure germania fibre is squares; pure alumina, circles.

4.3 Excited-state absorption

Excited-state absorption, or ESA, is defined as an absorption process which does not originate in the ground state of the laser ion, but from a higher lying excited state, most usually the upper laser level in the ion of interest. ESA can affect oscillators and amplifiers in two distinct ways: through parasitic absorption at the pump wavelength (pump ESA) or at the wavelength of the amplified signal or laser output (signal ESA). The effect of these two processes is different, although both will result in a reduction in device efficiency, and they will be considered separately.

4.3.1 Excited-state absorption at the pump wavelength

The detrimental effects of pump ESA will be illustrated with reference to the erbium system again, because pump ESA is generally of more importance for three-level laser systems such as erbium. Referring again to Figure 4.1, potential ESA transitions arise from the metastable $^4I_{13/2}$ level. Unfortunately, several of these transitions coincide with pump transitions from the ground state; the 800 nm pump wavelength is used as an illustration. The effect of pump ESA can be severe for an amplifier or oscillator pumped at these wavelengths. For the case of erbium in silica, pump ESA does not result in a permanent removal of population from the metastable level; the extremely high non-radiative decay rates from the closely spaced upper levels ensure that ions that become highly excited through pump ESA rapidly relax back to the metastable level. The net effect is that pump energy is being lost for no increase in useful inversion.

This will have a particularly detrimental effect on the efficiency of high-gain amplifiers, which will be pumped close to full inversion. This may be understood qualitatively by considering the pump propagation in such a system. For erbium in silica, the only levels that are appreciably populated are the ground-state and $^4I_{13/2}$ levels (Chapter 2), so the build up of population in other intermediate levels can be neglected. If small-signal operation is also assumed, when the pump propagation is independent of the signal, the expression for pump decay with distance is

$$\frac{dP}{dz} = (\sigma_A N_1 + \sigma_{ESA} N_2)P$$

$$= \left(1 + \frac{\sigma_{ESA}}{\sigma_A}\frac{N_2}{N_1}\right)P\sigma_A N_1. \quad (4.2)$$

The first term on the right-hand side of Equation (4.2) represents 'useful' absorption which is effective in pumping ions from the ground state to the upper laser level, while the second term represents the power loss due to ESA, which produces no overall change in the relative level populations. For comparable ground-state and excited-state absorption coefficients, it can be seen that most of the pump energy is dissipated uselessly in the case of a strongly inverted, highly pumped amplifier, and even small amounts of ESA are expected to have a significant effect. This is quantified in Figure 4.8, where results from a simple plane-wave small-signal model of pump and signal propagation are plotted for varying values of the excited-state to ground-state absorption ratio, using other parameters appropriate to an 800 nm pumped erbium amplifier, and optimum lengths for all pump powers. Clearly even small amounts of ESA have an important effect on amplifier performance. A reduction in efficiency is also expected for fibre oscillators in the presence of pump ESA, although the effect is not normally quite as dramatic in this case, as oscillators are not usually operated with the high gains associated with small-signal amplifiers; overall inversion levels will be much closer to the transparency inversion (inversion necessary to achieve an exact balance between gain and loss at the laser wavelength). A knowledge of the pump ESA coefficient is nevertheless an important factor in the design of an efficient oscillator. It is also necessary to take pump ESA into account if fluorescence quantum efficiency measurements are being made at a pump wavelength which suffers from ESA; the pump intensity must be kept low enough such that only a very small fraction of the ions are in the excited state. If this is not the case, some of the absorbed pump power will be due to ESA, and an incorrectly low value for fluorescence quantum efficiency will be deduced.

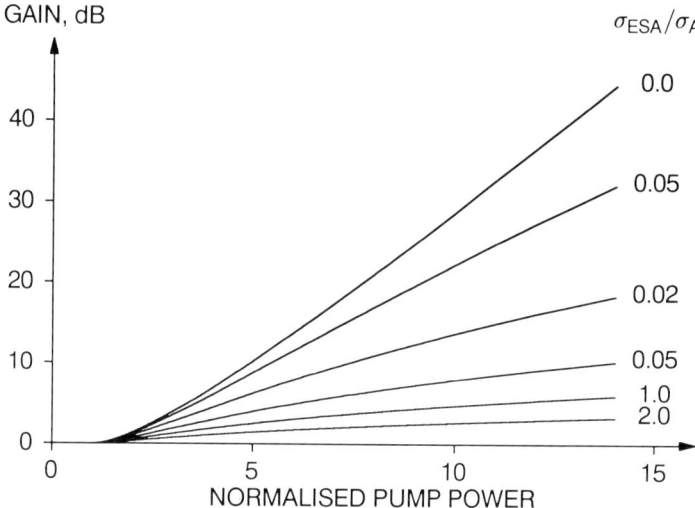

Figure 4.8 Calculated effect of pump ESA on small-signal gain performance of an erbium amplifier. Pump power is normalised to the transparency power, and the fibre length is optimised for each pump power.

To measure the ESA coefficient from the $^4I_{13/2}$ level of a range of erbium doped fibres, a colour-centre laser tuneable around 1.5 μm to excite ions from the $^4I_{15/2}$ ground state to the $^4I_{13/2}$ excited state was used, and the change in transmission that this produces monitored [7]. For excitation powers many times the saturation power, this transition becomes strongly saturated, and a reproducible, well-defined population is obtained in the excited state. This enables the relative contributions of ground-state bleaching and excited-state absorption to be assessed. Other excitation methods have been used [8], but pumping directly into the metastable level using a colour-centre laser has the advantage that no visible fluorescence or scatter is produced. The degree of inversion achieved in this way varies slightly from fibre to fibre, and is a function of pump wavelength, as discussed earlier (see Figure 4.7). For most fibres, pumping near 1.53 μm produces an inversion of ~ 0.5. This is confirmed by comparing the fluorescence from the first excited state under these conditions with that produced when the transition is totally inverted with a high-power argon laser.

The experimental arrangement for these measurements is depicted in Figure 4.9. The length of doped fibre was adjusted to be long enough to produce a few decibels absorption around the visible absorption lines, while still remaining short enough that the transition is strongly saturated over the whole length of the fibre with the available pump power. By using the computer controlled spectrometer and white light source to measure transmission spectra with and without the colour-centre pump, and then performing a cut

SPECTROSCOPY OF RARE-EARTH DOPED FIBRES 89

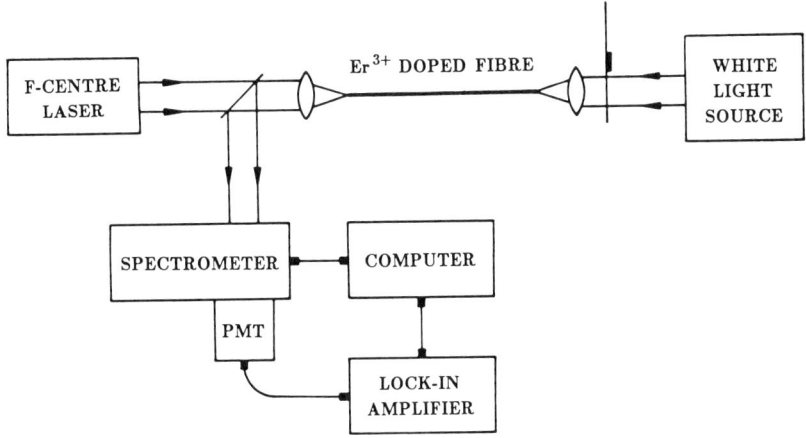

Figure 4.9 Experimental arrangement to measure pump ESA in erbium-doped fibres.

back to calibrate the system, values for the change in fibre transmission between the pumped and unpumped states can be obtained and hence the ground-state and excited-state absorption coefficients as a function of wavelength determined. A typical set of spectra is shown in Figures 4.10(a)–(c), for a SiO_2–GeO_2–P_2O_5 host. The ground-state and excited-state spectra are shown separately, together with the spectrum of the difference between these two coefficients. In this third graph, a negative-going feature corresponds to a wavelength for which the ground-state absorption exceeds the excited-state absorption, and hence corresponds to a favourable pump wavelength. A particular region of interest is around 800 nm, corresponding to wavelengths available from commercial AlGaAs lasers. Here it can be seen that a considerable overlap exists between the pump band and the excited-state absorption band. At other wavelengths – around 660 nm, for example – the situation is much more favourable, with ESA coefficients considerably less than ground-state coefficients. A number of fibres with differing compositions has been measured, and the results are presented in Tables 4.2 and 4.3. These results are in broad agreement with published work from other laboratories, for comparable fibre types [8,9] which is summarised in Table 4.4. It can be seen that increasing phosphorus, and the addition of aluminium, are both beneficial for the reduction of ESA around 800 nm. The situation is much more difficult to interpret in the 650 nm pump band because the ESA and ground-state absorption features are only overlapping in the wings, and slight changes to the measurement wavelength make large changes to this ratio. This is most clearly seen in the difference spectrum, Figure 4.10(c). Nevertheless, it is clear that an improvement in the excited-state to ground-state absorption ratio of 5- to 10-fold is achievable in this pump band compared to the 800 nm band. It is

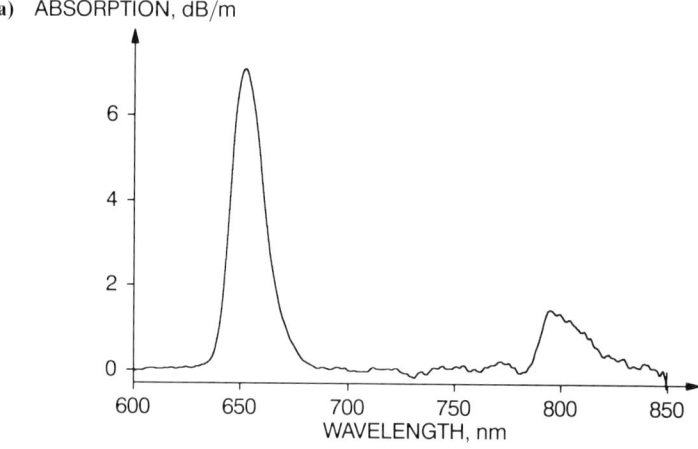
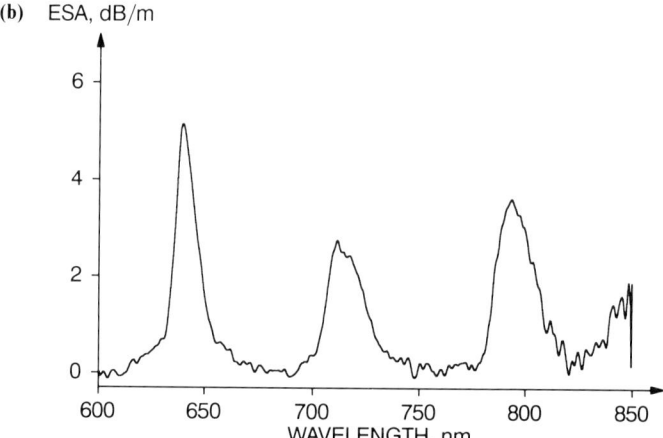
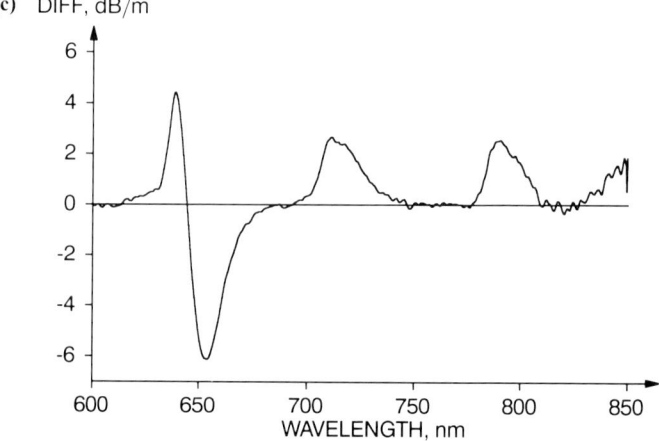

Figure 4.10 (a) Ground-state, (b) excited-state, and (c) difference spectra, for an aluminosilicate fibre.

Table 4.2 Fibre composition for erbium ESA studies.

Fibre	Material	Core co-dopants (ratio)	Er^{3+} concentration (ppm)	Core diameter (μm)	Δn
A	Silica	GeO_2	1600	5.3	0.0135
B	Silica	GeO_2–P_2O_5 (5:1)	500	7.5	0.0074
C	Silica	GeO_2–P_2O_5 (1:3)	600	6.5	0.008
D	Silica	GeO_2–P_2O_5 (1:5)	300	7.2	0.0105
E	Silica	GeO_2	500	3.1	0.033
F	Silica	GeO_2–P_2O_5–F (20:2:1)	300	10	0.006
G	Silica	Al_2O_3–P_2O_5	500	4.1	0.010
H	ZBLAN	ZBLANP	300	40	0.16

Table 4.3 ESA measurements on fibres in Table 4.2.

	Features around 650 nm					Features around 800 nm				
	GSA		ESA		ESA/GSA	GSA		ESA		ESA/GSA
Fibre	λ (nm)	Mag (dB)	λ (nm)	Mag (dB)	at 655 nm	λ (nm)	Mag (dB)	λ (nm)	Mag (dB)	at 810 nm
A	655	2.8	642	1.9	0.17	809	0.8	795	2.0	1.6
B	655	3.9	641	2.5	0.17	805	1.1	793	2.9	1.5
C	652	7.9	640	5.2	0.18	801	1.5	793	4.4	1.0
D	652	4.0	640	2.6	0.21	805	1.0	792	1.7	0.8
E	655	3.3	642	2.3	0.17	804	0.9	796	2.3	1.2
F	657	3.0	655	2.1	0.72	806	0.9	802	1.2	1.4
G	652	7.1	640	5.2	0.14	796	1.4	793	3.5	1.1
H						802	3.4	793	6.3	0.8

Table 4.4 ESA/GSA measurements on erbium-doped fibre from other labs.

Wavelength (nm)	Fibre composition				
	GeO_2	GeO_2–B_2O_3	GeO_2–P_2O_5	Al_2O_3	GeO_2
488	2.9		1.86	1.74	
514.5	0.95		0.55	0.5	
655	0.28	0.25	0.13	0.14	
810	2.0	2.0	1.0	1.0	1.4
980				0	
Reference	[8]	[8]	[8]	[8]	[9]

also clear from Figure 4.10 that a considerable improvement in the ESA:GSA ratio can be achieved by operating in the long wavelength wing of the 800 nm absorption. High gains have recently been achieved from erbium amplifiers pumped in this fashion [10]. Apparent from Table 4.4 is the complete lack of ESA at 980 nm, the pump wavelength for the $^4I_{11/2}$ level. This pump transition will be discussed in subsequent chapters, as it has been used to produce efficient oscillators [11, 12] and amplifiers [13–15], although, as commercial semiconductor lasers are not yet available in this wavelength region, other pump wavelengths remain of interest.

4.3.2 Excited-state absorption at the signal wavelength

If we now consider excited-state absorption at the signal wavelength, then we find that, although small amounts of signal ESA reduce amplifier gain only slightly, larger values can have a very serious effect. Considering the erbium in silica system again, a potential ESA transition from the $^4I_{13/2}$ metastable level to the $^4I_{9/2}$ excited state is illustrated in Figure 4.1. If σ_3 is non-zero, then the net stimulated emission coefficient is given by

$$\sigma_{\text{eff}} = \sigma_2 - \sigma_3 \tag{4.3}$$

and if the ESA coefficient exceeds the stimulated emission coefficient, the upward absorption rate will exceed the stimulated emission rate, and no gain is possible no matter how much pump power is available. Lower levels of ESA, while still allowing gain and laser action, will nevertheless affect device efficiency [16]. Pump probe measurements on the neodymium in silica system at 1.3 μm [17] have confirmed that ESA at the signal wavelength is responsible for the lack of success in constructing a 1.3 μm source in this system. Other hosts are much more favourable for this transition, as will be discussed in a subsequent chapter. We shall concentrate here on erbium.

Two methods to estimate the magnitude of signal ESA for the 1.5 μm transition will be described. The first relies on the measurement of the change in transmission of a fibre, at the signal wavelength, for pumped and unpumped cases. If we define the transmission through the test fibre for a particular signal wavelength as T_0, and the transmission at that same signal wavelength under pumped conditions as T_1, with N_0 ion density in the fibre, N_1 ions in the excited state under pumped conditions and overlap factor k between the signal beam and the ion distribution, then we can write down the following equations, for a fibre of length l, assuming no saturation at the signal wavelength

$$T_0 = \exp(-\sigma_1 N_0 k l) \tag{4.4}$$

and

$$T_1 = \exp(-\sigma_1 (N_0 - N_1)l + N_1(\sigma_2 - \sigma_3)l). \tag{4.5}$$

Taking logs, and subtracting (4.4) from (4.5), we get

$$\ln(T_1) - \ln(T_0) = k l N_1 (\sigma_1 + \sigma_2 - \sigma_3) = k'(\sigma_1 + \sigma_2 - \sigma_3). \tag{4.6}$$

Thus, the shape of this change is independent of both the degree of excitation and the degree of overlap between the signal mode and the ion distribution, as these are only scaling factors. The cross-sections σ_1 and σ_2 may be measured as described in (4.1) above, and the spectral shape of their sum compared with an experimentally measured difference of log transmissions. Any discrepancy in these shapes will indicate the presence of a non-zero σ_3. Applying this technique to a four-level transition such as the 1.3 μm neodymium system [17] is simpler, as $\sigma_1 = 0$, for a four-level system. In this case, pumping will produce either absorption or gain, depending on the relative magnitudes of σ_2 and σ_3. The point at which $\sigma_2 = \sigma_3$ is the wavelength at which the fibre remains transparent when pumped.

The second measurement method is complementary to the first, and utilises the fluorescence from the higher lying terminal state for the signal ESA to map out the shape of the signal ESA spectrum. The fluorescence from any level is proportional to the number density in that level, and so can be used as a monitor for that population. Thus, (Figure 4.1)

$$P_{\text{fluor}}(1.5\,\mu\text{m}) = k_1 N_1 \quad (4.7\text{a})$$

and

$$P_{\text{fluor}}(0.8\,\mu\text{m}) = k_2 N_2 \quad (4.7\text{b})$$

where k_1 and k_2 are constants containing lifetimes, geometrical factors, etc. If we now consider the transition from the $^4I_{13/2}$ level to the $^4I_{9/2}$ level, then (Section 2.5.2):

$$N_2 = k_3 \sigma_3 N_1 P_{\text{las}} \quad (4.8)$$

where k_3 is a constant, P_{las} is the power of the laser used to stimulate the transition, and we have made the assumption that the transition is unsaturated; that is $N_2 \ll N_1$. This is a good approximation for the present case due to the high non-radiative decay from the $^4I_{9/2}$ level. If we now substitute (4.7a) and (4.7b) in (4.8), and rearrange, we obtain an expression for σ_3:

$$\sigma_3 = \frac{k_1 P_{\text{fluor}}(0.8\,\mu\text{m})}{k_2 k_3 P_{\text{fluor}}(1.5\,\mu\text{m}) P_{\text{las}}}. \quad (4.9)$$

Therefore, if we measure fluorescence at 1.5 μm simultaneously with fluorescence at 0.8 μm, together with the laser power at the pump wavelength, we have a measure proportional to σ_3 from (4.9), and can map out the spectral variation of σ_3 as a function of pump wavelength.

In the first experiment, in which transmission changes are measured, an Ar^+ ion laser operating at 528.7 nm was used to pump a length of Er^{3+}-doped fibre, and two external cavity semiconductor diode lasers [18] with a combined tuning range of 1.49–1.64 μm were used to probe the transmission of the doped fibre. Approximately 100 mW of 528.7 nm radiation was launched into the fibre, this power being sufficient to invert the 1.5 μm laser transition almost totally. The transmission of the fibre in its pumped and unpumped states was then measured at a large number of wavelengths across the 1.5 μm band. The results of these measurements are displayed in Figure 4.11, where the difference between the log transmissions is shown as a function of wavelength for a typical SiO_2–Al_2O_3–P_2O_5 host. Also shown in Figure 4.11 is the sum of the absorption and emission cross-sections, normalised to the experimental data. From the close correspondence between the shapes of these two sets of data, it can be concluded that the cross-section for excited-state absorption at the signal wavelength is insignificant in comparison to σ_1 and σ_2 over the majority of the fluorescence band.

The experimental arrangement for the second measurement is shown in Figure 4.12. A short length of doped fibre is pumped in the 1.5 μm band with a tuneable colour-centre laser. 800 nm fluorescence from the $^4I_{9/2}$ level is detected by a photomultiplier, a germanium photodiode detects 1.5 μm fluorescence from the $^4I_{13/2}$ level, and a power meter measures the pump power exiting the fibre close to where the measurements are made. Measuring the fluorescence transversely in this fashion enables fluorescence to be collected from a short, well-defined length of fibre, over which the pump laser power will only change by a small amount. From (4.9), the ratio of the fluorescence at 0.8 μm to the product of the 1.5 μm fluorescence and the 1.5 μm pump power is proportional to the desired absorption cross-section σ_3 from $^4I_{13/2}$ to $^4I_{9/2}$. Figure 4.13 shows the results of this experiment for the same fibre as in Figure 4.11. A large peak is seen at 1.63 μm, while a relatively low level signal is present at shorter wavelengths. This result confirms that signal ESA is at a low level in the main erbium window between 1.53 and 1.56 μm, but

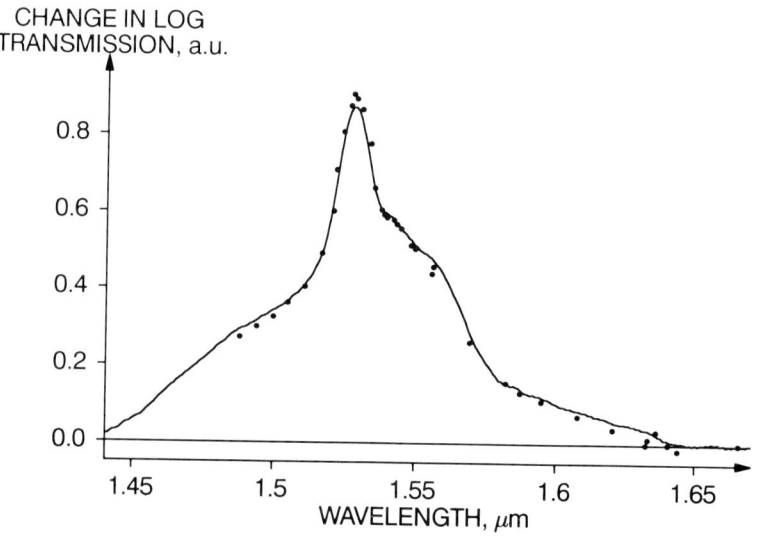

Figure 4.11 Normalised sum of absorption and emission cross-sections (solid line), and experimentally measured transmission changes with pumping (points) for an aluminosilicate fibre.

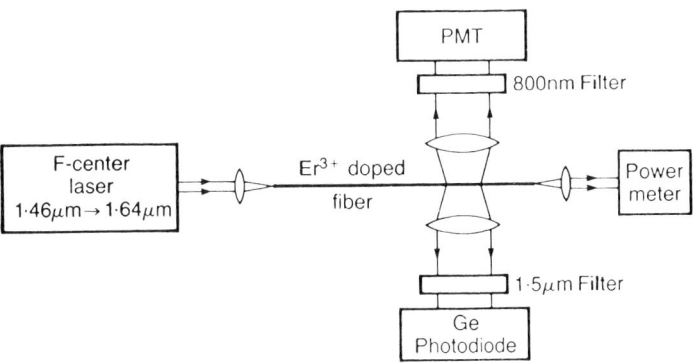

Figure 4.12 Experimental arrangement for measurement of spectral shape of signal ESA around 1.5 μm in erbium.

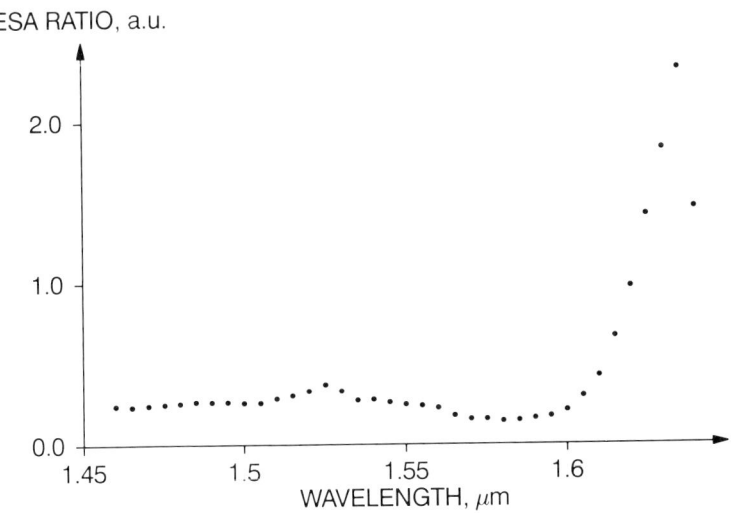

Figure 4.13 Spectral shape of ESA around 1.55 μm in erbium.

is becoming significant for wavelengths longer than 1.6 μm. An indication of this is also visible in Figure 4.11 in the 1.63 μm region, although the data are quite noisy here.

It is therefore concluded that ESA at the signal wavelength is not serious for erbium amplifiers in the 1.5–1.6 μm band, but may become a problem for oscillators working longer than 1.6 μm. The effect of this will be to limit their efficiency and maximum tuning range.

4.4 Fluorescence measurements

In the second part of this chapter we shall look at how detailed measurements of the decay of fluorescence from the upper laser level can provide information about the relative importance of ion–ion interactions such as quenching processes and up-conversion. These are undesirable processes for oscillators and amplifiers, and increase pump power requirements for a given performance by decreasing the effective lifetime of the upper laser level. Quenching and up-conversion have been discussed in Section 2.4.3.2, and are illustrated in Figures 2.8(a) and (b). The quenching process particularly affects neodymium, while the up-conversion process has an adverse effect on the 1.5 μm erbium transition. Because both processes typically have a probability that depends on the ion separation as the inverse sixth power, they will both have a strong dependence on doping density. In addition, we expect differing behaviours from the two processes as a function of excitation level. The resonant quenching process depends upon an interaction between a ground-state ion and an excited-state ion, and therefore its effect is expected to be relatively independent of excitation level for low to medium inversion levels. The up-conversion process, on the other hand, occurs between two excited ions and is negligible for extremely low excitation levels, becoming dominant at high levels of excitation if the basic ion density is high enough. This will manifest itself as an excitation dependent fluorescence decay rate, and produce strongly non-exponential decay curves, for high excitation levels, returning to exponential decay for low levels of excitation. To illustrate both these effects, and to quantify their magnitudes in a silica host, we have performed detailed fluorescence decay measurements on both Nd^{3+}- and Er^{3+}-doped fibres.

As both Nd^{3+} and Er^{3+} dopants have an absorption band near 800 nm, a modulated AlGaAs semiconductor laser makes an ideal excitation source. To increase the amount of pump available to excite the fibre, a laser array has been used, which allows up to ~ 100 mW to be launched into the test fibre. The fluorescence in the region of interest is detected transversely to the fibre, close to the input, by a germanium photodiode, together with an appropriate filter. Transverse detection is preferable to longitudinal detection, as stray pump levels are far lower, reabsorption effects are eliminated and, for high-concentration samples, fluorescence may be detected from highly excited regions only by suitably positioning the detector close to the pump input. If intensity dependent effects are under consideration, it is preferable to restrict the fluorescence detected to that emanating from a uniformly excited region. It is also important with high-concentration samples to use short lengths, and broken exit ends of the fibre, to remove the possibility of the build up of amplified spontaneous emission, or lasing, which would itself lead to an apparent lifetime shortening. The overall time resolution of this detection system is $\approx 1\,\mu s$, compared to typical fluorescence decay times of hundreds of microseconds. Decay curves are stored on a Tektronix 7854 digital storage

SPECTROSCOPY OF RARE-EARTH DOPED FIBRES 97

oscilloscope to allow subsequent analysis of the signal to determine decay rates and departure from exponentiality.

4.4.1 Neodymium

If we first consider the case of neodymium in silica, then we are interested in fluorescence from the metastable $^4F_{3/2}$ level (Figure 4.14) as this is the upper level for the 940 nm (to the $^4I_{9/2}$ ground state) and 1.06 μm (to the $^4I_{11/2}$ level) lasing transitions. We have studied a number of fibres, with SiO_2–Al_2O_3–P_2O_5 host glass, and up to 15 wt% of Nd^{3+} dopant [19].

Typical results are shown in Figures 4.15, 4.16, and 4.17 for several doping levels. Three traces are displayed for each dopant concentration for clarity. Firstly, the decay curve itself is shown, secondly, its natural logarithm, and finally, the differential of the natural logarithm. This method of data display makes any small departures from exponentiality easy to see, and shows the instantaneous decay rate at any time in the decay process. A purely exponential decay would have a constant slope of the natural logarithm against time, and a differential logarithm which is constant. Figure 4.15 is a typical result for a low (<0.5 wt%) concentration sample. The decay curve in this case is very close to a true exponential, with a decay rate of $2 \times 10^3 \, s^{-1}$, which corresponds to an equivalent lifetime of 500 μs, which is very close to Nd^{3+} in a pure silica host glass [20].

As the concentration is increased above 0.5 wt%, the decay rate becomes faster, as illustrated in Figure 4.16 for 4 wt% Nd^{3+}, and tabulated in Table 4.5

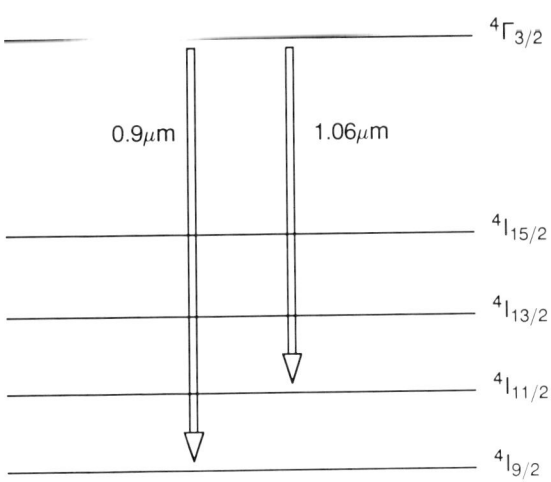

Figure 4.14 Partial energy level diagram for neodymium.

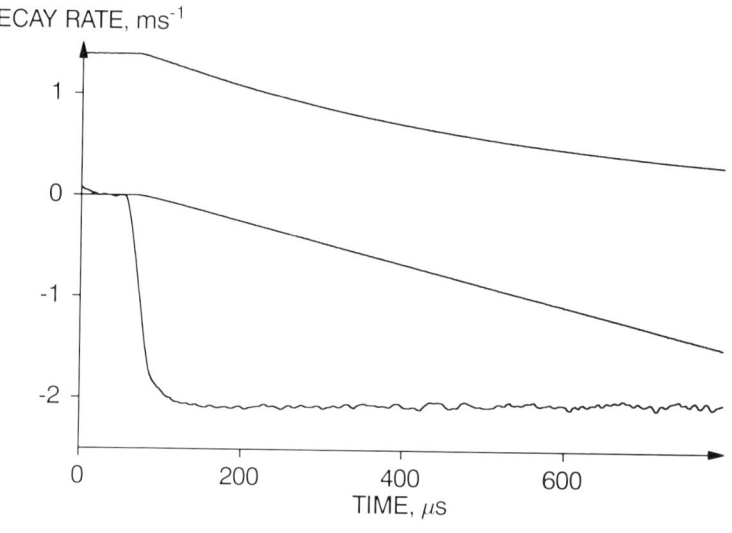

Figure 4.15 Fluorescence decay curves for 0.01 wt% neodymium.

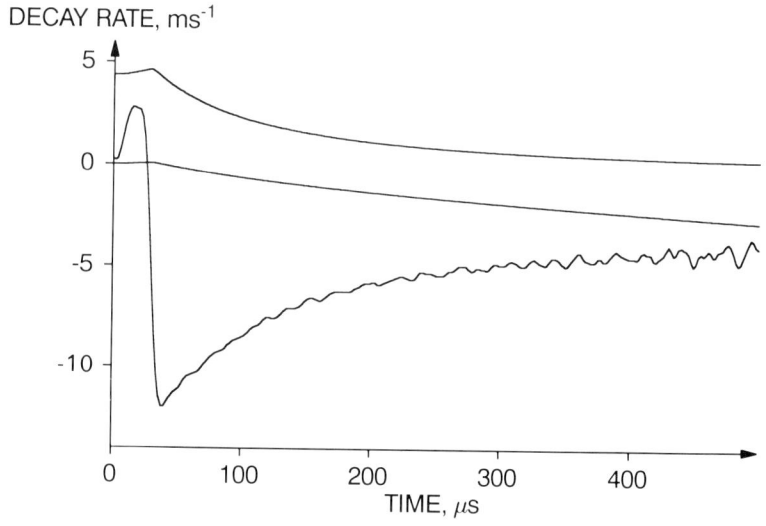

Figure 4.16 Fluorescence decay curves for 4 wt% neodymium.

for a range of concentrations. Also evident from Figure 4.16 is a non-exponential decay characteristic, with an initial decay rate ~ 4 times higher than the final rate. Similar behaviour has also been observed in bulk laser glasses [21]. This is believed to be a manifestation of the resonant quenching process discussed earlier. In this case, the relevant upper level is the $^4F_{3/2}$, and the most likely intermediate level is the $^4I_{15/2}$. The critical concentration of

Table 4.5 Fluorescence decay characteristics for a range of neodymium-doped fibres.

[Nd^{3+}] (wt%)	Cluster comp.	Decay rate (ms^{-1})	
		max.	min.
0.04	No	2.0	2.0
0.5	No	2.2	2.0
1.0	No	3.6	2.5
3.9	No	13	4
7	No	16	6
15	Yes	12	5

Nd^{3+} appears to be 0.5 wt%, which produces a 10% lifetime shortening at the beginning of the decay trace, while for the 1 wt% sample the initial decay rate is approaching twice that of the low-concentration samples. When the concentration is increased to 4 wt%, quenching becomes the dominant decay mechanism. The severe non-exponentiality that is also apparent on the decay traces is believed to reflect the distribution of ions within the glass host, with isolated excited ions living longer than those in close proximity to other ions. Up-conversion does not seem to play a role for this ion, as the decay traces do not depend on the excitation level to within experimental error, whereas an up-conversion process would increase in severity as excitation level increases. This result also suggests that local heating due to the absorbed pump power is sufficiently small to have little effect on the observed decay rates. The decay rate for these multimode fibre samples was also dependent to some extent upon launch conditions, indicating some variation in dopant concentration across the core. The figures in Table 4.5 cover the whole range of observed decay rates; typical variation of peak decay rate across a sample was ∼ 30%.

For the highest concentrations, two distinct decay regions are seen. Figure 4.17 shows the fluorescence from a 15 wt% sample, with an extremely rapid initial decay, which is unresolved by the experimental apparatus, followed by a non-exponential curve similar to that seen at lower concentrations. This sample was drawn from a preform that showed visible signs of clustering (see Chapter 3). Analysis revealed two distinct phases, one containing 40 wt% Nd^{3+}, while the background glass was ∼ 6 wt% Nd^{3+}. The high concentration region would be expected to yield an extremely rapid decay rate, while the medium concentration background glass should decay with a characteristic appropriate to a uniform fibre of that concentration. This agrees reasonably well with our measurements, suggesting that rare-earth concentrations in the clustered and unclustered regions are similar in preform and fibre.

To summarise, we have shown that for this range of fibres no deviation from exponential fluorescence decay, or increase in the fluorescence decay

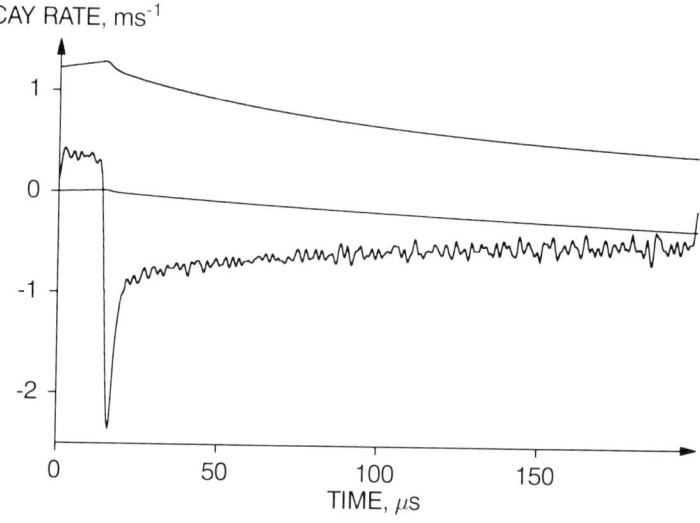

Figure 4.17 Fluorescence decay curves for 15 wt% neodymium.

rate, was observed for Nd^{3+} concentrations up to 0.5 wt%. Intermediate concentrations up to 7 wt% showed a single, non-exponential decay, with the initial decay rate exceeding the final rate by up to a factor of 4. At the highest concentrations of 15 wt%, phase separation was apparent in the samples, and the fluorescence decay curves had two distinct regions corresponding to the concentrations of the two phases. The neodymium concentration should therefore be kept below 0.5% if efficient devices are required.

4.4.2 Erbium

We now consider the case of erbium, in both SiO_2–Al_2O_3–P_2O_5 and SiO_2–GeO_2 hosts. We have carried out a similar series of measurements to those detailed above for the Nd^{3+}-doped fibre, for a range of Er^{3+} concentrations ranging from 0.02 to 18 wt%.

Figure 4.18 shows results which are typical of fibres with Er^{3+} concentration < 0.1 wt%. The decay is a smooth exponential, and has an equivalent lifetime of ~ 10 ms. This low-level lifetime is slightly host dependent, as shown by the decay rates in Table 4.6, varying from ~ 12 ms to ~ 10 ms. Decay traces are shown for two excitation levels from the laser array; the maximum input level corresponding to ~ 100 mW launched power, and the low-intensity trace ~ 10 mW. No significant changes in decay rate are seen with excitation level for these low-concentration samples.

Figure 4.19, on the other hand, shows the behaviour for an Er^{3+} concentration of 2.3 wt%. In this case, we see a strongly non-exponential decay

SPECTROSCOPY OF RARE-EARTH DOPED FIBRES

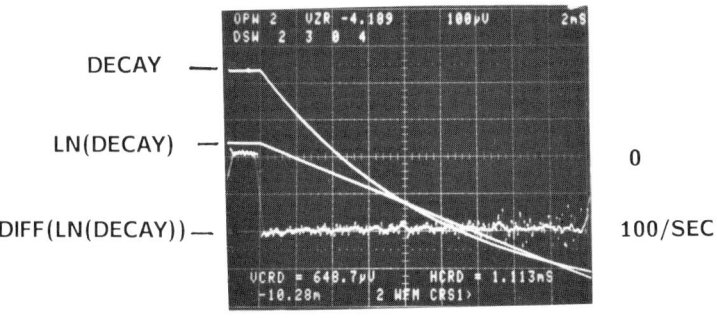

MAXIMUM INPUT

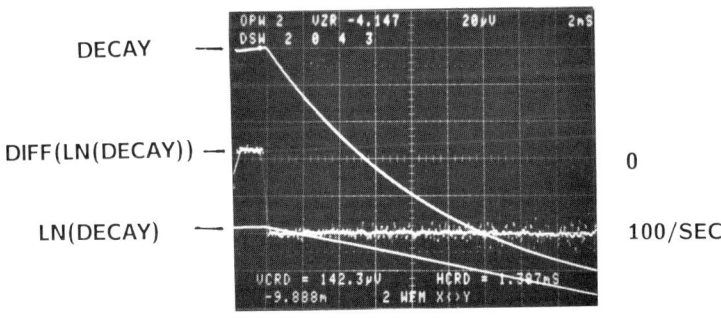

10% INPUT

Figure 4.18 Fluorescence decay for 0.07 wt% erbium. The zero for the decay and log decay traces is the bottom of the screen, while the zero for the differential trace is the centre of the screen.

for input powers of ~ 100 mW, returning to a nearly exponential decay late in the decay process, or for inputs reduced to 1% of maximum. Results for other concentrations display a similar behaviour, and are summarised in Table 4.6 for maximum excitation. These results are consistent with a cooperative up-conversion mechanism, as discussed earlier. Excited ions in the $^4I_{13/2}$ metastable level can transfer energy between themselves in such a way that one ion is excited to the $^4I_{9/2}$ level while the second ion returns to the ground state.

Table 4.6 Fluorescence decay characteristics for a range of erbium-doped fibres.

$[Er^{3+}]$ (wt%)	Host glass	Cluster comp.	Decay rate (s^{-1}) max.	Decay rate (s^{-1}) min.	Quantum effic. (a.u.)
0.02	Ge	No	93	93	–
0.07	Al	No	97	97	–
0.15	Ge	No	91	84	206
0.15	Al	No	115	102	202
0.6	Ge	No	110	87	–
0.2	Al	No	115	97	–
1.1	Ge	No	125	88	197
1.5	Al	No	240	100	267
2.3	Al	No	360	102	216
2.4	Ge	Yes	140	92	89
7.5	Al	No	1800	110	–
14	Al	Yes	700	116	–
18	Al	Yes	300	102	–

The $^4I_{9/2}$ level will then rapidly relax to the $^4I_{13/2}$ level by non-radiative decay. This produces a decay rate depending on the excited-state density, as observed.

Detection of emission at 0.8 μm from the $^4I_{9/2}$ level, from ions excited in the up-conversion process, would provide further evidence that this process was occurring. This emission has indeed been observed, and Figure 4.20 shows its temporal behaviour, after strongly exciting the 2.3 wt% fibre with 1.5 μm radiation from a colour-centre laser. The decay trace is extremely noisy due to the very low fluorescence quantum yield from this level, $< 10^{-4}$. Note that the decay characteristic shown here is not due to the fluorescence lifetime of the $^4I_{9/2}$ level which has been measured to be < 10 μs, but is instead indicative of a decaying pumping mechanism into the level, followed by rapid fluorescence or non-radiative decay. This time constant must therefore be connected with the lifetime of the pumping level. Up-conversion is expected to depend upon the square of the population in the excited state, and so, if this were the pumping mechanism for the $^4I_{9/2}$ level, this could be verified by simultaneously monitoring the fluorescence at 1.5 μm from the $^4I_{13/2}$ level to indicate the population density in that level. The square of this 1.5 μm emission is therefore also shown in Figure 4.20, and may be seen to have a similar temporal dependence to the 0.8 μm emission produced by the up-conversion. There seems little doubt then that the observed non-exponential decays, and increased decay rates, are due to this up-conversion mechanism.

At concentrations of 2.4 wt% in the germania host and 14 wt% in the alumina host, rapid decay components were observed, similar to those in high-concentration neodymium-doped fibres. The fibre preforms were visibly

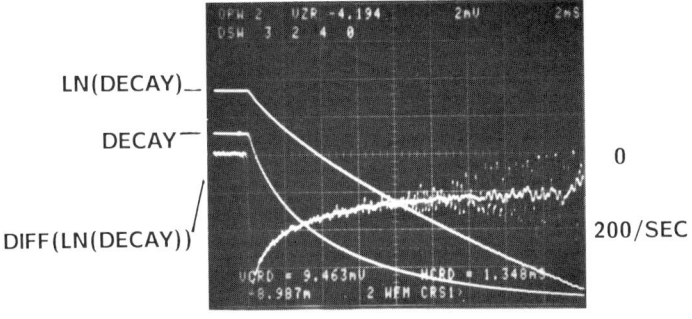

MAXIMUM INPUT

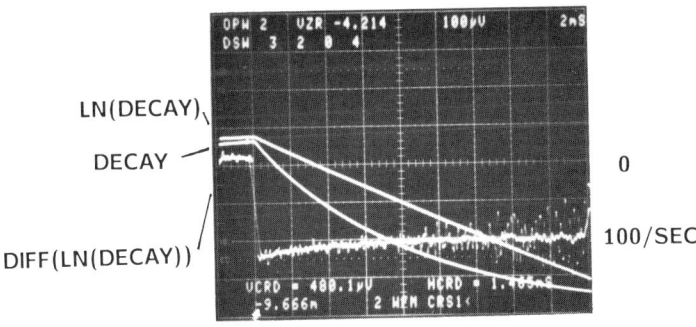

1% INPUT

Figure 4.19 Fluorescence decay for 2.3 wt% erbium. Scales as Figure 4.18.

clustered, and a fast decay rate would be expected from the high concentration of rare earth in the clusters, together with a decay characteristic of the background concentration of erbium. It may be seen from these results that nearly an order of magnitude more rare earth can be incorporated without clustering in the alumina host. This result is of significance if a short device is required, regardless of efficiency.

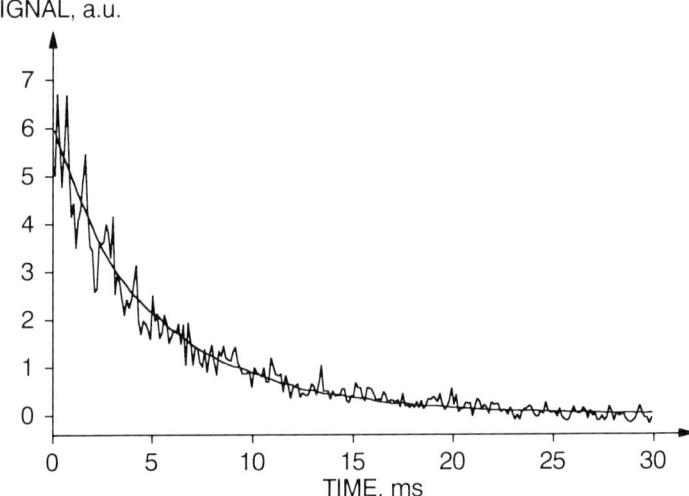

Figure 4.20 Temporal behaviour of the 0.8 μm fluorescence compared with the square of the 1.5 μm fluorescence from erbium. The 0.8 μm fluorescence is the noisy trace. The strong correlation indicates that up-conversion is the dominant mechanism for populating the $^4I_{9/2}$ level.

The relative quantum efficiency of several of the above fibres has also been measured, using an argon ion pump source; these results are also given in Table 4.6. Care was taken during these measurements to ensure that the fundamental quantum efficiency was being measured, rather than a value reduced by the presence of pump ESA, by taking data at a number of pump powers and extrapolating to zero power. A relatively constant value is obtained for all unclustered fibres, even at high concentrations, while the clustered fibre shows a strong reduction in quantum efficiency, as expected.

Erbium concentration should therefore be kept below ~ 0.1 wt% if undesirable up-conversion processes are to be avoided. Above this level, while the quantum efficiency remains high, up-conversion becomes a significant problem at high pumping levels and will result in amplifier inefficiency. Finally, clustering occurs for concentrations of 2.4 wt% in the germania host and 14 wt% in the alumina host, and concentrations should be kept below these levels if clustering is to be avoided.

4.5 Conclusions

In this chapter we have considered basic absorption and fluorescence cross-section measurements, and discussed how these may be used to predict the spectral response for a three-level amplifier such as erbium. Excited-state absorption has been measured for erbium fibres for both pump and signal wavelengths, and fluorescence measurements for neodymium and erbium have defined concentration limits in silica fibre to avoid quenching and up-conversion losses.

References

1. W.J. Miniscalco, L.J. Andrews, B.A. Thompson, T. Wei and B.T. Hall, *Proc. SPIE Conf. on Fiber Laser Sources and Amplifiers, Boston*, **1171** (1989) 93.
2. E. Desurvire and J.R. Simpson, *IEEE J. Light. Technol.* **7** (1989) 835.
3. K. Dybdal, N. Bjerre, J. Engholm Pedersen and C.C. Larsen, *Proc. SPIE Conf. on Fibre Laser Sources and Amplifiers, Boston* **1171** (1989) 209.
4. J.R. Armitage, *IEEE J. Quant. Electron.* **26** (1990) 423.
5. C.G. Atkins, J.F. Massicott, J.R. Armitage, R. Wyatt, B.J. Ainslie and S.P. Craig-Ryan, *Electron. Lett.* **25** (1989) 910.
6. R. Olshansky, *Electron. Lett.* **24** (1988) 1363.
7. C.G. Atkins, J.R. Armitage, R. Wyatt, B.J. Ainslie and S.P. Craig-Ryan, *Opt. Commun.* **73** (1989) 217.
8. R.I. Laming, S.B. Poole and E.J. Tarbox, *Opt. Lett.* **13** (1988) 1084.
9. L.J. Andrews, W.J. Miniscalco and T. Wei, *Proc First Int. School on Excited States of Transition Elements, Poland* (1988), p. 9.
10. K. Suzuki, Y. Kimura and M. Nakazawa, *Electron. Lett.* **26** (1990) 948.
11. R. Wyatt, *Electron. Lett.* **25** (1989) 1498.
12. W.L. Barnes, L. Reekie and D.N. Payne, *IOOC '89 Technical Digest*, Paper 20A3-3.
13. R.I. Laming, M.C. Farries, P.R. Morkel, L. Reekie, D.N. Payne, P.L. Scrivener, F. Fontana and A. Righetti, *Electron. Lett.* **25** (1989) 12.
14. M. Shimizu, M. Horiguchi, M. Yamada, M. Okayasu, T. Takeshita, I. Nishi, S. Uehara, J. Noda and E. Sugita, *Electron, Lett.* **26** (1990) 499.
15. J.F. Massicott, R. Wyatt, B.J. Ainslie, S.P. Craig-Ryan, *Electron. Lett.* **26** (1990) 1038.
16. J.R. Armitage, C.G. Atkins, R. Wyatt, B.J. Ainslie, S.P. Craig and D.P. Shepherd, 'Studies of excited state absorption at 1.5 μm in Er^{3+} doped silica fibres', *CLEO '89* (1989).
17. P.R. Morkel, M.C. Farries and S.B. Poole, *Opt. Commun.* **67** (1988) 349.
18. R. Wyatt, K.H. Cameron and M.R. Matthews, *Brit. Telecom Technol. J.*, **3** (1985) 5.
19. B.J. Ainslie, S.P. Craig, R. Wyatt and K. Moulding, *Mater. Lett.* **8** (1989) 204.
20. B.J. Ainslie, S.P. Craig, S.T Davey, D.J. Barber, J.R. Taylor and A.S.L. Gomes, *J. Mater. Sci. Lett.* **6** (1987) 1361.
21. D.C. Brown, *High Peak Power Nd:Glass Systems*, Springer, Berlin (1981), p. 7.

5 Components for fibre amplifiers and lasers

D.M. COOPER, T. FINEGAN, C.J. ROWE and I.J. WILKINSON

5.1 Introduction

The preceding chapters have described the basic physics underlying the operation of rare-earth doped fibres. As such they have many desirable properties for laboratory experiments. Their incorporation into practical systems introduces several problems: they must be stable, easy to use, physically small, and have low power requirements; and, ideally, they should consist of a box with two fibre connectors for input and output, an electrical power supply socket and an on/off switch. In this chapter we shall describe the ancillary components necessary to achieve such a device and discuss some of the problems involved in packaging an amplifier in order to turn it into a fully engineered device.

Before we give a more detailed description of the devices needed within fibre amplifiers and lasers, it would be helpful to describe the basic configurations. The simplest form of fibre laser is shown in Figure 5.1. Aside from the lasing cavity being a fibre rather than a crystalline rod, it is directly analogous to a ruby or a Nd:YAG laser. A basic fibre amplifier is also simple, as shown in Figure 5.2. In this case the input mirror is replaced by a dichroic (transmits one wavelength and reflects the other) beamsplitter to introduce both the pump and a signal into the fibre. The amplified signal is output from the far end of the fibre. More practical devices and configurations will be described later in the chapter.

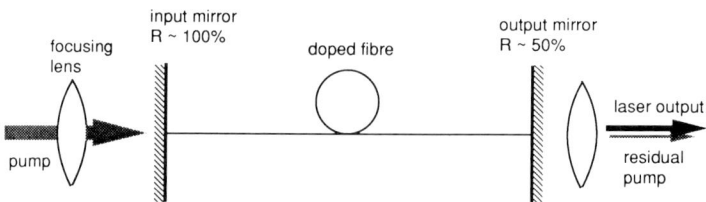

Figure 5.1 Block diagram of a doped fibre laser.

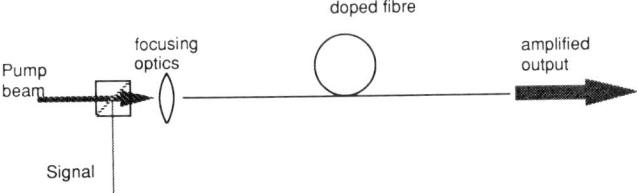

Figure 5.2 Block diagram of a rare-earth doped fibre amplifier.

5.2 Pump sources

Rapid advances have been made in the research and development of erbium-doped fibre amplifiers. Optical pumping of these devices has necessitated the discovery of laser sources, and three main absorption bands have been found at 807, 980 and 1480 nm. A number of possible pump sources are available, but only the most commonly used ones will be reviewed here.

5.2.1 Ti:sapphire lasers

In 1982 Moulton first demonstrated the titanium-doped sapphire laser [1] and, subsequently, this has been the subject of considerable investigation. This type of laser is now a most attractive candidate for an efficient, broadly tuneable solid-state laser operating over the wavelength range 700–1050 nm. At the heart of these devices lies a sapphire crystal (Al_2O_3) doped with titanium (Ti^{3+}) and the crystals are usually grown by conventional crystal growth techniques, i.e. Czochralski [2]. The Ti^{3+} ion substitutes for an Al^{3+} on a lattice site and the spectral characteristics are then dominated by the single 3d electron on the titanium ion. Absorption spectra exhibit a double humped peak centred on 499 nm and extending from 400 to 600 nm. The titanium transition responsible for this absorption is from the state $^2T_{2g} \rightarrow {}^2E_g$. A relatively weak absorption band is also observed in the wavelength range 650–1600 nm and is thought to arise from $Ti^{3+} \rightarrow Ti^{4+}$ pairs in the crystal. Optical pumping of the band at 400–600 nm produces a broad fluorescence band extending from around 600 to 1000 nm, with a peak at approximately 750 nm. The absorption band centred on 490 nm makes the use of an argon ion laser ideal for CW pumping of this system. Sapphire is also an excellent laser material having both a high optical damage threshold and good thermal conductivity. This allows high-power argon ion laser radiation to be focused directly onto the crystal without causing damage. The laser cavity is formed using external mirrors and the wavelength tuning is accomplished with a wavelength dispersive element, i.e. birefringent filter. Typical commercial lasers [3], when pumped with 20 W of argon ion radiation, deliver a maximum of 3 W output at 800 nm and a tuning range of 700–1000 nm.

5.2.2 0.98 μm InGaAs–GaAs lasers

5.2.2.1 Fabrication. A second source for use at 0.98 μm is a gallium arsenide based semiconductor laser diode [4, 5]. Conventional gallium arsenide laser emission is at a shorter wavelength and the only way to achieve emission at 0.98 μm is to use a single, strained, quantum well for the 'active layer'. Quantum-well lasers use a thin layer, a 'well' typically < 10 nm wide, of narrow bandgap material surround by a 'barrier' of wider bandgap material. Since the well width is comparable to the DeBroglie wavelength, the minimum energy state of the carriers lies above the bottom of the well [6]. Thus, the lasing wavelength is determined only by the well width and bandgap difference between the well and barriers. If the well is also strained this can change the band structure and consequently the lasing wavelength. A schematic diagram of the structure used in these devices is shown in Figure 5.3, where the single quantum well is surrounded by a graded index region of AlGaAs. This is followed by a region of wider bandgap AlGaAs and finally surrounded with GaAs. Ridge waveguide lasers have been fabricated using this structure and produced output powers of up to 100 mW/facet.

5.2.2.2 Reliability. Some initial lifetesting of these devices has been undertaken and the results are promising [4]. Devices have been tested at 50°C, with output powers of up to 30 mW, and almost half of the devices are still operating after 3000 hours. Some optical damage of the facet has been reported, but this is claimed to result from bandgap changes at the facet due to the change from biaxial to uniaxial strain. This produces absorption and is responsible for enhanced oxidation. It is reported that facet coating of the devices relieves this problem [4].

5.2.3 1.48 μm InGaAsP–InP lasers

Long-wavelength lasers have been developed as sources for optical transmission systems. The requirements for high output powers in the silica transmission windows has increased, as the quest for longer distance transmission experiments has proceeded. A more recent application for high

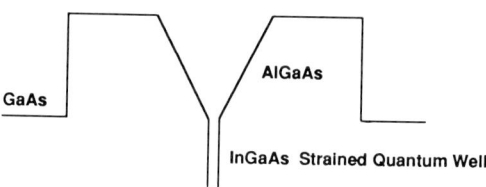

Figure 5.3 Schematic diagram of structure of 0.98 μm semiconductor laser.

optical powers at 1.48 μm has arisen with the advent of erbium fibre amplifiers. The currently required levels of output power are possible using InP-based lasers. Long-wavelength lasers of this type have a higher catastrophic optical damage threshold than GaAs-based lasers.

5.2.3.1 *Device structure.* The usual device structure used for 1.48 μm high-power lasers is the buried heterostructure (BH) Fabry–Perot laser. In this structure, the narrower bandgap active region is surrounded by wider bandgap InP layers, schematically illustrated in Figure 5.4. Surrounding the active layer are InP-based p–n–p–n structures to prevent leakage current flow. There are a number of laser designs that can be used to achieve this aim. Successfully fabricated 1.48 μm lasers include the V-groove inner stripe on p-substrate (VIPS) laser [7], the double-channel planar buried heterostructure (DCPBH) laser [8] and a buried heterostructure device fabricated using only metal organic vapour phase epitaxy (MOVPE) [9].

The first of these devices uses a conventional bulk active layer with reduced thickness in order to reduce the internal losses. The remaining devices use a multiple quantum-well (MQW) active layer to achieve this aim. In MQW devices, the differential gain for a unit thickness of material in the wells is increased. Consequently, the effective active layer volume required to reach the lasing threshold is reduced and, as a result, the internal losses are also decreased.

5.2.3.2 *Cavity length dependence.* When the lasers are driven to produce high output powers, the injected current can be up to 20–30 times the threshold current. Semiconductor lasers have internal loss mechanisms with typical values in the range 10–40 cm^{-1}, and the internal conversion efficiency of electron hole pairs to photons is around 90% at 25°C. Both of these processes cause a considerable amount of energy to be dissipated in the active region, resulting in an increase in the active layer temperature. As a result of this temperature rise, the threshold current increases and the internal efficiency falls further. Therefore, both of these effects fuel further rises in the active layer temperature. The maximum output power of a device is limited,

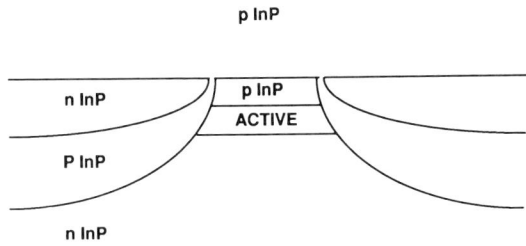

Figure 5.4 Schematic diagram of buried heterostructure laser.

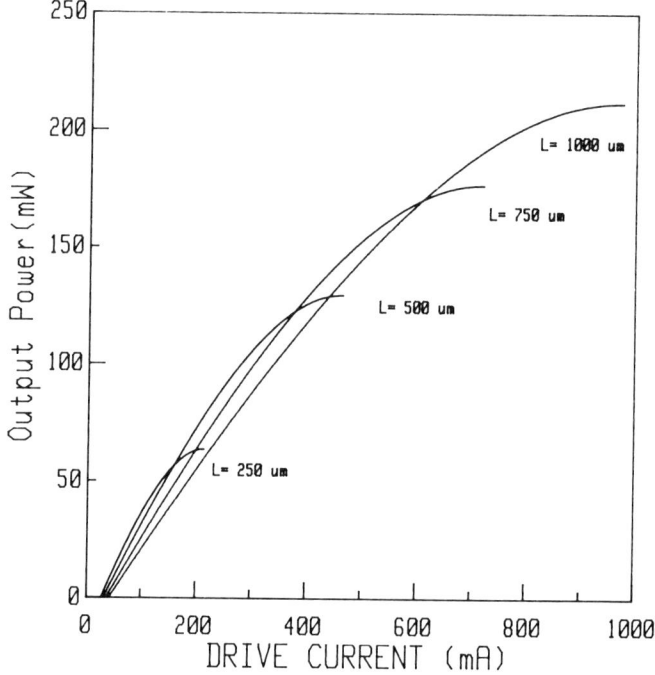

Figure 5.5 The calculated light current curve for four device lengths 250, 500, 750 and 1000 μm.

therefore, and the active layer temperature at this maximum operating point is typically calculated to be in the range 110–130°C. Heat is conducted away from the active layer, through the surrounding InP, to the contact, which is usually soldered to a heatsink with AuSn eutectic solder. For high-power lasers, it is important to reduce the thermal resistance in the device. This can be done by increasing the device length, but as that also has the effect of reducing the external differential efficiency, there is therefore an optimum device length for maximum output power. Figure 5.5 shows the calculated light against current curves for four different lengths of laser 250, 500, 750 and 1000 μm. As the device length is increased, the external differential efficiency is seen to fall but the maximum output power is increased. These characteristics have been confirmed experimentally at 1.3 μm [10].

5.2.3.3 *Facet coating.* Semiconductor lasers use the natural cleavage plains of the crystal to provide high-quality laser facets and the reflectivity of these is around 30%. In a device with two uncoated, cleaved facets, equal amounts of power are emitted from both ends. All of this output power can be made to emanate from one facet if the laser diode facets are coated to change the reflectivity. To achieve this, one facet is coated to increase its reflectivity to

COMPONENTS FOR FIBRE AMPLIFIERS AND LASERS   111

> 90%, while the other facet has its reflectivity reduced to the 5–10% range. The ratio of output power from the two facets is given by

$$\frac{P_1}{P_2} = \frac{(1-R_1)}{(1-R_2)}\sqrt{\frac{R_2}{R_1}} \qquad (5.1)$$

where R_1 and R_2 are the facet reflectivities. A high reflectivity can be achieved by coating one facet with alternating layers of alumina (Al_2O_3 low refractive index) and silicon (Si high refractive index) with thicknesses equal to one-half of the lasing wavelength. The alternating low and high refractive index layers cause large reflections. The half wavelength thicknesses produce constructive interference in the reflected waves, resulting in a high reflectivity. The reflectivity of the other facet can be lowered by coating with approximately a quarter wavelength thickness of alumina, which has an intermediate refractive index between the InP and the air. Changing this thickness allows precise control of the reflectivity. As this reflectivity is reduced, the threshold current in the device starts to rise and so the temperature and alternative leakage mechanisms increase. Consequently, there is an optimum value of reflectivity for this front coating. Figure 5.6 shows the calculated maximum output power against the front facet reflectivity for two 1000 μm long devices, one with

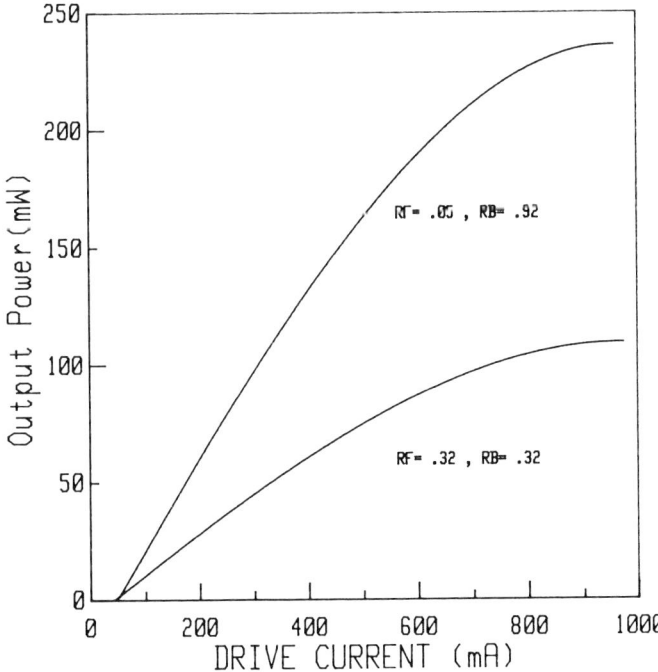

Figure 5.6 The calculated light current curve for an uncoated and a high/low-coated device.

uncoated facets and the other with one high-reflectivity coated facet (95%) and one low-reflectivity coated facet (5%). The maximum output power reported at 1.48 μm is 190 mW from a 1000 μm long device with a high/low coating [8], but more routinely achievable powers are around 100 mW.

5.2.3.4 *Reliability.* The reliability of 1.48 μm pump lasers is of great importance if diode-pumped erbium fibre amplifiers are to be used in telecommunications networks. Some testing has been performed on these diodes [7] and shows degradation rates of 3% per kilohour at 200 mW output power and $-40°C$. Although there is little published work on 1.48 μm lasers, there has been considerably more investigation of high-output power lasers at 1.3 μm [10] and 1.55 μm. Results from 1.48 μm lasers are expected to show similar characteristics to those at 1.3 μm. Tests, at a constant temperature and output power, have shown that the degradation rate of 1.3 μm lasers is dependent on the ratio of the power output during the test to the maximum output power of the laser [10]. From this finding, the lifetest conditions are chosen such that the diode is tested at only 75% of its maximum output power. Lasers, 500 μm long with a high and low coating, have been tested at 25°C at 100–130 mW and show degradation rates of 2% per kilohour.

5.3 Pump/signal multiplexing components

5.3.1 *Fibre couplers*

Fibre couplers provide beamsplitting and combining mechanisms 'in-fibre', avoiding diffraction losses associated with bulk and micro-optics. A variety of devices are available commercially and the technology is mature.

When two or more fibres are brought into close proximity, a transfer of energy can occur. This cross-talk arises because the fields of a fibre extend to infinity through the cladding. These fields can then interact with the fields of neighbouring fibres. In a normal single-mode fibre, the transfer of energy is negligible, since there is little overlap of fields. It is, however, possible to increase the field overlap to couple large amounts of power between fibres. If the coupled fibres are identical, complete power transfer can occur. Power transfer is well described by coupled-mode theory [11]. Two common techniques of making fibre couplers are described here.

5.3.1.1 *Polished couplers.* If the cladding of a fibre is removed close to the core, the evanescent field can be accessed. Two such fibres mated together exchange energy. The normal method of fabricating such a device is to mount the fibre in a slot in a quartz block [12]. The slot has a convex curvature to provide mechanical stability and to allow control of the interaction length. The block and fibre are polished until the core of the fibre is within a few

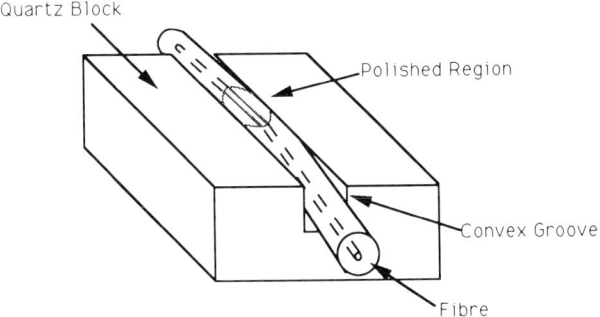

Figure 5.7 A polished half coupler block.

microns of the surface (Figure 5.7). Two polished half-coupler blocks are then placed together using index oil for good optical contact. If light is launched into one fibre, the amount of coupled power in an output port is

$$P = P_0 \cos^2(c_0 L) \tag{5.2}$$

where L is the effective interaction length, P_0 is the launched power and c_0 is the coupling coefficient. If the curvature of the fibres is taken into account, it can be shown [12] that

$$c_0 L = \int_{-\infty}^{+\infty} c(z)\, dz \tag{5.3}$$

with

$$c = \frac{\lambda}{2\pi n_1} \cdot \frac{u^2}{a^2 V^2} \cdot \frac{K_0[v(h/a)]}{K_1^2(v)} \tag{5.4}$$

where n_1 is the core refractive index, a is the fibre core radius, h is the distance between fibre axis, K_v are modified Bessel functions of the second kind and u and V are the normal fibre parameters.

The device is bidirectional. If the coupling ratio from port 1–4 is K (Figure 5.8), then the coupling ratio of ports 4–1, 2–3 and 3–2 is also K. The coupling ratio from 1–3, 3–1, 2–4 and 4–2 is $1 - K$. Coupling ratios from 0 to 100% can be obtained by altering the core–core distance, h, and the interaction length; and h can be controlled by fine polishing and L by moving the substrates relative to each other using micrometers. This ability to control the interaction length is very important since it allows an assembled device to be tuned, which makes such couplers useful components in constructing loop cavities for lasers (Figure 5.9). The coupler can be designed to allow efficient

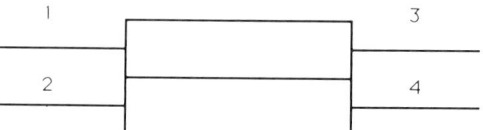

Figure 5.8 Port configuration of a fibre coupler.

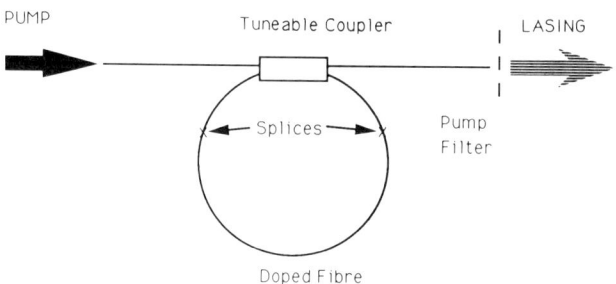

Figure 5.9 Block diagram of a rare-earth doped fibre laser featuring a ring cavity.

coupling at the pump wavelength with a high finesse at the lasing wavelength. Insertion losses at the lasing wavelength are as low as 0.005 dB.

These couplers have two major disadvantages, however, which must be taken into account. Firstly, polishing the cladding is a time-consuming and skilled job which leads to expensive devices. Secondly, the coupling ratio is sensitive to environmental factors. One reason for this is the use of index oil between the coupler halves. Significant changes in the oil's refractive index can occur for modest changes in temperature. Changes in the coupling ratio of $6\% \, °C^{-1}$ have been reported. Cemented devices have been reported where this sensitivity has been reduced to $0.2\% \, °C^{-1}$, but these are no longer tuneable [13].

5.3.1.2 Fused couplers. The fused-taper technique has become the most popular method of producing couplers, being less time consuming and cheaper to perform than polishing. Two fibres have their coatings removed along a short section of a few centimetres. The fibres are then twisted together, heated and drawn while the throughput and coupled powers are constantly monitored. Temperatures of $> 1500°C$ are required and this is normally achieved using gas burners. The fibres are tapered until the desired coupling ratio is achieved [14]. The coupling mechanism in these devices is slightly different to that for the polished couplers. As the fibre is tapered, the decreasing core size leads to an increase in the mode field diameter of the guided mode. Eventually, the mode is no longer confined by the core (the effects of which are usually ignored in theoretical studies) and the cladding/external medium interface becomes the guide. Coupling occurs through a beating process between the lowest order symmetric and antisymmetric modes of the composite waveguide [15]. The coupling coefficient

$$C = \frac{\beta^i_{11} - \beta^i_{12}}{2} \qquad (5.5)$$

can be obtained numerically providing the propagation constants of the two modes of the composite waveguide, β^i_{11} and β^i_{12}, are known. The figure-of-eight geometry of the waveguide makes this calculation complex and, as a result, most attempts to model the properties of these couplers use simplified cross-sections. Rectangular and elliptical sections are the most common [16, 17].

The tapered region is extremely fragile and must therefore be attached to a substrate. Since the coupling ratio is sensitive to bends, twists and tensioning of the fused region and the refractive index of the medium surrounding the taper, careful thought must be given to the packaging method. Some manufacturers pot the tapered region in a low-index silicone resin for additional support [18]. Once the taper is mounted the coupling ratio is fixed; however, tuneable devices have been reported by some researchers using controlled distortions [19] or twists [20] in the tapered region.

Coupling ratios from 0 to 100% can be obtained with excess losses of < 0.05 dB for commercially available off-the-shelf devices. The port configuration is identical to Figure 5.8 and the device is bidirectional. The environmental stability of fused devices is extremely good, with variations of only 0.2% over a temperature range of 100°C [21]. The manufacturing techniques are relatively straightforward, leading to a very cheap product.

The fused-taper technique has also been applied to fluorozirconate glasses [22], which make excellent hosts for rare-earth ions. Duplicating the process in fluoride glass, however, presents three problems to be overcome:

(i) the glass transition temperature is only ~260°C;
(ii) the viscosity is low and drops rapidly with temperature;
(iii) recrystallisation occurs at temperatures very close to the glass transition temperature.

These problems nevertheless have been overcome with coupling ratios of 40% observed in multimode fibres.

5.3.2 *Fused-fibre wavelength division multiplexers*

The coupling of light described earlier is wavelength dependent. Coupling coefficients are stronger for longer wavelengths (since their mode field diameter is greater). Figure 5.10 plots diagrammatically the beating of power between two fibres at two wavelengths. By designing the coupled region

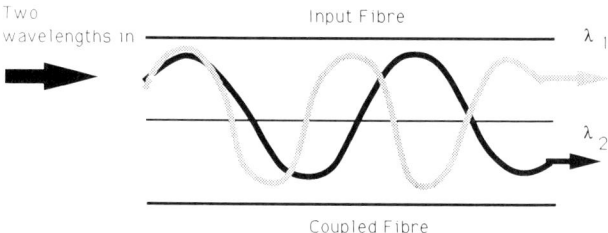

Figure 5.10 Principle of a fibre wavelength division multiplexer.

appropriately, the two wavelengths can be separated. Conversely, two wavelengths entering the tapered region on separate fibres will appear on a single output fibre. Such a device is known as a wavelength division multiplexer (WDM), which allows pump and signal wavelengths of fibre amplifiers (Figure 5.11(a)) to be combined with very low insertion losses. Early attempts were made to fabricate WDMs using the polished half-coupler block method described in Section 5.3.1 [23]. The simplest method, however, is to use the fused-taper technique [24].

5.3.2.1 *Fused-taper WDMs.* The manufacturing process is essentially the same as that described for the fused coupler, but in this case the fibres are drawn until the desired degree of wavelength selectivity is obtained. This is usually achieved by monitoring the coupled power at two wavelengths simultaneously. Power entering the fused region can oscillate many times between the two composite fibres. One complete oscillation is termed a beatlength. The nature of the coupling mechanism leads to an approximately periodic, raised cosine, spectral response (Figure 5.11(a)). The spacing between neighbouring pass and stopbands follows the relationship

$$\text{Channel spacing} \propto \frac{1}{\text{No. of beatlengths}}. \tag{5.6}$$

The constant of proportionality is determined by the size and shape of the fused region. In order to obtain narrow channel separations it is necessary to taper the fibres through many beatlengths. Typically, to obtain a 60 nm channel spacing (for a 1480–1540 nm WDM, for example), six to nine beatlengths are required, depending on manufacturing conditions. Subsequently, long interaction lengths are needed. These long interaction lengths present new problems, however, since they lead to a polarisation-dependent behaviour [25, 26]. The figure-of-eight shaped geometry of the fused region causes form birefringence, hence, orthogonal polarisation components of each mode will have slightly different propagation constants. It is therefore necessary to consider coupling coefficients for both x and y polarised states, C_x and C_y, where x and y are orthogonal. For randomly

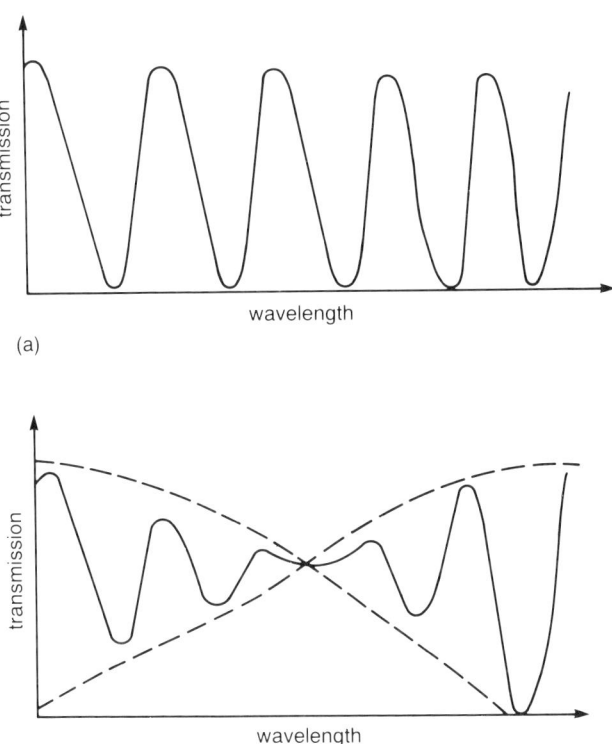

Figure 5.11 (a) Spectral response of a fused fibre WDM. (b) Spectral response of a polarisation-sensitive fused fibre WDM featuring slow modulation.

polarised illumination, the power in an output port is then

$$P = \frac{P_0}{2}[1 + \cos(C_x - C_y)L \cdot \cos(C_x + C_y)L] \tag{5.7}$$

i.e. the spectral response features a slow modulation (Figure 5.11(b)). This results in increased insertion loss at the passband wavelengths and reduced isolation between neighbouring channels. The insertion loss and isolation will also vary as the input state of polarisation (SOP) is rotated. Figure 5.12 shows the spectral response of one output port of a WDM changing as a linear input SOP is rotated.

These problems can be overcome, however. One method is to pot the taper in a dielectric resin to reduce form birefringence [27]. This can reduce polarisation sensitivity to less than 0.5 dB variation in passband insertion loss. It does, however, have one major disadvantage – the dielectric resin has a refractive index that is temperature sensitive, and this in turn will lead to the

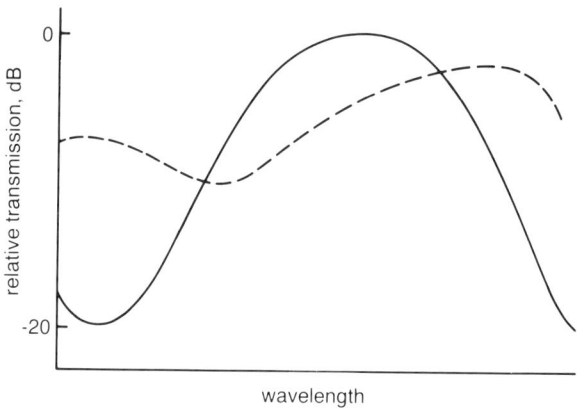

Figure 5.12 Spectral response of a polarisation-sensitive WDM at two different input states of polarisation.

WDM having a temperature-sensitive performance. Passbands can shift by several nanometres per degree Celsius. Temperature sensitivities of this order rule out systems applications of these devices.

An alternative method of reducing the polarisation sensitivity has been developed at BTRL [28]. In this method, a number of twists are applied to the fused region after manufacture. The high twist rate induces circular birefringence thus preventing polarisation mode coupling in the fused region. The method of applying high twist rates to preserve polarisation has previously been used to fabricate long lengths of polarisation-preserving fibre

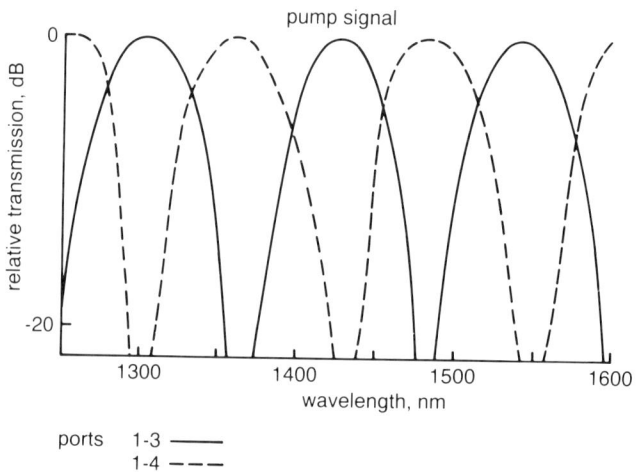

Figure 5.13 Spectral response of a fused fibre WDM for a 1480 nm pumped erbium fibre amplifier.

[29]. Twisted WDMs have < 0.1 dB variation in insertion loss with input SOP. They are also temperature stable, displaying a similar environmental performance to that of standard fused-fibre couplers. The WDMs can be fabricated with < 0.1 dB excess loss and 25 dB peak isolation between channels (measured with a 5 nm wide source). Figure 5.13 shows the measured spectral response of ports 1–3 and 1–4 for a 1480–1536 nm WDM for use in a 1480 nm pumped erbium fibre amplifier systems. The periodic nature of the coupling has been used in this instance to provide simultaneously a passband at 1300 nm, allowing this fibre window to be used as well. Erbium-doped fibre is transparent at 1300 nm.

This fused-taper method can be used to produce low loss WDMs with channel spacings from > 500 nm to < 5 nm. Very large channel spacings, e.g. for a 980–1540 nm WDM, can only be achieved with low loss if the fibre is single-moded at both wavelengths.

5.3.3 *Pump rejection filters*

Optical filters are used in systems and instruments where it is necessary to isolate precisely a discrete wavelength region of the electromagnetic spectrum. The filter may pass, reflect, absorb or attenuate the wavelength region of interest – the particular function is determined by the application.

In networks and instruments employing optical amplifiers it could be necessary to provide attenuation of the optical pump wavelength region and pass the amplified wavelength region. This would be particularly important in a system in which the magnitude of the received power from the pump laser and the system laser wavelength are comparable. In practice, the pump power is absorbed by the amplifier, but if only a small proportion gets to the system receiver the signal-to-noise ratio (SNR) is reduced. The required SNR varies according to the particular application.

In the following subsections the manufacture of optical filters is briefly described followed by a review of the desired optical performance and packaging options.

5.3.3.1 *Filter production.* Optical filters with optimum performance can be produced using uniform dielectric thin-film coatings constructed as alternating layers of high and low refractive index materials. There are a variety of materials that can be used for these coatings, the choice of which can impinge on the overall stability and the optical performance of the filter. Figure 5.14 shows the basic layer structure of the dielectric filter. Cavities within the filter structure reflect or transmit specific regions of the electromagnetic spectrum. The optical performance can be varied by changing any of the following factors: number of layers in a basic period; number of repeats of this period (cavity number); and the placement of half-wave spacers in the cavities. The desired performance of the filter is mathematically

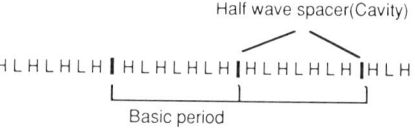

H - high refractive index material   L - low refractive index material

Figure 5.14 Basic layer structure of a thin-film dielectric filter.

modelled and the layer structure is determined theoretically before depositing the material layers onto the substrate [30].

5.3.3.2 Filter characteristics. The spectral response of a filter (filter profile) is specified to determine its use in a particular system. In practice, for optical amplifier applications the desired response may be to absorb the 1480 nm wavelength region (pump wavelength) and to pass the remainder of the spectrum with negligible loss.

Dielectric filters generally reflect the stopband (non-transmitted region) and very steep roll-off profiles are difficult to achieve due to slight material variations. In Figure 5.15 we show the spectrum of a high-pass filter. This filter blocks the pump wavelength but passes the 1500 nm system wavelength region. The majority of the pump power is reflected, a small amount is absorbed and about 0.1% may be transmitted in this filter. Similarly, a low-pass filter could be used to reject the pump wavelength but pass the 1300 nm region. Filter technology is sufficiently developed that bandpass filters at virtually any wavelength in the usable wavelength region could be manufactured [31]; the choice depends on the application.

5.3.3.3 Filter packaging. In order for the filter substrate described above to be used in a system it must either be placed directly onto a receiver/detector or be fibre pigtailed. We briefly describe here some of the options for packaging

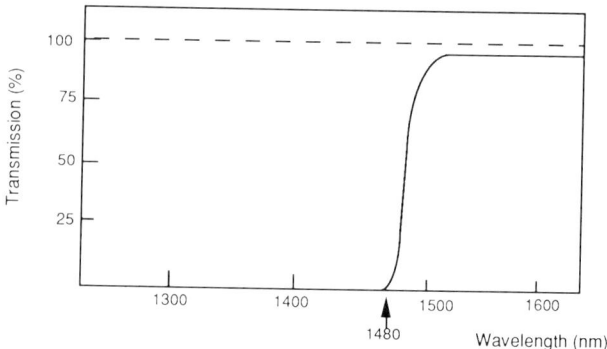

Figure 5.15 'High pass' filter transmission spectrum.

COMPONENTS FOR FIBRE AMPLIFIERS AND LASERS 121

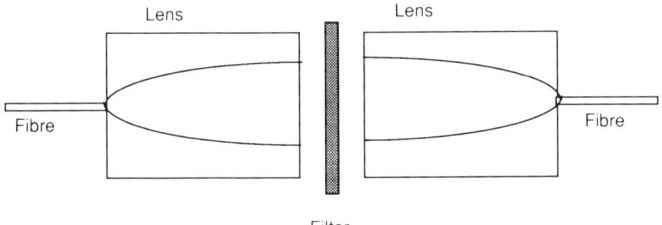

Figure 5.16 Filter package using GRIN lenses to produce a parallel beam.

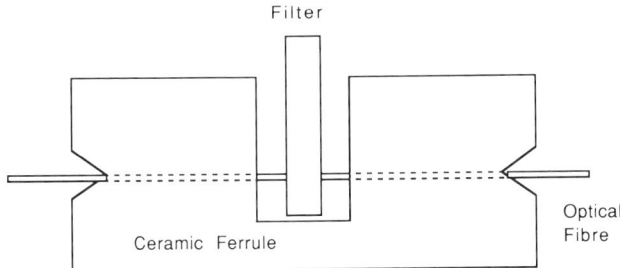

Figure 5.17 Filter package using a ceramic ferrule to provide the fibre alignment.

filters in a single-mode fibre in/out configuration. In Figure 5.16 lenses are used to collimate the beam from a single-mode fibre. The filter substrate is placed in the gap between the lenses to produce a low-loss device, however, the precision alignment required could result in an expensive option. A more simplistic approach is to use a very thin filter substrate ($< 100\,\mu$m) and butt the two fibres directly to it [32]. The loss in this option is controlled by the fibre offset (filter thickness). Again, a low-loss device is produced but with a potential cost benefit. Figure 5.17 shows a ceramic ferrule which is used for the low-cost precision fibre alignment in this device.

5.3.3.4 *Filter reflections.* As the reflected power from an optical filter could cause problems to optical amplifiers, it is important that these are minimised. Lenses and fibre ends could be antireflection coated or the filter substrate tilted so that reflected light is not collected by the input fibre. Tilting the filters can cause the wavelength characteristics to shift dramatically [33], a 6-degree tilt should be sufficient to avoid reflections from the filter re-entering the fibre.

5.3.3.5 *Filter reliability.* The long-term reliability of filters within systems and amplifiers must be determined. The performance of the filter substrates can depend upon the particular choice of deposited materials. A combination of materials must be used which are environmentally hardened against scratching, temperature and humidity but can still achieve the required optical performance [34].

Problems can occur with large variations in temperature. The layer thicknesses alter with a resultant variation in centre wavelength and spectral response. In some instances humidity can introduce stress into the filter layers and affect the overall performance. These environmental effects can be reduced by hermetically sealing the components and by suitable design of the materials and structures. Temperature variations can be limited to within 0.05 nm/°C.

5.3.4 *Dielectric filter WDMs*

We have already discussed the importance of the WDM in networks, and in this and the following subsection two further WDM configurations are discussed to supplement the fused-fibre device outlined in Section 5.3.2.

The optical filters described in the previous section are the basic building blocks for a WDM in which the filter defines the channels of the device. Multiplexing the 1480 nm pump wavelength of an amplifier with the 1500 nm system wavelength region can be achieved using the high-pass filter response of Figure 5.14. The physical arrangement of this WDM is shown in Figure 5.18. Two GRIN lenses collimate and focus the input light onto the filter; the 1500 nm wavelength region is transmitted into the output fibre, while the 1480 nm pump wavelength is reflected into the common output fibre. This type of two-channel multiplexer has been produced with an excess loss < 0.5 dB and > 20 dB rejection between the channels [35]. As the variation of the filter performance with temperature can significantly change the WDM windows if they are too narrow, this type of WDM has generally broad windows. The cost of this type of device is very high in comparison with the fused-fibre device because of the precision alignment needed to produce the low excess loss.

A limitation of this filter-based WDM is that although it is possible to multiplex more than two channels, the complexity increases significantly. More recent developments in planar glass substrate technology could provide the relaxation of the alignment required to make this type of device cost effective. 'Y' junction waveguides are fabricated in a glass substrate and a thin

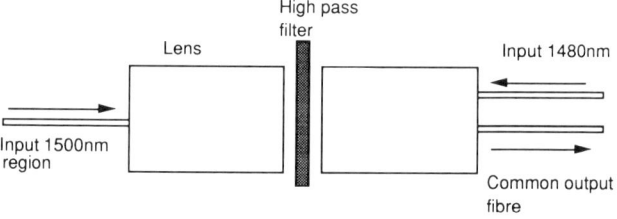

Figure 5.18 Filter-based WDM configuration.

COMPONENTS FOR FIBRE AMPLIFIERS AND LASERS

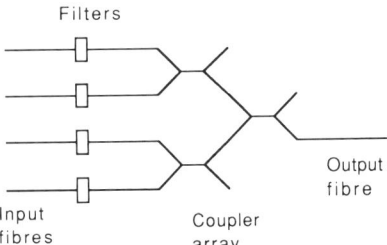

Figure 5.19 WDM produced by combining fused fibre and filter technology.

filter slice is then placed in a narrow micro-machined channel at the waveguide junction [36].

The WDMs discussed in Section 5.3.2 could also be combined with optical filters on the outputs (Figure 5.19) in order to improve the cross-talk levels. The filters must have spectral responses very similar to the fused-fibre WDM channels and must also be wide enough to remove potential temperature drifting misalignment problems [37]. Throughput losses for a four-channel device have been demonstrated to be less than 4 dB with very good channel spacing, which makes this a potentially very low cost and high-performance WDM.

5.3.5 Grating-based WDMs

When a diffraction grating is illuminated by a broad wavelength source, it reflects or refracts light at different angles according to its wavelength. Most diffraction gratings used within WDMs have groove sizes very much greater than the wavelength of light used [38].

One advantage of the grating-based WDM is that gratings multiplex all input wavelengths simultaneously and loss or complexity is not affected by an increase in the number of channels. However, when the number of channels is increased the focusing optics and the placement of the input and output fibres become critical to maintain uniformity of losses between the channels. The most common diffraction grating WDM schematic is the Littrow configuration [38, 39]. In this arrangement a GRIN rod lens is directly attached to an angled grating (Figure 5.20) to make a compact and rugged device. The number of channels and the operating wavelength of the WDM are determined by the diffraction grating spatial frequency, the input and output fibre core size, the spacing between the fibre cores, and the lens characteristics. A larger lens may accommodate more output fibres (more channels) at the expense of increased losses due to chromatic dispersion in the lens. Increasing the number of grooves per millimetre in the grating can result in a larger linear dispersion, but the device may become more sensitive to polarisation [38].

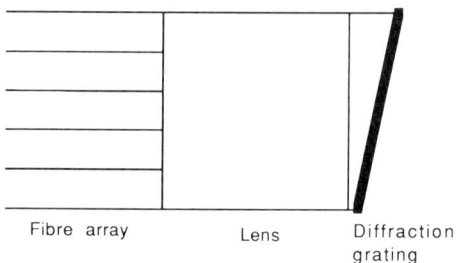

Figure 5.20 Grating-based WDM configuration.

As fibre alignment in this device is very important, silicon-machined vee-grooves are frequently used to give the required consistent micrometre accuracy.

Fabrication of the diffraction grating is by a photolithographic technique and is described in detail by Linglesheim and Aytac [40]. The blade angle of the grating is chosen to give the maximum efficiency versus wavelength.

The basic Littrow multiplexer has a relatively high excess loss (> 3 dB) and a large channel spacing (12 times the channel width). The channel spacing can be reduced by etching the fibres so that the core-to-cladding ratio is higher, and significant improvements have been reported by Lipson et al. [41]. The channel spacing has also been reduced by using a $LiNbO_3$ waveguide concentrator; this again results in an increase in the effective core-to-cladding ratio [42].

Excess losses of these devices are higher than the fused-fibre WDM but the channel rejection is much improved. The complexity of production of this WDM results in an expensive device, the long-term reliability of this device is very good and there is only a small wavelength variation with temperature (0.007 nm/°C).

A comparison of pump/signal multiplexing technologies is given in Table 5.1.

Table 5.1 Comparison of pump/signal multiplexing technologies

	Fused fibre	Dielectric	Grating
Insertion loss (dB)	< 0.2 dB	< 1.0 dB	< 3.0 dB
Polarisation sensitivity (dB)	< 0.2 dB	None	None
Cost	Low	Medium	Medium
Temperature stability, (nm/°C)	0.005	0.05	0.007
Manufacture	Simple	Skilled	Skilled
Spectral response	Broad raised Cosine	Broad square	Narrow
Approx. package size (mm)	10 × 10 × 80	8 × 8 × 60	10 × 10 × 80
Port configuration	2 IN, 1 OUT	2 IN, 1 OUT	2 IN, 1 OUT

COMPONENTS FOR FIBRE AMPLIFIERS AND LASERS

5.4 Laser cavity elements

5.4.1 Introduction

Fibre amplifiers are converted into lasers by creating some form of resonant cavity, such as causing a reflection at each end of the doped fibre. The feedback causes the amplifier to oscillate. Various means for introducing this feedback will be described along with their advantages and disadvantages.

5.4.2 Dielectric stack mirrors

The simplest way to create a fibre laser is to attach a mirror to each end of the amplifying fibre, as shown in Figure 5.21(a) [43]. This arrangement is entirely analogous to that found in a standard solid-state laser, such as a Nd:YAG. A separate WDM component is not required. Dielectric stack mirrors can be fabricated with a suitably high transmission at the pump wavelength and a high reflection at the lasing wavelength.

The theory of dielectric mirrors and filters has been described above. The main constraint on performance is not optical performance but resistance to high-power damage. The pump, of up to several watts, is focused onto the surface of the mirror. The mirror coatings need to be able to resist power densities $> 10^{10}$ W/m^2.

Slightly different configurations are possible which eliminate this limitation. An example is shown in Figure 5.21(b) which removes the danger of mirror damage [44, 45].

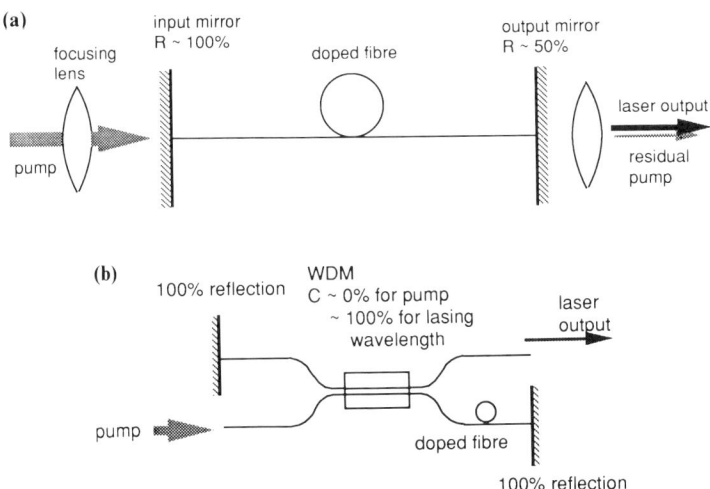

Figure 5.21 (a) Simple fibre laser; (b) alternative configuration.

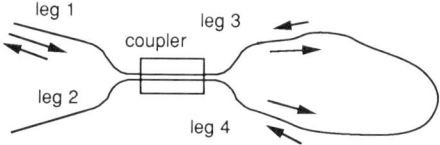

Figure 5.22 Fibre loop mirror.

5.4.3 *Loop reflectors*

Mirrors as described above enable the construction of simple fibre lasers. Unfortunately they negate one of the main advantages of fibre lasers, namely their compatibility with fibre systems. Mirrors require the use of bulk or micro-optics to interface with fibres, entailing losses and instabilities. Ideally, what is required is a device that acts like a mirror but is of an all-fibre construction. Two such devices exist: the first, which we shall now describe, is the loop mirror [46].

This device simply consists of the two legs of a fused-fibre coupler connected together, as shown in Figure 5.22. It is the fibre equivalent of the Sagnac interferometer. At wavelengths at which the splitting ratio of the coupler is 50%, a strong reflection is produced. It is assumed that light travels towards the coupler on leg 1 and is split equally between legs 3 and 4. Fifty per cent of the light travels in each direction around the loop. Light that couples across the waveguides undergoes a $\pi/2$ phase change. The transmitted light intensity in leg 2 is therefore the sum of a clockwise field of arbitrary phase ϕ and the anticlockwise field of phase $\phi - \pi$. As both are of equal intensity, the resultant transmitted intensity is zero and hence, by conservation of energy, all the power is reflected back into leg 1. Note that although this is an interferometric effect it is insensitive to environmental effects because the two light paths are identical. Of course this is the ideal case and ignores effects that would appear in a real device such as unequal splitting ratio, excess loss, fibre birefringence, etc.

For a typical coupler the reflection peak is fairly broad, several tens of nanometres. This is similar to a mirror and so the performance of fibre lasers constructed with these devices is similar to those using mirrors. The coupler can be overpulled to produce a more wavelength-selective device giving a reflection wherever the coupler response is 50:50 [47]. The linewidth of the laser is reduced as the coupler becomes more wavelength sensitive, but this cannot be continued indefinitely. As explained above, couplers have a cyclic wavelength response; if the coupler is pulled too far then the loop mirror will have two or more strong reflections within the gain spectrum of the fibre. Lasing will occur at several wavelengths. To produce narrower linewidth lasers, devices are required with narrow, non-cyclic responses.

5.4.3.1 *Bulk optic gratings.* One such device is the diffraction grating used as shown in Figure 5.23. Feedback for the laser exists only over a narrow

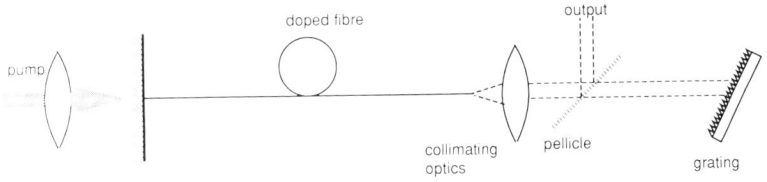

Figure 5.23 Fibre laser incorporating bulk grating.

bandwidth. In addition, changing the angle of the grating gives a degree of tuneability.

The bandwidth of the reflection is narrow, typically 1 nm. Lasing only occurs at the very peak of the response so that linewidths are very much narrower still, < 0.1 nm. Varying the angle of the grating tunes the wavelength of operation. Tuneable fibre lasers have been constructed with up to 80 nm of tuning range in a Nd-doped fibre [48]. This is limited by the available gain bandwidth of the fibre.

5.4.3.2 Fibre gratings. Like dielectric mirrors, diffraction gratings do provide some very beneficial properties but they suffer from similar problems. The major problem is the loss due to the coupling optics. Again an all-fibre device exists which provides the same function. This is the fibre grating, as shown in Figure 5.24, which produces a narrow linewidth Bragg reflection.

The basis of the device is a polished fibre half-coupler, as described in an earlier section. A diffraction grating is introduced into the evanescent field. The perturbation of the guiding field causes a strong reflection at a wavelength given by

$$\lambda_B = 2n_e \Lambda \tag{5.8}$$

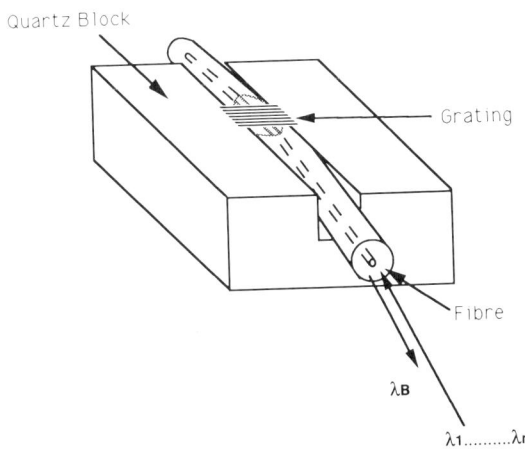

Figure 5.24 Fibre gratings.

where λ_B is the Bragg wavelength, Λ is the grating period and n_e is the effective guiding index. Longer wavelengths are unaffected by this perturbation. Shorter wavelengths are coupled out of the fibre into the air and the substrate at angles given by

$$\beta - K = k_{s,c} \sin \theta_{s,c} \tag{5.9}$$

where β is the propagation constant, $K = 2\pi/\Lambda$ where Λ is the grating period, $k_{s,c} = n_{s,c} k_0$, $k_0 = 2\pi/\lambda$, λ the free-space wavelength, $n_{s,c}$ the refractive index of the substrate (silica block) and superstrate (air) and $\theta_{s,c}$ the angles to the normal [49].

There are several ways of forming a grating on the surface of a polished half-coupler. The simplest method is to contact a normal metal grating onto the surface [50]. Unfortunately, the reflection is polarisation dependent due to the metal and the grating must be aligned to the waveguide. Another method is to form a photoresist grating on the surface [51], but the most robust method is to etch the grating into the surface of the fibre itself [52, 53].

The grating is formed by a holographic method. First, a thin layer of photoresist is spun onto the surface of the block. This is then exposed by a two-beam interference setup such that the period of the fringe pattern formed is that of the required Bragg grating. When developed, the parts of the photoresist exposed by a dark fringe are dissolved leaving behind a grating of the correct pitch. This is then transferred into the block (and fibre) by a dry-etching process such as ion beam milling or reactive ion etching. Finally a thin layer of alumina and silica is deposited on the surface to enhance the grating efficiency. A layer of oil can be used instead of the silica. The final reflectivity can be above 90% with a bandwidth of <0.8 nm [53].

The device has been used to define one end of a fibre laser cavity with a consequent narrowing of the laser response. For a fairly long laser cavity the linewidth is of the order of 0.06 nm [54]. This consists of several tens or hundreds of lines corresponding to the cavity Fabry–Perot modes. Reducing the length of the laser reduces the number of modes that exist within the gain bandwidth. If the fibre length is sufficiently short then only one mode can oscillate, producing a single longitudinal mode laser equivalent to the semiconductor DFB. Such a laser has been constructed by this method with a bandwidth of 1.3 MHz; however, this laser has little or no tuning capability [55].

A variant of the metal grating approach produces a tuneable device. In this case the external grating is etched into a metal or a deposited silicon layer. Instead of a series of parallel lines a fan-shaped pattern is used [56, 57]. The spreading angle is very small so that over a small area the lines are essentially parallel. As the grating is moved sideways over the block, the grating period changes so altering the wavelength of the reflection peak. Tuning ranges of 80 nm have been achieved and used to tune external cavity semiconductor lasers.

5.4.4 Ring cavities

All the components described so far have all been used to form some variant of the familiar Fabry–Perot resonator. Another configuration that has been used with fibre lasers is the ring resonators shown in Figure 5.25(a) and (b) [58]. The first configuration is, at least in theory, the simplest means of implementing a ring resonator. In practice, a splice is needed within the ring with its inherent extra loss. The other configuration is better as it can be made on a single length of fibre. Although at first sight this appears to be similar to the loop mirror, it is a resonant structure that involves energy storage within the ring. The mirror is a non-resonant interferometer. Resonance occurs when the loop length is an integral number of wavelengths. The free spectral range (FSR) is given by

$$\text{FSR} = \frac{c}{nL} \tag{5.10}$$

where c is the speed of light, n is the guiding index of the fibre and L is the loop length. For the second type of ring the finesse factor F is given by

$$F = \frac{\pi\sqrt{\kappa_r}}{1 - \kappa_r} \tag{5.11}$$

in which

$$\kappa_r = (1 - \gamma_0)e^{-2\alpha_0 L} \tag{5.12}$$

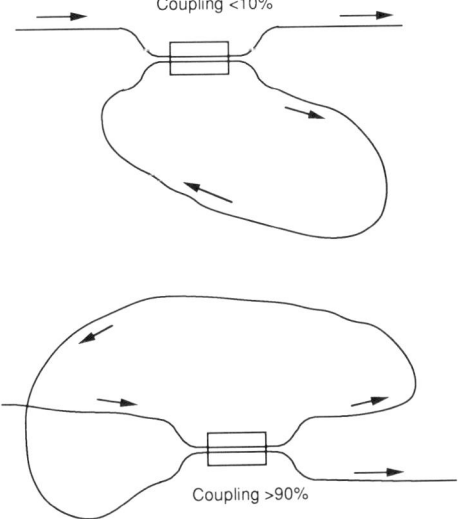

Figure 5.25 Loop resonator.

where γ_0 is the coupler fractional loss and α_0 is the amplitude attenuation coefficient of the fibre.

Its main feature is the ease with which it can be implemented, needing only one splitter (which ideally is also a WDM) and possibly some form of polarisation control within the loop. Only one device is needed to form a very simple laser [48].

5.5 Packaged fibre amplifiers

The preceding sections of this chapter have described the necessary components required to build both fibre amplifiers and lasers. To date amplifier applications are more numerous. Complete and self-contained amplifiers (only needing an electrical supply) are appearing on the market (June 1990). In this final section we would, firstly, like to highlight some of the other problems involved in packaging. Secondly, the performance and characteristics of a production-packaged amplifier are described.

5.5.1 *Component assembly*

5.5.1.1 *Reflections.* It is essential that all steps are taken to minimise reflections in the amplifier (and indeed throughout any system incorporating an amplifier) since they can easily lead to lasing. Reflections can arise by various means, some of which are:

- unterminated fibre ends
- the facets of the pump laser
- poor fusion splices
- connectors
- the reflectance of any dielectric stack type filters.

Most of these problems can be overcome with care, for example, the use of a high return loss, super-polished FC/PC connectors, etc.

5.5.1.2 *Splicing.* Large splice losses can arise when the doped fibre is spliced to B-type fibre. Such splices are necessary when connecting the doped fibre to both the WDM component and B-type fibre at the other end of the amplifier. The losses arise due to large mode field mismatches between the two fibre types. B-type fibre has a core diameter typically of $\sim 9\,\mu m$ while the core of doped fibre is much smaller, $\sim 5\,\mu m$ [59]. The small core size is chosen to increase pump-power densities in the fibre. A joint made using a standard fusion splice leads to $\sim 3.0\,dB$ loss which will obviously reduce the input pump power and the gain of the amplifier. The mode field mismatch can be reduced, however, by tapering the splice [60]. Excess losses of $\sim 0.7\,dB$ can be achieved by tapering the junction of a B-type/doped fibre splice.

COMPONENTS FOR FIBRE AMPLIFIERS AND LASERS

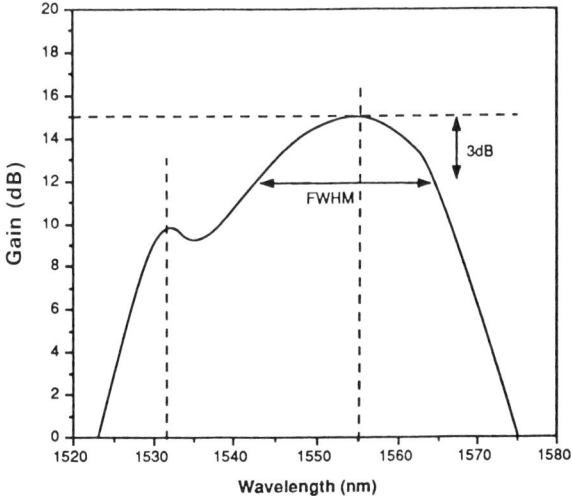

Figure 5.26 Wavelength response of packaged fibre amplifier.

5.5.2 *Amplifier performance*

Very recently, modular erbium-doped fibre amplifiers have become available commercially, and their performance is reviewed in this section. The EFA3000 series, launched by BT&D [61], has 15 dB peak gain and a 3 dB gain bandwidth of 25 nm. The fibre has been optimised to give broadband gain in the range 1530–1560 nm using a Al_2O_3–GeO_2–SiO_2 core with an erbium concentration of ~ 300 ppm. The centre wavelength is 1550 nm (Figure 5.26). The peak gain and bandwidth can be adjusted by optimising the fibre length and composition. Shorter lengths of doped fibre give increased gain bandwidths of 41 nm but with reduced gain. The saturation input power is -12 dBm and the noise figure 7 dB. The package has dimensions of $\sim 244 \times 250 \times 127$ mm and is designed for operation in the range 10–30°C and storage from 0 to 40°C.

References

1. P.F. Moulton, *Opt. News* **8** (1982) 9.
2. P. Lacovara, L. Esterowitz and M. Kokta, *IEEE J. Quant. Electron.* **QE-21** (1985) 1614.
3. Spectra Physics, Titanium Sapphire Lasers, Model 3900.
4. M. Okayasu, M. Fukuda, T. Takeshita, O. Kogure, T. Hirono and S. Uehara *OFC 1990, San Francisco (USA)*, Paper PD29.
5. A. Larsson, S. Forouhar, J. Cody and R.J. Lang, *OFC 1990, San Francisco (USA)*, Paper PD30.
6. M.J. Adams, A.G. Steventon, W.J. Devlin and I.D. Henning, 'Semiconductor lasers for long wavelength optical fibre communications systems', *IEE Materials and Devices*, Series 4, Peter Peregrinus Ltd.

7. Y. Kawai and T. Yamada, *Electron. Lett.* **26** (1990) 53.
8. I. Mito, H. Yamazaki, H. Yamada, T. Sasaki, S. Takano, Y. Aoki and M. Kitamura, *7th IOOC, Kobe, Japan* (1989), Paper 20PDB-12.
9. D.M. Cooper, C.P. Seltzer, M. Aylett, D.J. Elton, M. Harlow, H. Wickes and D.L. Murrell, *Electron. Lett.* **25** (1989) 1635.
10. S. Oshiba, A. Matoba, M. Kawahara and Y. Kawai in *IEEE J. Quant. Electron.* **QE-23** (1987) 738.
11. A.W. Snyder, *J. Opt. Soc. Amer.* **62** (1972) 1267.
12. M.J.F. Digonnet and H.J. Shaw, *IEEE J. Quant. Electron.* **QE-18** (1982) 746.
13. J.D. Beasley, D.R. Moore and D.W. Stowe, *Proc. SPIE* **417** (1983) 36.
14. C.A. Villarruel and R.P. Moeller, *Electron. Lett.* **17** (1981) 243.
15. J. Bures, S. Lacroix and J. Lapierre, *Appl. Opt.* **22** (1983) 1918
16. F.P. Payne, C.D. Hussey and M.S. Yataki, *Electron. Lett.* **21** (1985) 461.
17. K.S. Chiang, *Electron. Lett.* **22** (1986) 1221.
18. A. Fielding, *Opt. Laser Technol.* (June 1986) 145.
19. B.S. Kawasaki, M. Kawachi, K.O. Hill and D.C. Johnson, *J. Light. Technol.* **LT-1** (1983) 176.
20. T.A. Birks, *Appl. Opt.* **28** (1989) 4226.
21. T. Bricheno and A. Fielding, *Electron. Lett.* **20** (1984) 230.
22. C.J. Rowe, M.W. Moore, I.J. Wilkinson and N.E. Achurch, *Electron. Lett.* **26** (1990) 316.
23. M.J.F. Digonnet and H.J. Shaw, *Appl. Opt.* **22** (1983) 484.
24. C.M. Lawson, P.M. Kopera, T.Y. Hsu and V.J. Tekippe, *Electron. Lett.* **20** (1984) 963.
25. M.S. Yataki, D.N. Payne and M.P. Varnham, *Electron. Lett.* **21** (1985) 248.
26. T. Bricheno and V. Baker, *Electron. Lett.* **21** (1985) 251.
27. S.P. Shipley and P.F. Trwoga, *IEE Colloq. Digest* No. 90 (1988).
28. I.J. Wilkinson and C.J. Rowe, *Electron. Lett.* **26** (1990) 382.
29. R. Ulrich and A. Simon, *Appl. Opt.* **18** (1979) 2241.
30. D. Gibson, P. Lissberger, I. Salter and D. Sparks, *Opt. Acta* **29** (1982) 221.
31. J. Minowa and Y. Fujii, *Electron. Lett.* **21** (1985) 915.
32. M. Reeve et al, *Brit. Telecom Technol.* **7**. No. 2 (1989) 89.
33. H. Yanagawa, T. Ochiai, H. Hayakawa and H. Miyazawa, *Electron. Lett.* **24** (1988) 312.
34. T. Christmas and D. Richmond, *Opt. Laser Technol.* **9** (1977) 109.
35. J. Lipson and G. Harvey, *J. Light. Technol.* **LT-1** (1983) 387.
36. M. Seki et al, *Electron Lett.* **23** (1987) 948.
37. T. Finegan, A. Hunwicks and I. Wilkinson, *Proc. Broadband FOC/LAN, San Francisco* (1989), p. 34.
38. R. Erdmann, C. Perry and C. Parmenter, *Proc. SPIE* (1982) 12.
39. K. Kobayashi and M. Seki, *IEEE J. Quant. Elect.* **QE-16** (1980) 11.
40. T. Linglesheim and S. Aytac, *Proc. SPIE* **503**, *Application theory* (1984) 15.
41. J. Lipson et al, *J. Light. Tech.* **LT-3** (1985) 16.
42. J. Lipson et al, *J. Light. Tech.* **LT-3** (1985) 1159.
43. R.J. Mears, L. Reekie, S.B. Poole, D.N. Payne, *Electron. Lett.* **17** (1985) 709.
44. P. Urquhart, *Appl. Opt.* **26** (1987) 456.
45. M. Brierley and P. Urquhart, *Appl. Opt.* **26** (1987) 4841.
46. D.B. Mortimore, *J. Light. Technol.* **LT-6** (1988) 1217.
47. C.A. Millar, I.D. Millar, D.B. Mortimore, B.J. Ainslie and P. Urquhart, *IEE Proc.* Part J, **135** (1988) 303.
48. L. Reekie, R.J. Mears, S.B. Poole and D.N. Payne, *J. Light. Technol.* **LT-4** (1988) 956.
49. A. Yariv, *IEEE J. Quant. Electron.* **QE-9** (1973) 919.
50. W.J. Sorin and H.J. Shaw, *J. Light Technol.* **LT-3** (1985) 1041.
51. P.St-J. Russell and R. Ulrich, *Opt. Lett.* **10** (1985) 291.
52. I. Bennion, D.C.J. Reid, C.J. Rowe and W.J. Stewart, *Electron. Lett.* **22** (1986) 341.
53. C.J. Rowe, I. Bennion and D.C.J. Reid, *IEEE Proc.* Part J, **134** (1987) 197.
54. I.M. Jauncey, L. Reekie, R.J. Mears, D.N. Payne, C.J. Rowe, D.C.J. Reid, I. Bennion and C. Edge, *Electron. Lett.* **22** (1986) 987.
55. I.M. Jauncey, L. Reekie, J.E. Townsend, D.N. Payne and C.J. Rowe, *Electron. Lett.* **24** (1988) 24.
56. M.S. Whalen, D.M. Tennant, R.C. Alferness, U. Koren and R. Bosworth, *Electron. Lett.* **22** (1986) 1307.

57. W.V. Sorin and S.A. Newton in *Electron. Lett.* **23** (1987) 390.
58. L.F. Stokes, M. Chodorow, H.J. Shaw, *Opt. Lett.* **7** (1982) 288.
59. C.G. Atkins, J.F. Massicott, J.R. Armitage, R. Wyatt, B.J. Ainslie and S.P. Craig-Ryan, *Electron. Lett.* **25** (1989) 910.
60. D.B. Mortimore and J.V. Wright, *Electron. Lett.* **22** (1986) 318.
61. BT&D Technologies Ltd. Preliminary Product Sheet.

6 Silica fibre amplifiers and systems

D.M. SPIRIT

6.1 Introduction

In this chapter, optical amplifiers are introduced by considering their application to the enhancement of optical transmission system performance. The emphasis is on erbium-doped silica fibre amplifiers for use in the 1550 nm transmission window of standard single-mode silica fibre. Non-linear silica fibre and semiconductor laser amplifiers are discussed in sufficient depth to permit the comparison of amplifier characteristics in various applications. The final section will summarise some of the recent demonstrations involving fibre amplifiers and will serve to act as a focus for the systems issues raised in earlier parts of the chapter.

In general terms, the maximum loss capability of an unrepeatered optical transmission system is set by the difference between the transmitter laser launch power and the receiver sensitivity. In order to design a system with a large loss capability, whether a long-span point-to-point link or a wavelength division multiplexed (WDM) broadcasting system, the requirements are for high launch powers and highly sensitive receivers. For any given information rate, the receiver sensitivity may be significantly improved by the use of avalanche photodiodes (rather than PIN diodes) or, if the extra complexity is justified, by the use of coherent detection techniques. However, shot noise sets a fundamental limit beyond which the sensitivity may not be further improved.

At the transmitter end of the system, safety considerations left aside, the launch power cannot be increased indefinitely. Fundamental limits to the transmitter power are set by non-linear optical effects in the fibre such as self-stimulated Raman scattering (SRS) and self-stimulated Brillouin scattering (SBS) which cause the transfer of optical power to other wavelengths and backwards travelling waves [1]. Both these effects can cause severe degradation of system performance and are described in Section 6.5. Typical mean powers for the onset of SRS are of the order of watts – a level not likely to be approached for semiconductor lasers operating in the 1300 and 1550 nm transmission windows in the immediate future. For SBS, the mean threshold power is strongly dependent upon the laser linewidth and therefore on the modulation format, but in the worst case of an almost monochromatic beam, the threshold is typically a few milliwatts [2], which is modest by today's standards. The highest power (> 50 mW) semiconductor lasers available at present have spectra which contain many longitudinal modes and occupy many nanometers of wavelength. It is difficult to modulate directly such high power lasers at high data rates. Also, for long-distance point-to-point transmission, dispersion of the pulses produced from these broad linewidth lasers sets a limit on the length of system that can be operated. Alternatively, in WDM broadcasting systems, individual channels would use up a large proportion of the available bandwidth of the fibre. Hence, increasing the power of the laser transmitter by using devices with broad output spectra may limit the performance of an optical system.

In order to achieve maximum performance for both long-span point-to-point and high-capacity WDM broadcasting systems the transmitter is commonly a narrow linewidth distributed-feedback (DFB) laser. These devices are currently limited to output powers of a few milliwatts. At the highest data rates, modulation is imposed on the optical carrier in external modulators rather than direct modulation of the bias current which causes a modulation of the laser spectrum during a pulse (chirp). In intensity-modulated direct detection systems, dispersion of chirped pulses sets another limit to the maximum transmission span. The external modulator is usually a lithium niobate integrated optical device. The combined propagation losses of the waveguides in this material and coupling from fibre into the device and back into fibre again are typically 5–8 dB. The optical power in the fibre after the modulator will therefore be at least 10 dB (usually more, since the laser spectrum always has a finite width) below the SBS threshold in the transmission fibre. Hence it may be possible to increase the transmitter power by at least this amount before observing system degradation due to non-linear optical effects.

Active optical devices can therefore be used to boost the output power of a transmitter laser to just below the non-linear threshold. This provides an increased loss capability in any system, without necessarily requiring a special transmitter laser or a highly sensitive receiver.

6.2 Applications of optical amplifiers

(a) *Power amplification.* In Section 6.1 it has been shown how an optical amplifier could be used to increase the loss capability of an optical transmission system. In this case, the amplifier is required to have a good saturation performance, as the input signal level to the amplifier will be relatively high (typically of the order of 1 mW, 0 dBm).

(b) *Linear gain stage.* Current medium- and long-span optical transmission systems use opto-electronic repeaters at intervals along the length of the system to provide periodic regeneration of the signal to compensate for both fibre loss and dispersion. The repeaters essentially consist of a receiver, electronic amplifiers, a digital regenerator and a laser transmitter. The electronic circuitry in such a repeater is designed for use at a specific data rate, so that upgrading such a system would require the replacement of components in each repeater. While for terrestrial systems this may be feasible, for repeatered undersea systems the cost would be prohibitive. A linear optical gain stage would only require a CW bias current supply (plus supervisory equipment) and the performance of such an amplifier would be independent of data rate. Hence, optical amplifiers could be used as a direct replacement for opto-electronic repeaters in systems where regeneration of optical signals is not required (medium-length systems of a few hundred kilometres). In principle, this would allow the system to be upgraded to an increased data rate, bidirectional or WDM operation, simply by changing the terminal equipment – the transmission link can be thought of as being transparent to the data. For long-span systems (e.g. trans-Atlantic cables), while a chain of amplifiers could compensate for the loss of the system, the spontaneous emission from the amplifiers builds up along the system, setting an upper limit to the number of amplifiers that can be cascaded. Opto-electronic regenerators may still be required to prevent this build up of noise and also to restore the optical pulse distortion arising from fibre dispersion.

(c) *Optical switch.* All types of optical amplifier require the supply of electrical current to a laser device to drive the amplification mechanism. Removal of this bias not only removes the optical gain but, in some cases, the amplifier becomes opaque to the signal wavelengths. Hence, such a device can be used as an electrically controlled optical switch.

(d) *Pre-amplifier receiver.* In this application, the amplifier is used to amplify signals that may be below the sensitivity limit of a receiver. The amplifier is placed directly in front of a receiver. The output from the amplifier consists of the amplified signal and also a certain amount of amplified spontaneous emission (ASE). Since the input signal power to the amplifier will be small, the total power due to ASE can be a substantial proportion of (or even more than) the amplified signal power. Normally, the ASE is spread over a much broader optical spectrum than the signal, so that the use of narrowband, low-loss optical filters can reject most of ASE, which could otherwise dramatically reduce the sensitivity of the receiver.

All types of optical amplifier use the single-pass gain obtained by stimulated emission of photons by a signal laser from a population inversion in an optically active device. There are three main classes of optical amplifiers: rare-earth doped fibre amplifiers, non-linear fibre amplifiers [3] and semiconductor laser amplifiers [4]. Some of these classes are more suited to certain applications than others. Fibre amplifiers exhibit a number of advantages over semiconductor laser devices in that (a) the coupling losses between the fibre amplifier and the transmission fibre are low, (b) they show greater tolerance to signal wavelength and (c) the amplification mechanism is insensitive to the state of polarisation of the input signal. For current optical transmission systems, the main signal wavelengths are in the region of 1300 nm and 1550 nm, corresponding to the lowest loss windows in single-mode silica-based fibres. Although the 1300 nm window is associated with higher loss than the 1550 nm window, standard step index telecommunications fibre exhibits zero dispersion around 1300 nm, the transmission window which is most commonly used for installed systems to date. However, in the remainder of this chapter the emphasis will be on optical amplifiers for the 1550 nm window, where ultimately higher line rates and longer transmission spans are feasible. The next three sections will give a description of each class in isolation and will be followed by a section which will make comparisons between the types of amplifier.

6.3 Rare-earth doped fibre amplifiers

In Chapter 3 of this book, it has been shown how various rare-earth materials may be incorporated into single-mode optical fibres to provide gain media for fibre lasers. By removing optical feedback from the ends of the doped fibre, amplifiers can be made which use the single-pass gain available from the population inversion established by optical pumping of the dopant.

A rare-earth doped fibre amplifier consists of three basic components (Figure 6.1): a length of suitable rare-earth doped fibre, a pump laser, and a dichroic coupler to combine the pump and signal wavelengths. In some applications it is necessary to reject any unused pump light, so an optical filter may be used at the opposite end of the amplifier. This could be achieved by using a second dichroic coupler, in which case a second laser could be used to pump the fibre from the other end. This would allow the fibre length to be increased, implying increased gain from the amplifier. Alternatively, the

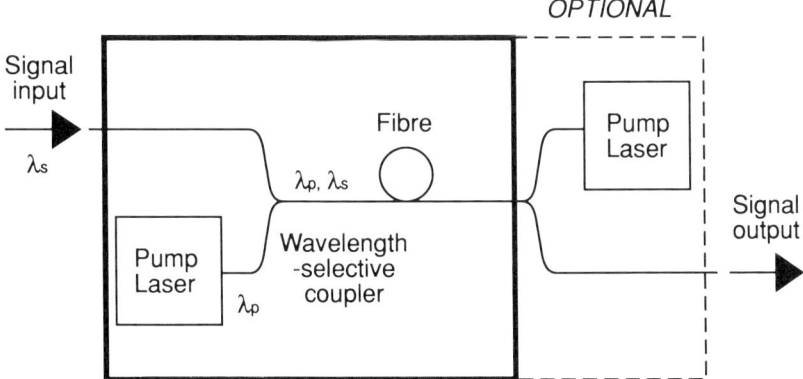

Figure 6.1 Schematic of fibre amplifier.

second coupler provides a monitor port for the transmitted pump laser, which could be used to provide the input to an automatic gain control mechanism. In all cases, any insufficiently pumped fibre exhibits high loss at the signal wavelength. The use of insufficient pump power or too long a fibre can lead to part of the amplifier providing loss rather than gain. Alternatively, increasing pump power arbitrarily or using too short a fibre leads to inefficient use of the pump power. There is therefore an optimum fibre length that has a strong dependence on pump power [5] and also on the signal input power.

The theory governing rare-earth doped fibre amplification is based on the dynamics of photon interactions with three-level systems [6, 7]. The treatment outlined here follows the results of reference [8]. The fibre gain coefficient per unit length at the signal wavelength λ is given by

$$g(\lambda) = 2\pi\sigma_e(\lambda) \int \rho(r)\psi(r) \left[N_2(r) - \frac{\sigma_a(\lambda)}{\sigma_e(\lambda)} N_1(r) \right] r\, dr \qquad (6.1)$$

where $\rho(r)$ is the erbium distribution function, $\sigma_{a,e}(\lambda)$ are the absorption and emission cross-sections, $N_{1,2}$ are the ground and upper state populations and $\psi(r)$ is the optical beam profile. The populations of the upper and lower states are governed by the differential equation

$$\frac{dN_1}{dt} = -\frac{dN_2}{dt} = -\frac{P_p \psi(r) \sigma_a(\lambda_p)}{h\nu_p} N_1 + \frac{N_2}{\tau} \qquad (6.2)$$

where P_p is the optical pump power and τ is the lifetime of the upper state. In the undepleted steady state, the solution to Equation (6.2) is given by $N_1 = 1/(1 + \alpha)$ and $N_2 = \alpha/(1 + \alpha)$, where $\alpha = P_p \psi(r) \sigma_a(\lambda_p) \tau / h\nu_p$. Assuming a Gaussian form $\psi(r) = \exp(-r^2/\omega^2)/\pi\omega^2$ for the field distribution, where ω is the mode field radius, and a uniform erbium distribution $\rho(r) = \rho_0$ in the fibre core, zero otherwise, Equation (6.1) may be integrated to give

$$g(P_p, \lambda) = \rho_0 \sigma_e(\lambda) \left\{ 1 - \eta + \frac{P^*}{P_p} \left[1 + \frac{\sigma_a(\lambda)}{\sigma_e(\lambda)} \right] \ln\left[\frac{P^* + \eta P_p}{P^* + P_p} \right] \right\} \qquad (6.3)$$

where $P^* = h\nu_p \pi \omega^2 / \sigma_a(\lambda_p) \tau$ and $\eta = \exp(-a^2/\omega^2)$, where a is the fibre core radius. Good agreement between experimental measurements and results from this model have been obtained [8].

Erbium has an emission line conveniently placed in the 1550 nm window at 1536 nm, and it is this example that will form the basis for most of this section. For operation in the 1300 nm transmission window, the neodymium emission line at 1.32 μm offers some prospects for the future. However, in silica host glasses, neodymium suffers from signal excited-state absorption (see later), which causes the fluorescence (and therefore the gain) to peak at around 1370 nm, which is uncomfortably close to the OH absorption peak of silica fibres. Excited-state absorption is of reduced significance in fluoride glasses, and recently a small signal gain of 10 dB has been reported at 1330 nm signal wavelength [9]. Amplification in rare-earth doped fluoride fibre will be discussed in detail in Chapter 8.

A schematic of the energy level diagram of erbium is shown in Figure 6.2 [10]. The 1536 nm spectral line is clearly marked as arising from the $^4I_{13/2} \rightarrow\ ^4I_{15/2}$ transition. The population inversion is achieved by optically pumping the erbium ions from the ground state to some higher lying state from which the ions relax to the upper state of the lasing or amplifying transition. The fluorescence decay time of the $^4I_{13/2}$ state is approximately 14 ms [11]. The reliability requirements for transmission system components dictate that any pump laser must be a semiconductor device. The transitions that can be pumped with semiconductor lasers at 807, 980 and 1480 nm are shown in Figure 6.2. Comparison between these three wavelengths has shown that 980 and 1480 nm pumping yield similar efficiencies, much greater than that obtained with 807 nm pumping [12]. Since these are ionic absorption lines, the tolerance on pump wavelength is generally of the order of only a few nanometres. Other wavelengths in the visible part of the spectrum are also suitable for pumping this system.

The first pump wavelength to be investigated was 807 nm, as high power AlGaAs lasers have been available at this wavelength for some time. In the first experiments, 6 dB net gain was obtained from 8 m of fibre doped with a few ppm of erbium for 15 mW launched pump power from an injection-locked diode array [13]. The limitation of using this pump wavelength is that as the population of the $^4I_{13/2}$ excited state increases, further absorption occurs at the same wavelength from this state to a higher lying state. This phenomenon is known as excited-state absorption (ESA) and is the major limitation to the

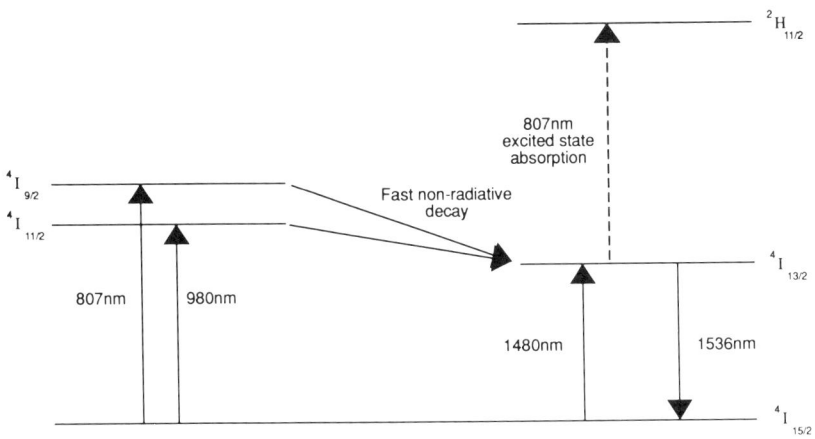

Figure 6.2 Schematic energy level diagram for erbium ions showing absorption lines which can be accessed with semiconductor lasers.

efficiency of erbium fibre amplifiers pumped at 807 nm [14]. Current systems research efforts concentrate on pumping at wavelengths of 980 and 1480 nm, where ESA is absent.

The most recent pump wavelength to be investigated with semiconductor, dye and titanium-sapphire laser pump sources is 980 nm [15]. This pumping scheme offers the highest efficiency of the three cases considered here due to the lack of ESA and the highest ground-state absorption cross-section. Until very recently, no semiconductor pump lasers were available at this wavelength, as it is difficult to process materials of a suitable bandgap. The other consideration of pumping at this wavelength is that to obtain the best performance from the amplifier, the fibre must be single mode at both 980 and 1536 nm. However, the high efficiencies promised by pumping at this wavelength (implying a low bias current requirement for the pump laser) are of great interest, particularly in submarine applications. The maximum efficiency reported to date is 4.9 dB/mW for 5 mW pump power (24.5 dB gain) at 980 nm from a strained-layer InGaAs quantum layer laser, requiring only 60 mA bias current [16]. This result was obtained from a compact packaged amplifier module which exhibited a maximum gain of 33 dB, requiring an optical pump power of 10 mW, an electrical bias to the laser of 80 mA and total electrical power consumption of only 175 mW.

The variation of gain with pump wavelength in the region of 980 nm has been studied using a tuneable Ti:sapphire laser [17]. It was shown that the pump power level at which the erbium-doped fibre became transparent (and therefore the point at which the amplification mechanism would be most efficient) was a minimum of 8.2 mW for a wavelength of 981.5 ± 0.5 nm. A ± 1 nm change in pump wavelength was shown to degrade the slope efficiency (gain/pump power) by 10%. A maximum gain of 48 dB was obtained for a launched pump power of approximately 45 mW. Figure 6.3 shows the variation of gain with pump power at various signal wavelengths and for 5 mW (-53 dBm). The saturation of gain at the highest pump power arose from the high (2 mW) levels of ASE produced in the fibre itself.

The 1480 nm pump wavelength arises from the high energy tail of the $^4I_{15/2} \rightarrow {}^4I_{13/2}$ absorption itself. In this case, the ions may be pumped over a range of at least 50 nm around 1480 nm. The tolerance on the pump wavelength has been shown to be at least 20 nm for a reduction in differential gain coefficient of 10% [18]. One practical implication of this is that multimode lasers (of many nanometres bandwidth) may be used as pump sources, rather than single-mode lasers which are necessary to obtain the highest efficiencies for 980 and 807 nm pumping. This wavelength range is easily accessible with colour-centre lasers and high-power ($>$ 50 mW ex-facet) commercial semiconductor lasers, offering the advantage that the mode field distribution is similar for both the pump and signal wavelengths. The maximum gain reported to date with semiconductor pumps in the region of 1480 nm is 46.5 dB for a launched optical pump power of 133 mW [19].

SILICA FIBRE AMPLIFIERS AND SYSTEMS 141

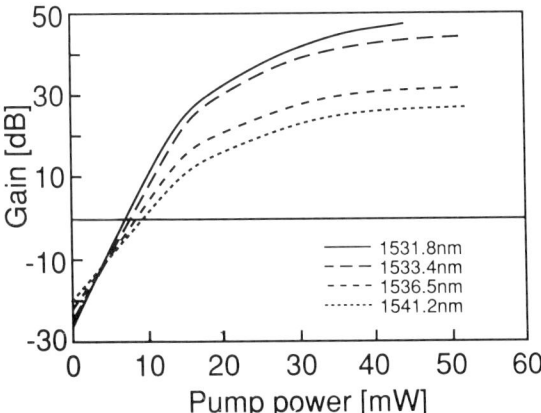

Figure 6.3 Variation of gain of erbium-doped fibre amplifier with pump power at 975.5 nm for signal wavelengths as shown (after [17]).

The spectral width of the 1536 nm transition in bulk erbium is very limited. However, perturbation of erbium ions by the glass host is sufficient to cause substantial broadening. In germania-doped silica fibres, 3 dB bandwidths of up to 10 nm have been reported [20]. The inclusion of alumina in the fibre core causes further broadening, particularly to longer wavelengths. Figure 6.4 shows the variation of gain spectrum with pump power for an erbium fibre, co-doped with germania and alumina, pumped at 1490 nm [21]. The fibre was doped with 1100 ppm erbium, was 10 m long and had a core diameter of 7 μm. The fibre has an unpumped loss of over 10 dB m at 1531 nm. It is clear that the

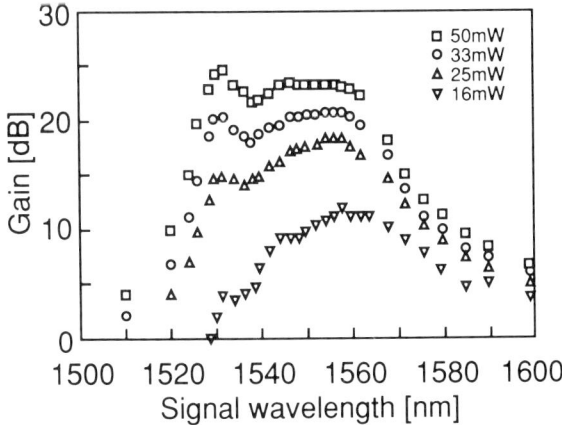

Figure 6.4 Gain spectrum of alumina co-doped erbium fibre amplifier for pump powers indicated at 1480 nm (after [21]).

amplifier exhibits small-signal gains in the region of 20–25 dB for 50 mW pump power and that the gain spectrum is fairly flat from 1530 to 1560 nm. Over 10 dB gain was obtained between 1520 and 1580 nm.

Further modification of the erbium gain spectrum may be obtained by the inclusion of the rare-earth ions into optical fibres made from fluoride glasses [22]. Since the erbium ions are now immersed in a totally different environment, the gain spectrum is rather different than that obtained in silica-based fibres, offering the potential for extended wavelength operation of erbium-doped fibre amplifiers. This will be examined in more detail in Chapter 8.

Gain saturation in rare-earth doped fibre amplifiers occurs when the amplified signal power (or the total ASE, as in the example above) approaches the pump power in the fibre, causing depletion of the population inversion. Figure 6.5 shows the variation of gain of an erbium-doped fibre amplifier with signal output power at four different pump powers [8]. At each pump power the fibre length has been optimised for maximum gain, and the values of fibre length and saturated output power for 3 dB gain compression are given as an inset to the figure. In each case, the saturated output power is around 15–30% of the pump power. Operation in severe gain compression approached 100% quantum efficiency, yielding a maximum output power of +15 dBm for a pump power of +17.3 dBm. In this regime, the output powers which can be obtained are limited mainly by the pump power.

The 14 ms fluorescence lifetime of the upper state of the amplifying transition in erbium has many important implications. In all types of optical amplifier, electrical noise on the bias supply to the semiconductor optical devices will introduce a degree of modulation on the gain of the amplifier. In the case of erbium-doped fibres, the relatively long lifetime of the upper state leads to significant modulation of the gain only for pump noise components

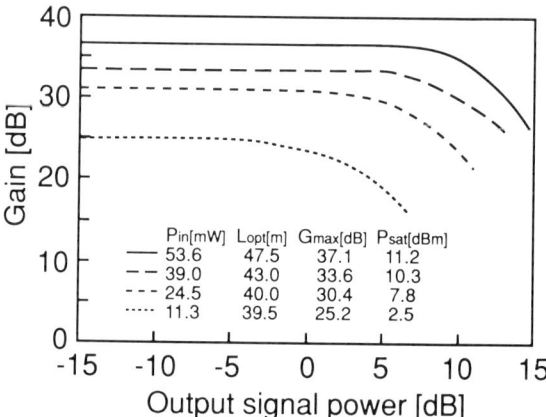

Figure 6.5 Variation of gain of erbium-doped fibre amplifier with amplified output power for pump powers as shown (after [8]).

below 100 kHz [11]. The electrical bias supplies for the pump lasers need to be noise-free only at relatively low frequencies, leading to relatively simple drive circuitry. Also, the propagation of a signal pulse through a rare-earth doped fibre can cause a modification of the population inversion and therefore a modulation of the amplifier gain. This effect is particularly marked for pulses that are sufficiently intense to saturate the gain. Severe pulse shaping can occur as a result of this process [23].

In multi-wavelength transmission, the modulation of the gain by one signal laser can cause cross-talk effects on the other signals, even if the amplifier is still operating in the small-signal regime. It has been shown that the intermodulation distortion induced by this process is negligible for frequency components above 100 kHz, another consequence of the long fluorescence lifetime of the $^4I_{13/2}$ state [24]. Cross-talk effects can become more pronounced when the amplifier is driven strongly into gain saturation, but again these effects are negligible for high-frequency components. Slow speed ($< 1\,\mu s$) automatic gain control mechanisms could be used to compensate for the reduction in gain caused by saturating signals in systems where cross-talk reduction is likely to be important [23]. It has been shown that intermodulation distortion and saturation-induced cross-talk have only a small effect on the performance of erbium fibre amplifiers used in high-speed WDM systems [25].

The random orientation of erbium ions in the fibre core leads to the averaging out of polarisation effects along the length of the fibre. The gain in erbium-doped fibres has been shown to be insensitive to the state of polarisation of the signal [26, 27], removing the need for polarisation control at the input to the amplifier. However, polarisation-insensitive dichroic couplers must be used in order to create an amplifier with no polarisation sensitivity.

Optical amplification, whether in fibre or semiconductor laser devices, introduces extra noise components into transmission systems due to contributions arising primarily from the amplifier spontaneous emission. Signal-spontaneous beat noise tends to be the dominant noise mechanism at high signal input powers, while shot noise and spontaneous-spontaneous beat noise terms dominate at low input power. Rate equation modelling has shown that an ideal erbium-doped fibre amplifier, optimised for maximum gain, would exhibit a noise figure of 3.2 dB, a degradation of only 0.2 dB over the value expected for an amplifier with a complete population inversion [28]. Figure 6.6 shows the variation of gain and noise figure with fibre length for this ideal amplifier operating with -30 dBm signal input power, assuming a pump power of 100 mW. It is clear from the figure that the noise performance of a counter-pumped amplifier can be significantly worse than that of a co-pumped amplifier, since in the former case, the population inversion at the input of the amplifier is relatively low. Noise figures as low as 5.2 dB have been reported for a co-propagating pre-amplifier pumped at 1490 nm [29].

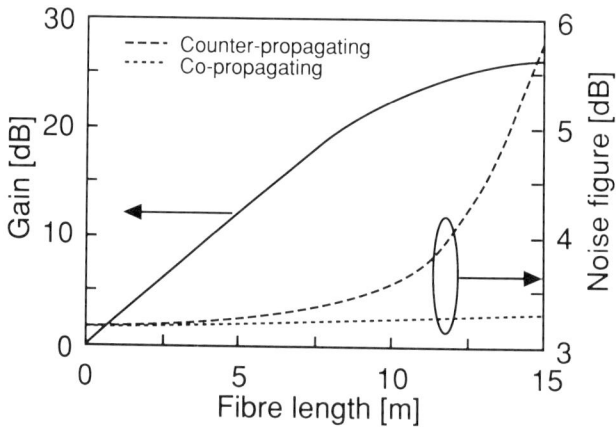

Figure 6.6 Variation of gain and noise figure with erbium-doped fibre length for a pump power of 100 mW and signal power of 1 μW. (Redrawn and adapted from [28].)

All the erbium-doped fibres that have been used in systems demonstrations are relatively short, typically between a few metres long with dopant levels of ≈ 1000 ppm and 200 m long with dopant concentrations of < 100 ppm. For certain applications, particularly non-linear transmission ([30] and Section 6.7.4), it can be more useful to have the gain distributed over many kilometres of fibre which can be achieved using much lower dopant concentrations. A 10 km length of fibre with an erbium dopant concentration of 10–100 ppb has been fabricated which became almost transparent when pumped with 70 mW power at 1480 nm [31].

6.4 Non-linear fibre amplifiers

Non-linear optical processes have already been presented in the introduction as fundamental limits to the power levels used in optical transmission systems. However, the same processes can also be used to advantage as amplification mechanisms. In this section, Raman and Brillouin scattering mechanisms will be described, showing how each process can be used to amplify signals beams in a fibre [3]. The main emphasis will be on Raman amplification, as Brillouin amplification has only a limited intrinsic bandwidth (< 100 MHz at 1550 nm).

6.4.1 Raman amplification

Figure 6.7 is a schematic of the vibrational energy levels of the molecular core material in any optical fibre. A number of other processes can occur as well as absorption. If the photon energy is not equal to the energy of a vibrational

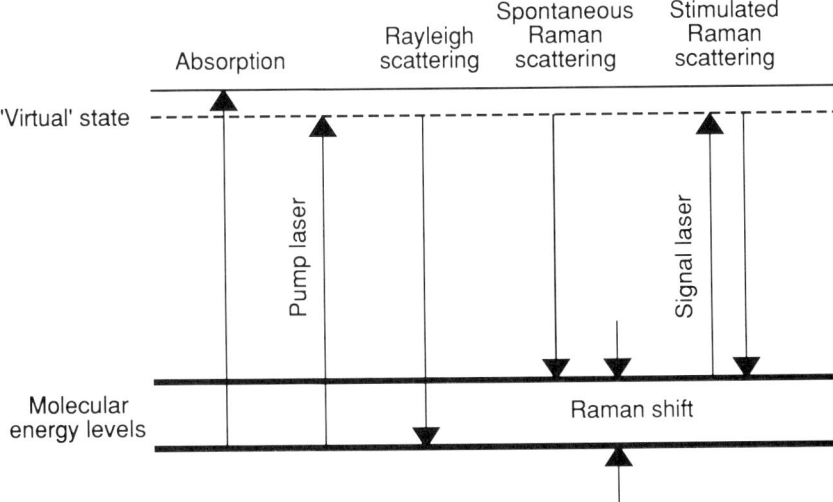

Figure 6.7 Molecular vibrational energy level diagram illustrating Raman scattering mechanisms.

transition of the molecule, then a 'virtual' excited state is created (shown dashed), which immediately ($\ll$ ps) re-radiates to the ground state by the emission of a photon at the same wavelength. This process is known as Rayleigh scattering, and forms one of the fundamental loss mechanisms in optical fibres. There is, however, a small probability that the virtual excited state will emit a photon of lower energy (longer wavelength) and leave the molecule in a different final state. This process is known as spontaneous Raman scattering. The difference in frequency between the exciting photon and the spontaneous emission is independent of wavelength and is known as the Raman shift (more generally for non-linear processes as the Stokes shift).

If sufficient power is coupled into an optical fibre, the level of population inversion between the virtual state and the final state can reach a sufficiently high value for amplification of the spontaneous emission to occur through stimulated Raman scattering (SRS). The threshold power for SRS is of the order of watts for single-mode optical fibres of kilometre lengths. Above threshold, considerable fractions of the optical power can be transferred to longer wavelengths. The SRS threshold is one of the fundamental limits to optical fibre transmission, but is unlikely to be a serious practical limit for semiconductor laser systems in the near future.

Raman amplification of a signal laser beam can occur through stimulated Raman scattering if a fibre is pumped at the appropriate wavelength. The vibrational spectra of the core material defines the Raman shift, and therefore if the wavelength of the signal laser beam is known, then the optimum pump wavelength can be calculated – i.e. Raman amplification can occur at any wavelength as long as an appropriate pump laser is available.

The implementation of a non-linear fibre amplifier is very similar to that of the rare-earth doped fibre amplifier. There are again three basic components: pump laser, wavelength selective coupler and fibre (Figure 6.1). The major difference here is that the fibre can be standard single-mode optical fibre, although for certain applications it may be desirable to use special fibre designs to enhance the amplifier performance.

Raman scattering has a spectrum that is created by the superposition of many transitions. Clearly, the incorporation of different molecular species into the fibre core will cause a modification of the Raman cross-section. In particular, the addition of germania into the core can cause significant enhancement to the Raman scattering cross-section [32]. Standard telecommunications optical fibres contain approximately 4 mol% germania, and so the Raman scattering cross-section (and, hence, maximum Raman gain) is a maximum for Raman shifts of 420–450 cm^{-1}. In the 1300 and 1550 nm transmission windows, this is equivalent to wavelength shifts of approximately 80 and 100 nm respectively. The full width half maximum of the cross-section spectrum is around 40 nm at 1550 nm.

In the absence of gain saturation, the gain of a Raman amplifier is given by

$$\text{Raman gain [dB]} = \frac{(10\log_{10} e) KgL_e P}{A} \qquad (6.4)$$

where g is the Raman gain coefficient (which is proportional to the Raman scattering cross-section), P is the mean optical pump power, A is the fibre spot area, implying that the amplification is proportional to pump intensity, and K is a parameter designed to account for the polarisation dependence of the fibre: $K = 1$ for polarisation-maintaining fibre, $K = \frac{1}{2}$ otherwise. The pump power is attenuated linearly in the fibre through the normal loss mechanisms, and so an effective length, L_e, is defined which takes account of this. For a fibre of length L and attenuation α_p at the pump wavelength,

$$L_e = \frac{1 - \exp(-\alpha_p L)}{\alpha_p}. \qquad (6.5)$$

In the limiting case of an infinitely long fibre, $L_e \to 1/\alpha_p$, so for standard single-mode optical fibre with lowest losses in the region of 0.2 dB/km, the maximum effective length is of the order of 15 km. With standard optical fibres, the gains that may be achieved in the 1550 nm transmission window for 100 mW pump power and 100 km fibre are of the order of 2.5 dB for step-index fibre and 5 dB for dispersion-shifted fibre [33]. The difference between the two fibres arises mainly because the dispersion-shifted fibre contains more germania in the fibre core. In both cases, 95% of the limiting gain is obtained in the first 50 km of fibre. Using 100 mW pump power at the appropriate wavelength from each end of 100 km of dispersion-shifted fibre would reduce the loss of the link from 20 dB to only 10 dB at 1550 nm.

Special designs can dramatically increase the Raman gain obtainable from optical fibres. Reference to Equation (6.4) shows that the gain for any given pump power may be increased by reducing the core area and increasing the Raman gain coefficient. Both of these may be obtained by increasing the level of germania doping in the fibre core, but this is always at the expense of increasing the loss of the fibre (and hence decreasing the effective length). The two extremes of fibre designs for Raman amplifiers are summarised in Figure 6.8, which shows the development of optical signal power with fibre length in unpumped and pumped Raman fibres.

In the case of standard fibres of low loss and low levels of germania doping, the Raman gain appears directly as an improvement in the system budget. In the opposite extreme of highly doped fibres with small spot sizes, the fibre alone is much more lossy than standard fibre. However, when subjected to the same Raman pump power, the Raman gain obtained is much greater than for standard fibres. There is an optimum fibre length for which the net gain (Raman gain – Linear loss at signal wavelength) is a maximum. A Raman power amplifier can be made by selecting a short (few kilometres) length of special fibre of optimum length given the available pump power, and then the installed transmission link would consist of standard telecommunications

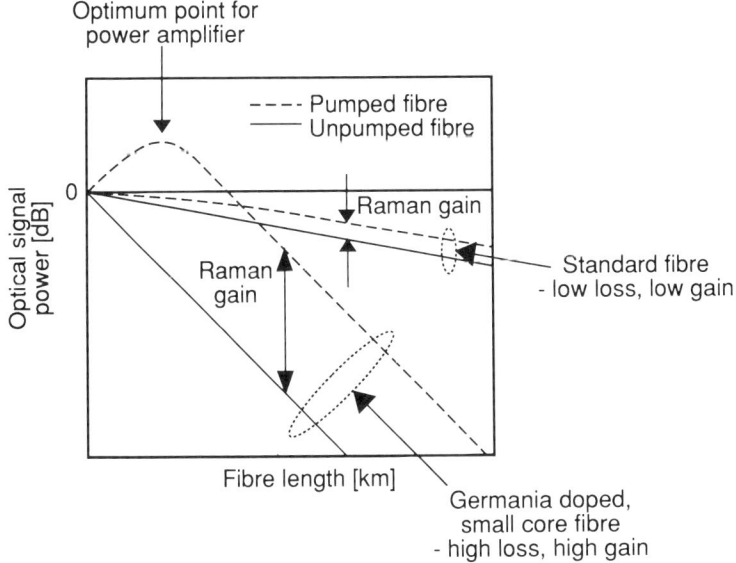

Figure 6.8 Variation of optical signal power with fibre length for both pumped and unpumped Raman fibre amplifiers, indicating the distinction between standard transmission fibre and small core, germania-doped fibres (after [34]).

fibre. Gain saturation occurs when the amplified signal beam becomes of comparable power to the pump power at the same point in the fibre. High (> 50%) conversion efficiencies from the pump beam to the amplified signal beam have been obtained [34]. Measured parameters for this high-gain Raman fibre are losses of 0.7 dB/km at 1480 nm, 0.5 dB/km at 1580 nm and germania doping level of 15 mol%, resulting in a mode field radius of 2.7 μm at 1500 nm. For 200 mW pump power, the optimum net gain of the amplifier is 8.5 dB, corresponding to a Raman gain of 13 dB. The 9 km length of fibre has a 3 dB gain bandwidth of approximately 25 nm.

6.4.2 Brillouin amplification

The main disadvantage of Raman amplification is the high powers required to obtain useful levels of gain, even in special fibres. The reason for this is the small magnitude of the Raman gain coefficient. At the opposite extreme is amplification by stimulated Brillouin scattering (SBS), where the gain coefficient g is several orders of magnitude greater. Mathematically, the description of SBS is identical to that of SRS, with the insertion of the appropriate gain coefficient, so that the equations above still hold. However, in SBS, the pump photons couple to acoustic waves (or phonons). Spontaneous Brillouin scattering occurs when a pump photon couples with a phonon, creating a photon of lower energy. The Brillouin shift is equal to the frequency of sound waves in the core material, which is approximately 11 GHz (or 0.1 nm at 1550 nm). Phase-matching conditions are satisfied only when the pump and scattered photons are travelling in opposite directions in the fibre. This is one significant difference between SBS and SRS which is bidirectional. The threshold value for stimulated scattering has been measured to be of the order of only 5 mW for a narrow linewidth laser beam propagating in 13 km of standard fibre [2].

Amplification can be obtained by pumping the fibre below the SBS threshold and simultaneously injecting a signal laser beam from the opposite end of the fibre. Gains of the order of 25 dB have been obtained for pump powers of only 5 mW [35]. At this power level, the threshold for SBS was reached and so the gain could not be increased any further. The bandwidth available in a Brillouin amplifier arises from the thermal distribution of the phonons in the fibre core material, and is less than 100 MHz (0.001 nm). Despite the high efficiencies that can be obtained, the use of SBS in transmission systems is extremely limited as a result of the small bandwidth. Modification of the Brillouin gain spectrum is possible by using fibre with different core materials [36] or by using a pump laser of finite linewidth greater than the intrinsic Brillouin gain spectrum [37]. In both cases, it is difficult to see how the useful gain bandwidth can be greater than the Brillouin shift (0.1 nm). The most efficient use of the mechanism requires the use of narrowband ($\ll$ 100 MHz) lasers, separated by 11 GHz to an accuracy of only a few megahertz. Only for certain specialised applications is this complexity justified [35, 38].

6.5 Semiconductor laser amplifiers

The optical spectrum of a multimode semiconductor laser is a superposition of the material gain and the spectrum of the Fabry–Perot (FP) cavity formed by the cleaved ends ($\approx 30\%$ reflectivity). Operating the device at a bias current below the lasing threshold can create a significant population inversion in the optically active region. The injection of light into a device operated in this way stimulates the emission of photons and hence leads to single-pass amplification of the signal laser beam. Gains well in excess of 20 dB may be obtained from FP laser amplifiers [39]. However, the FP mode spacing is typically of the order of 1 nm. Between the modes, the device can exhibit loss, and so accurate alignment of the signal laser wavelength to the resonances is essential.

In order to obtain broadband operation of semiconductor laser amplifiers, the facet reflectivity must be reduced either by polishing facets at Brewster's angle or, more commonly, by depositing antireflection coatings (usually single layers of SiO_x) on both facets of a standard semiconductor laser. Ideally, a perfect coating would produce a device with no ripple on the gain spectrum caused by FP modes, and in this case the device is known as a travelling wave (TW) laser amplifier. In practice, materials and growth limitations always leave some residual reflectivity – a few times 10^{-4} can be obtained routinely. The remainder of this section will consider these antireflection coated near-travelling wave amplifiers in more detail.

The optical gain spectrum of a near-TW amplifier is dependent upon the bandgap of the material used (which broadly defines the wavelength at which maximum gain is obtained), the bias current supplied to the device (which defines the broad envelope of the gain) and the residual reflectivity of the facets (which governs the fine structure on the overall envelope). A typical device is 500 μm long with residual facet reflectivities of 8×10^{-4} [39]. Operating at 70% of the threshold bias current, the device exhibits a peak internal small-signal gain of around 24 dB with a gain ripple of 3 dB. The 3 dB bandwidth of the envelope of maximum gain is around 45 nm. Coupling of signal light between single-mode fibre and the amplifier can routinely be achieved with losses of 3 dB per facet, and so the peak fibre-to-fibre gain of the amplifier is around 18 dB. For ordinary device structures, the saturated output power is of the order of a few milliwatts, although special devices have shown saturated output powers of over 100 mW [40].

Since the optically active region of a laser amplifier is rectangular rather than square or circular in cross-section, near-TW amplifiers can exhibit several decibels of polarisation sensitivity. However, new device structures have yielded amplifiers with less than 1 dB polarisation dependence [41], although these structures may not necessarily have, for example, good saturation performance. It should be clear that in order to obtain the maximum gain from a near-TW amplifier the polarisation of the input signal should be controlled. Also, the signal wavelengths should be coincident with the peaks of the residual FP resonances, implying active temperature control of the amplifier and signal laser. These requirements imply extra complexity in the active devices of the transmission system.

The speed at which intermodulation distortion and saturation-induced cross-talk can occur in semiconductor laser amplifiers is governed by the carrier recombination time. This is of the order of only 1 ns, implying that cross-talk can be a serious problem for high-speed multi-channel systems and also that signal patterning can occur even on single-channel digital transmission systems operating at several gigabits per second.

Despite the stringent requirements on the alignment of the signal laser to the cavity modes, FP amplifiers can be useful for optical processing. Operation in non-linear and bistable regimes has been obtained. Devices which exhibit hysteresis can allow pulse shaping of the type required by an all-optical regenerator. The reshaping of a triangular pulse train to an almost square output pulse train by the appropriate operation of an uncoated FP amplifier has been observed [42]. The input signal level required for this demonstration was of the order of only -30 dB m. The degree of non-linearity is dependent on the carrier density and therefore is limited in speed to the carrier recombination time of 2 ns. Hence although the switching speed is fast (≈ 10 ps), the bistability disappears for data rates in excess of 500 Mbit/s. The major limitation of FP amplifiers in this application is the limited range of wavelengths over which bistability can be obtained. Similar pulse shaping and bistable behaviour has more recently been reported for near-TW amplifiers where the optical waveguide contains an absorptive section as well as a region which exhibits optical gain [43].

6.6 Comparison between fibre and semiconductor laser optical amplifiers

6.6.1 *Amplifier characteristics*

Although the performance of each type of amplifier is specific to, for example, pumping conditions, an attempt has been made to give parameters for a typical amplifier of each type. These data are summarised in Table 6.1 and will form a basis for the discussion in this section. Brillouin fibre amplifiers will be ignored in the discussion, since although they require low electrical bias currents and can produce high gains, the bandwidth of well below 1 nm severely limits their use in transmission systems.

The most important property of an optical amplifier is the net gain. For semiconductor laser amplifiers and erbium fibre amplifiers, the typical

SILICA FIBRE AMPLIFIERS AND SYSTEMS

Table 6.1 Comparison of general amplifier characteristics in 1550 nm transmission window

Property	Laser	Erbium fibre	Raman fibre	Brillouin fibre
Unsaturated device gain	> 20 dB	> 20 dB	5–10 dB	20 dB
Optical pump power	N/A*	20–50 mW	100–200 mW	< 10 mW
Optical pump wavelength	N/A	807 nm, 980 nm 1460–1500 nm	Stokes shift below signal λ	
Electrical bias current	50 mA	> 100 mA	> 500 mA	< 50 mA
Wavelength of operation	Any	1530–1560 nm	Any, but subject to pump λ	
Stokes shift	N/A	N/A	100 nm	0.1 nm
Bandwidth	20–50 nm	10–40 nm	20–40 nm	0.001 nm
Coupling loss	5–6 dB	< 1 dB	< 1 dB	< 1 dB
Polarisation sensitivity	< few dB	0 dB	0 dB	0 dB
Saturated output (−3 dB)	< few mW	few mW	Limited only by pump power	
Directions	Bi-directional	Bi-directional	Bi-directional	Unidirectional
Noise	Low	Low	Very low	Very low
Cross-talk	Severe in gain compression	Only below 100 kHz	Low	N/A
Intermodulation distortion	Yes	Only below 100 kHz	Low	N/A

*N/A – not applicable

unsaturated gains are greater than 20 dB, while for Raman fibre amplifiers the gain is restricted to lower values by the stringent pump power requirements. However, coupling losses in the case of the semiconductor laser amplifier are of the order of 2.5–3 dB per facet, and so the net amplifier gain is reduced by 5–6 dB. For both types of fibre amplifiers, coupling losses can be reduced to below 0.5 dB at each end by using fusion splicing (even for special small-core single-mode fibres).

The pump requirements and gain spectrum of each class of amplifier are very different. The semiconductor laser amplifier is the least demanding, needing only an electrical bias supply at levels of around 50 mA. The wavelength of optimum gain can be selected by growth of the appropriate semiconductor material. However, there will always be some fine structure on the gain spectrum arising from the residual FP resonances, spaced by approximately 1 nm for a 500 μm long device.

Both fibre amplifiers require pump lasers of high output power which implies a higher bias current supply to the devices. Raman fibre amplifiers need pump powers of at least 100 mW in the fibre before significant useful gain is obtained, whereas erbium fibre amplifiers can yield gains of 20 dB for pump powers of as little as 10 mW. The pump lasers and gain spectrum for fibre amplifiers must be at or around specific wavelengths. It has been shown above that erbium fibre amplifiers require semiconductor pump lasers at one of 807, 980 or 1460–1500 nm and that the gain spectrum can extend across the range

1525–1565 nm by using alumina co-doped fibres. For Raman fibre amplifiers, the pump laser wavelength is usually 100 nm (Raman shift) below that of the signal. However, as the current availability of commercial high power (> 70 mW) semiconductor lasers is restricted to wavelengths between 1470 and 1550 nm, in practice the signal wavelength must be longer than 1570 nm to obtain optimised Raman gain from an amplifier. The 20–40 nm bandwidth of Raman fibre amplifiers is comparable to that of both erbium fibre amplifiers and semiconductor laser amplifiers. In most cases, the selection of the signal wavelength defines which amplification mechanism(s) may be applied.

The polarisation dependence of semiconductor laser amplifiers can be reduced to levels of 1 dB or less by using special device structures. However, a device designed for low polarisation sensitivity will probably not be optimum in other respects. The fibre amplification mechanisms (except for the special case of birefringent fibres) are polarisation insensitive, although care must be taken to ensure that any dichroic devices used to combine the pump and signal wavelengths introduce no polarisation dependence into the amplifier.

The saturated output power at 3 dB gain compression is high for the Raman fibre amplifier, mainly because the pump power requirement is greatest in this case. For both semiconductor laser and erbium fibre amplifiers the saturated output power is typically a few milliwatts. However, for erbium fibre amplifiers output powers well above a few milliwatts can be obtained in saturation as the limit on the output power is set mainly by the optical pump power supplied to the fibre. Driving any of the amplifiers into gain saturation has the added advantage that the population inversion is depleted, implying a reduction in the level of ASE from the amplifier.

Intermodulation distortion and saturation-induced cross-talk in WDM systems can occur in all three classes of amplifier, but on different timescales. For semiconductor laser amplifiers, the carrier recombination time of $\approx$ 1 ns sets a limit to their suitability for high-speed systems. The speed of erbium fibre amplifiers is governed by the 14 ms lifetime of the $^4I_{13/2}$ state, and so cross-talk between channels can be negligible for high-speed data with suitable coding. For Raman fibre amplifiers, the lifetime of the virtual state is extremely short and the gain is distributed over many kilometres or tens of kilometres of fibre. This implies that cross-talk due to gain saturation is negligible. However, if the power in any one signal channel becomes large (> tens of milliwatts), then that signal can become a *pump* for Raman amplification on all other signals at longer wavelengths. This is known as Raman-induced cross-talk, and forms a fundamental limit to high-power multichannel transmission.

6.6.2 *Suitability for specific applications*

(a) *Power amplifier.* All classes of amplifier could be used as power amplifiers, although the performance of each would be different. In all cases, the saturation limit is set by the bias supplied to the amplifier, whether electrical or optical.

The optical pump power required to drive Raman fibre amplifiers is the highest, implying the highest saturated output powers. Gain saturation induced cross-talk will be most severe in high speed WDM semiconductor laser power amplifiers.

(b) *Linear gain stage.* All the three types of amplifier are suitable for use as linear gain stages in both single-channel and WDM transmission systems. Intermodulation distortion is likely to be a considerable problem in WDM semiconductor laser amplifier systems, but not in the case of fibre amplifiers. The semiconductor laser amplifier (< 1 mm long) and erbium fibre amplifier (tens of metres long) can both be used as discrete amplifiers. However, Raman gain, which requires many kilometres of fibre, is most likely to be used to provide distributed gain in the transmission fibre itself.

(c) *Pre-amplifier.* The relatively low gains obtainable from Raman fibre amplifiers imply that they would not be used in this application. However, both semiconductor laser amplifiers and erbium fibre amplifiers can be used as optical pre-amplifiers for receivers. The semiconductor laser amplifier has the slight disadvantage that residual FP resonances can give small changes in sensitivity of the receiver for small changes in the signal wavelength.

(d) *Optical switch.* For each type of amplifier, the gain can be removed by turning off the electrical bias supply to the laser chips. Only limited extinction ratios would be obtained by removing the pump light from Raman fibre amplifiers which would not be suitable in this application. Without the bias, the semiconductor laser amplifier and erbium fibre amplifier become strongly absorbing, and are suitable for use in this application. However, these two switches have very different response times. The semiconductor laser amplifier responds on nanosecond timescales (carrier recombination time ≈ 1 ns) and so can be used as a high-speed switch. Contrast ratios in excess of 30 dB have been obtained for near-TW semiconductor laser amplifier optical switches [44]. The 14 ms fluorescence lifetime in erbium fibre amplifiers implies that the population inversion decays relatively slowly, and therefore that it is only possible to make slow switches with this type of amplifier (Section 6.7).

(e) *Non-linear processing applications.* The semiconductor laser amplifier can be used in a variety of non-linear applications including pulse reshaping, allowing the prospect of all-optical regeneration. Novel structures such as multi-electrode devices will increase the range of possibilities for optical processing using semiconductor laser amplifiers. The nature of the devices offers the prospect of integration of semiconductor laser amplifiers with other devices to create optical processors on a single chip. Both erbium and Raman fibre amplifiers are unsuited to non-linear processing applications and integration onto optical processing chips – in their present configuration they can only be used as gain stages.

6.7 Advanced system demonstrations

This section will present a discussion of some of the systems that have been demonstrated involving erbium-doped fibre amplification.

6.7.1 Long-span high-capacity transmission

Table 6.2 summarises the long-span, multi-gigabit per second systems that have been demonstrated to date, and gives some indication of the ways in which the amplifiers have been used – as a power booster, in-line repeater and pre-amplifier. As yet there are no systems results using amplifiers pumped by 980 nm semiconductor devices.

The direct detection, intensity-modulated results from NTT concentrate on unrepeatered transmission using both power and pre-amplification. The signal output from the power amplifiers was in the range $+8$ to $+15\,\mathrm{dBm}$ in these demonstrations, and the pump lasers were all $1.48\,\mu\mathrm{m}$ semiconductor devices. The longest unrepeatered transmission reported to date is 301 km long, operating at a data rate of 1.8 Gbit/s [48]. This 1552 nm system involved a combination of discrete amplification in erbium-doped fibres and Raman gain in the system fibre. Raman amplification was also used to provide some pre-amplification in the transmission fibre which was backwards pumped by

Table 6.2 Summary of recent long-span high-capacity transmission demonstration using erbium-doped fibre amplification. Intensity-modulated direct detection was used in all cases except where stated otherwise.

Distance (km)	Data rate (Gbit/s)	Power amp.	In-line amp.	Pre-amp.	Laboratory	Reference
218	1.2	No	1	No	KDD	45
904	1.2	Yes	11	No	KDD	46
250	1.8	Yes	No	Yes	NTT	47
310	1.8	Yes	Remote	Raman	NTT	48
710	2.4	Yes	9	No	KDD	49
2200	2.5	No	25	No	NTT, coherent	50
146	5	No	1	No	SEL	51
201	5	Yes	No	Yes	NTT	52
161	10	Yes	No	Yes	NTT	53
216	10	No	1	Yes	NTT	53
505	10	Yes	4	Yes	NTT	54
151	11	No	1	No	Bellcore	55
260	11	No	2	No	Pirelli & Bellcore	56
200	11	Yes	1	No	Pirelli & Bellcore	56
84	12	No	No	Yes	Fujitsu	57
100	12	No	1	No	Fujitsu	57

1.45–1.49 µm laser diodes. The in-line erbium fibre amplifier, situated 34 km from the receiver, was remotely pumped using the residual power left after transmission of the Raman pump beams through 34 km, and provided 9.9 dB gain. The 1450 nm end of the pump spectrum is optimal for providing Raman gain in the system, although the absorption cross-section in erbium is low. The opposite applies to the 1490 nm pump wavelength.

All the other experiments involve at least one in-line repeater, and introduce several important issues for the future installation of fully engineered long-distance transmission. The remainder of this discussion will concentrate on two examples which both used dispersion-shifted transmission fibre: 904 km, 1.2 Gbit/s intensity-modulated transmission with 12 cascaded amplifiers [46], and a 2200 km, 2.4 Gbit/s coherent FSK system with 25 in-line repeater amplifiers [50]. In both cases, the amplifiers are pumped at 1480 nm from one end only, so that unabsorbed pump power can enter the transmission system, which may well cause some degradation of the system performance. For a pump wavelength of 980 nm or below, the difference in fibre attenuation between the pump and signal wavelengths may well give sufficient rejection of the unused pump beams. However, for a 1480 nm pump wavelength, optical filtering is necessary to reject the unwanted pump, either as an edge filter to remove the pump alone [46] or as an optical bandpass filter to reject unabsorbed pump and broadband ASE from the amplifier [50]. The use of optical isolators in conjunction with erbium-doped fibre amplifiers can be necessary to prevent reflective feedback from other parts of the system causing the amplifier to act as a laser. Isolators may also be required to prevent Brillouin backscattered light from the amplified signal laser beating with the signal which can introduce extra noise components [56]. The use of narrow linewidth bandpass filters and isolators may serve to significantly improve the performance of a particular system, but also prevent later upgrades to the system by introducing more signal wavelengths or bidirectional transmission.

The 2200 km coherent system used amplifiers that contained an isolator at each end of the erbium-doped fibre (so that the fibre is pumped through an isolator) and a 1 nm or 3 nm filter at the output. The mean input power to each amplifier was -18.4 dBm and the mean gain was 17.7 dB. After transmission over 2200 km, the accumulated excess noise from the amplifiers introduced a penalty of only 4.2 dB. In comparison, the 904 km direct detection demonstration used only one isolator per amplifier on the output, with no optical filtering except for an edge filter to reject unabsorbed pump light at the receiver. The performance of the amplifiers was similar to those in the 2200 km coherent system. Noise accumulation around the gain peak in the chain of amplifiers led to a penalty of only 0.6 dB after 904 km transmission. Accumulation of the noise, which was observed to exhibit an almost linear dependence on the number of repeaters, sets an upper limit to the number of erbium-doped fibre amplifiers that can be cascaded, in a similar way to semiconductor laser amplifiers [58].

6.7.2 Pre-amplifier receiver

The long-span transmission experiments summarised in Table 6.2 include a number of systems in which pre-amplifiers yielded receiver sensitivity improvements of up to 14 dB at 5 Gbit/s [53] and 9.5 dB at 12 Gbit/s [57]. The ASE from the amplifier introduces excess noise at the receiver which has been studied in more detail both with and without narrowband optical filtering to reduce the spontaneous noise processes. Without filtering, the spontaneous noise bandwidth is determined by the amplifier gain spectrum, and a 22 dB amplifier gain resulted in a sensitivity improvement of 10.5 dB [59]. This pre-amplified receiver was shown to have a sensitivity of -46 dB m at 140 Mbit/s. With optical filtering the detected level ASE can be greatly reduced, significantly improving the receiver performance. A pre-amplifier receiver including a 0.6 nm grating filter has been demonstrated in an ASK system at 1.8 Gbit/s with a sensitivity of -43 dB m, or 215 photons/bit [60]. This was achieved using an amplifier with a measured gain of 35 dB gain and noise figure of 4.1 dB, and is the best sensitivity (photons/bit) reported to date at any data rate in an optical pre-amplified receiver.

6.7.3 Multichannel systems

This is perhaps the most diverse applications area, covering both long-span digital transmission and digital and analogue broadcasting in the subscriber loop. The use of WDM for long-span transmission has been demonstrated over 459 km using six cascaded erbium-doped fibre amplifier [61]. The four signal laser beams were each separated by 2 nm around 1550 nm, and were digitally modulated at 2.4 Gbit/s. A 1 nm optical bandpass filter was used to select a single channel at the receiver, but no filters were used in the individual amplifiers. A penalty of below 1.5 dB was detected on each of the four channels. Analogue AM and FM video distribution has been demonstrated over distances in excess of 250 km where the modulation is imposed on a single laser, intended as a direct replacement for co-axial cable TV systems [62]. Six cascaded amplifiers, each with an output power of 5–10 dB m, were used in this system, demonstrating only a small degradation in carrier-to-noise ratio over 480 km.

In the local broadcasting applications, a variety of systems have been demonstrated. Sixteen optical signals intensity modulated at 622 Mbit/s and separated by 5 GHz have been simultaneously amplified by 15–19 dB in a single amplifier [63]. A penalty on each channel of 1.1 dB was observed, which was associated with SNR degradation due to ASE rather than cross-talk in the amplifier. Sixteen optical signals, ten with analogue and six with digital modulation, spanning 34 nm across the erbium fibre gain spectrum, have been simultaneously amplified by 18–24 dB in a splitting network [64]. This system demonstrated the transmission of 100 FM TV channels and six 622 Mbit/s channels over 9 km to 4096 receivers. Finally, sixteen 155 Mbit/s coherent channels spaced by 10 GHz have been amplified by an in-line repeater amplifier to enable transmission over 102 km with a penalty below 1 dB [65].

6.7.4 Non-linear transmission

The propagation of high-intensity optical pulses in fibre can allow access to non-linear optical effects. In particular, the Kerr non-linearity can lead to pulse compression that can be used to offset the linear dispersion of the fibre. In a lossless transmission medium, selection of the appropriate signal peak power can allow the transmission of a pulse train over very long distances without distortion. In a real, lossy fibre, periodic discrete (semiconductor laser or erbium fibre) amplifiers or distributed gain (Raman) in the transmission fibre itself must be used to maintain the signal level as close to the critical power (referred to as the $N = 1$ soliton threshold) as possible. Transmission of a 55 ps pulse train many times through a 42 km fibre loop for a distance of 6000 km with low distortion has been demonstrated using colour-centre lasers as signal and Raman pump sources [66]. The interaction of soliton pulses of different wavelengths in erbium-doped fibre transmission systems has also been investigated, and showed that gain saturation and cross-talk effects are similar to those found for standard intensity-modulated transmission [30]. Practical systems, however, require semiconductor laser devices, and mode-locked lasers and gain-switched DFB lasers have emerged as signal sources for non-linear transmission. In both cases, integrated optic devices have been used to impose modulated data on the signal.

The first demonstration of an error-free, all-semiconductor laser soliton transmission system used a gain-switched DFB at 1550 nm wavelength, producing 16 ps pulses after amplification and compression in 3.7 km fibre [67]. After the external modulator, power amplification in an erbium-doped fibre was followed by Raman amplification in the 23 km transmission fibre. More recently, error-free transmission has been obtained at 5 Gbit/s over a distance of 250 km using a total of eleven erbium-doped fibre amplifiers to periodically restore the peak signal power above the $N = 1$ soliton threshold [68].

Mode-locked semiconductor lasers can generate trains of short (< 100 ps) pulses at gigahertz repetition rates. Transmission of 4 Gbit/s has been achieved over 136 km of standard step-index fibre at 1559 nm using four erbium amplifiers, two to boost the transmitter launch power and two as in-line repeaters [69]. Using an unmodulated 10 GHz repetition rate mode-locked signal laser (equivalent to a 20 Gbit/s 1010... sequence), transmission has been demonstrated over 100 km dispersion-shifted fibre using both a single in-line erbium amplifier and Raman gain in the system fibre [70]. The 20 ps input pulses to the system were narrowed to 16 ps after 100 km fibre in the non-linear regime. When the input power level was significantly reduced, the pulses broadened through linear dispersion to at least 35 ps, causing the pulses to overlap, which would lead to total system failure.

6.7.5 Routeing switch

As mentioned in Section 6.6.2, erbium-doped fibre amplifiers can only be used as slow optical switches due to the 14 ms lifetime of the $^4I_{13/2}$ state. However, relatively slow switches operating on this timescale would be useful for routeing signals around a network. An all-fibre 2 × 2 optical routeing switch has been demonstrated which uses erbium-doped fibres as on–off switches [71]. The two optical inputs are split by 1 × 2 couplers into four paths, each of which is followed by an erbium fibre amplifier. The output from the amplifiers is combined by 2 × 1 couplers in such a way that either input can be switched to either output port by pumping the appropriate amplifier. Alternatively, one or both inputs can be routed to both output ports to enable routeing diversity functions. Each amplifier is pumped at 980 nm, exhibiting 7 dB gain when pumped (sufficient to compensate for the 6 dB loss of the two couplers) and greater than 20 dB attenuation when unpumped.

6.7.6 OTDR enhancement

The performance of optical time domain reflectometers (OTDRs) is limited by the transmitter laser and the receiver sensitivity in a manner similar to the limits to system performance described in Section 6.1. Optical post-amplification at the output of the OTDR not only causes increased output power from the transmitter, but also improves the receiver sensitivity by acting as a pre-amplifier for the backscattered light from the system under test. The gain of the amplifier therefore appears directly as an increase in the dynamic range of the OTDR. This has been demonstrated with an OTDR modified to operate with a 1536 nm DFB laser transmitter and a 1 nm optical bandpass filter between the OTDR and the erbium amplifier to reduce the mean level of ASE incident on the receiver [72]. Using 125 ns pulses, the OTDR alone had a dynamic range of 15 dB. Modifications to the OTDR, including the insertion loss of a non-optimal filter, reduced this to 7.5 dB, with the 18 dB fibre amplifier resulting in an increase in one-way measurement range to 25.5 dB, an improvement of over 10 dB compared with the instrument alone. Not only does this demonstrate the bidirectional nature of the amplification mechanism, but it offers the prospect of high-resolution fault location over systems with large loss budgets.

6.8 Conclusions

Optical amplifiers, particularly semiconductor laser-pumped erbium-doped fibre amplifiers, offer great potential for use in future transmission systems.

The properties of high-gain, high-saturated output power, substantial bandwidth and low polarisation sensitivity allows the use of amplifiers in a wide variety of applications. In particular, fibre amplifiers will provide high-gain, low-noise linear amplification for transmission, while semiconductor laser amplifiers will find applications in non-linear optical processing.

References

1. R.G. Smith, *Appl. Opt.*, **11** (1972) 2489.
2. D. Cotter, *Electron. Lett.* **18** (1982) 495.
3. R.H. Stolen, *Proc. IEEE* **68** (1980) 1232.
4. M.J. O'Mahony, *J. Light. Technol.* **LT-6** (1988) 531.
5. E. Desurvire, J.R. Simpson and P.C. Becker, *Opt. Lett.* **11** (1987) 888.
6. J.R. Armitage, *Appl. Opt.* **27** (1988) 4831.
7. E. Desurvire and J.R. Simpson, *J. Light. Technol.* **7** (1989) 835.
8. E. Desurvire, C.R. Giles, J.R. Simpson and J.L. Zyskind, *Opt. Lett.* **14** (1989) 1266.
9. Y. Miyajima, T. Komukai, T. Sugawa and Y. Katsuyama, *Proc. OFC '90* (1990), Paper PD16.
10. G.H. Diecke and H.M. Crosswhite, *Appl. Opt.* **2** (1963) 675.
11. R.I. Laming, R.L. Reekie and D.N. Payne, *Electron. Lett.* **25** (1989) 455.
12. J.R. Armitage, *IEEE J. Quant. Electron.* (in press 1990).
13. T.J. Whitley, *Electron. Lett.* **24** (1988) 1537.
14. R.I. Laming, M.C. Farries, P.R. Morkel, L. Reekie, D.N. Payne, P.L. Scrivener, F. Fontana and A. Righetti, *Electron. Lett.* **25** (1989) 12.
15. R.S. Vodhanel, R.I. Laming, V. Shah, L. Curtis, D.P. Bour, W.L. Barnes, J.D. Minelly, E.J. Tarbox and F.J. Favire, *Electron. Lett.* **25** (1989) 1386.
16. M. Shimizu, M. Horiguchi, M. Yamada, I. Nishi, S. Uehara, J. Noda and E. Sugita, *Proc. OFC '90* (1990), Paper PD17.
17. W.J. Miniscalco, B.A. Thompson, E. Eichen and T. Wei, *Proc. OFC '90* (1990), Paper FA2.
18. Y. Kimura, K. Suzuki and M. Nakazawa, *Electron. Lett.* **25** (1989) 1656.
19. J.L. Zyskind, C.R. Giles, E. Desurvire and J.R. Simpson, *Proc. OFC '90* (1990), Paper FA4.
20. R.J. Mears, L. Reekie, I.M. Jauncey and D.N. Payne, *Electron. Lett.* **23** (1987) 1026.
21. C.G. Atkins, J.F. Massicott, J.R. Armitage, R. Wyatt, B.J. Ainslie and S.P. Craig-Ryan, *Electron. Lett.* **25** (1989) 910.
22. C.A. Millar, M.C. Brierley and P.W. France, *Proc. ECOC '88* (1988) 66.
23. C.R. Giles, E. Desurvire and J.R. Simpson, *Opt. Lett.* **14** (1989) 880.
24. M.J. Pettitt, A. Hadjifotiou, and R.A. Baker, *Electron. Lett.* **25** (1989) 416.
25. E. Desurvire, C.R. Giles and J.R. Simpson, *Proc. OFC '89* (1989), Paper TUG7.
26. C.R. Giles, E. Desurvire, J.R. Talman, J.R. Simpson and P.C. Becker, *J. Light. Technol.* **7** (1989) 651.
27. M. Suyama, K. Nakamura, S. Kashiwa and H. Kuwahara, *Proc. IOOC '89* (1989), Paper 20A4-3.
28. R. Olshansky, *Electron. Lett.* **24** (1988) 1363.
29. C.R. Giles, E. Desurvire, J.L. Zyskind and J.R. Simpson, *Proc. IOOC '89* (1989), Paper 20PDA-5.
30. M. Nakazawa, Y. Kimura, K. Suzuki and H. Kubota, *J. Appl. Phys.* **66** (1989) 2803.
31. J.R. Simpson, L.F. Mollenauer, K.S. Kranz, P.J. Lemaire, N.A. Olsson, H.T. Shang and P.C. Becker, *Proc. OFC '90* (1990), Paper PD19.
32. S.T. Davey, D.L. Williams, D.M. Spirit and B.J. Ainslie, *Proc. SPIE* **1191** (1989), Paper 19.
33. D.M. Spirit and L.C. Blank, *Electron. Lett.* **25** (1989) 1687.
34. D.M. Spirit, S.T. Davey, D.L. Williams and L.C. Blank, *IEE Proc.*, Part J, Special issue on *Optical Amplifiers for Communication* (in press 1990).
35. C.G. Atkins, D. Cotter, D.W. Smith and R. Wyatt, *Electron. Lett.* **22** (1986) 556.
36. N. Shibata, R.G. Waarts and R.P. Braun, *Opt. Lett.* **12** (1987) 269.
37. N.A. Olsson and J.P. van der Ziel, *J. Light. Technol.* **LT-5** (1987) 147.
38. R.W. Tkach and A.R. Chraplyvy, *Proc. OFC '89* (1989), Paper THG2.
39. M.J. O'Mahony, I.W. Marshall and H.J. Westlake, *Brit. Telecom. Technol. J.* **5** (1987), No. 3.

40. D.M. Cooper, M. Bagley, L.D. Westbrook, D.J. Elton, H.J. Wickes, M.J. Harlow, M.R. Aylett and W.J. Devlin, *Proc. OFC '90* (1990), Paper PD32.
41. S. Cole, D.M. Cooper, W.J. Devlin, A.D. Ellis, D.J. Elton, J.J. Isaac, G. Sherlock, P.C. Spurdens and W.A. Stallard, *Electron. Lett.* **25** (1989) 314.
42. H.J. Westlake, M.J. Adams and M.J. O'Mahony, *Electron. Lett.* **22** (1986) 541.
43. I.W. Marshall, M.J. O'Mahony, D.M. Cooper, P.J. Fiddyment, J.C. Regnault and W.J. Devlin, *Appl. Phys. Lett.* **17** (1988) 1577.
44. M. Ikeda, *Electron. Lett.* **21** (1985) 252.
45. N. Edagawa, K. Mochizuki and H. Wakabayashi, *Electron. Lett.* **25** (1989) 363.
46. N. Edagawa, Y. Yoshida, Y.H. Taga, S. Yamamoto, K. Mochizuki and H. Wakabayashi, *Proc. ECOC '89* (1989), Paper PDA-8.
47. K. Hagimoto, K. Iwatsuki, A. Takada, M. Nakazawa, M. Saruwatari, K. Aida, K. Nakagawa and M. Horiguchi, *Electron. Lett.* **25** (1989) 662.
48. K. Aida, S. Nishi, Y. Sato, K. Hagimoto and K. Nakagawa, *Proc ECOC '89* (1989), Paper PDA-7.
49. N. Edagawa, Y. Yoshida, Y.H. Taga, S. Yamamoto and H. Wakabayashi, *Proc. OFC '90* (1990), Paper PD3.
50. S Saito, T. Imai, T. Sugie, N. Ohkawa, Y. Ichihashi and T. Ito, *Proc. OFC '90* (1990), Paper PD2.
51. B. Wedding, T. Pfeiffer and M. Wittman, *Proc. ECOC '89* (1989), Paper TuA5-7.
52. K. Nakagawa, K. Hagimoto and S. Nishi, *Proc. OFC '90* (1990), Paper WC2.
53. K. Hagimoto, Y. Miyagawa, A. Takada, K. Kawano and Y. Tohmori, *Proc. ECOC '89* (1989), Paper TuA5-5.
54. K. Hagimoto, Y. Miyagawa, Y. Miyamoto, M. Ohhashi, M. Ohhata, K. Aida and K. Nakagawa, *Proc. IOOC '89* (1989), paper 20PDA-6.
55. M.Z. Iqbal, J.L. Gimlett, M.M. Choy, A. Yi-Yan, M.J. Andrejco, L. Curtis, M.A. Saifi, C. Lin and N.K. Cheung, *Proc. IOOC '89* (1989), Paper 20PDA-7.
56. A. Righetti, F. Fontana, G. Delrosso, G. Grasso, M.Z. Iqbal, J.L. Gimlett, R.D. Standley, J. Young, N.K. Cheung and E.J. Tarbox, *Proc. ECOC '89* (1989), Paper PDA-10.
57. H. Nishimoto, I. Yokota, M. Suyama, M. Seino, T. Horimatsu, H. Kuwahara and T. Touge, *Proc. IOOC '89* (1989), Paper 20PDA-8.
58. T. Mukai, Y. Yamamoto and T. Kimura, *IEE Trans. Microwave Theory and Techniques* **MTT-30** (1982) 1548.
59. M.J. Pettitt, R.A. Baker and A. Hadjifotiou, *Electron. Lett.* **25** (1989) 273.
60. C.R. Giles, E. Desurvire, and J.R. Simpson, *Opt. Lett.* **14** (1989) 880.
61. H. Taga, Y. Yoshida, N. Edagawa, S. Yamamoto, and H. Wakabayashi, *Proc. OFC '90* (1990), Paper PD9.
62. K. Kikushima, E. Yoneda, and K. Aoyama, *Proc. OFC '90* (1990), Paper PD22.
63. H. Toba, K. Inoue, N. Shibata, K. Nosu, K. Iwatsuki, N. Takato, and M. Shimizu, *Electron. Lett.* **25** (1989) 885.
64. W.I. Way, S.S. Wagner, M.M. Choy, C. Lin, R.C. Menendez, H. Thome, A. Yi-Yan, A.C. Von Lehman, R.E. Spicer, M. Andrejco, M.A. Saifi and H. Lemberg, *Proc. ECOC '89* (1989), Paper PDA-9.
65. R. Welter, R.I. Laming, W.B. Sessa, R.S. Vodhanel, M.W. Maeda and R.E. Wagner, *Electron. Lett.* **25** (1989) 1333.
66. L.F. Mollenauer and K. Smith, *Opt. Lett.* **13** (1988) 675.
67. K. Iwatsuki, S. Nishi, M. Saruwatari and M. Shimizu, *Proc. IOOC '89* (1989), Paper 20PDA-1.
68. M. Nakazawa, K. Suzuki, E. Yamada and Y. Kimura, *Proc. OFC '90* (1990), Paper PD5.
69. N.A. Olsson, A. Andrekson, P.C. Becker, J.R. Simpson, T. Tanbun-Ek, R.A. Logan, H. Presby and K. Wecht, *Proc. OFC '90* (1990), Paper PD4.
70. I.W. Marshall, D.M. Spirit, G.N. Brown and L.C. Blank, *Proc. OFC '90* (1990), Paper PD6.
71. E. Eichen, W.J. Miniscalco, J. McCabe and T. Wei, *Proc. OFC '90* (1990), Paper PD20.
72. L.C. Blank and D.M. Spirit, *Electron. Lett.* **25** (1989) 1693.

7 Silica fibre laser oscillators

D.C. HANNA and A.C. TROPPER

7.1 Introduction

Prior to the recent resurgence of interest in rare-earth doped fibre lasers and amplifiers, silica glass had received relatively little attention as a host glass for laser application. This was in part due to its inconveniently high melting temperature and in part due to the fact that pure silica offers only a small solubility for rare-earth dopants. In 1982 Namikawa et al. [1] reported on the preparation of bulk samples of Nd-doped SiO_2 glass by plasma torch CVD, in which a Nd_2O_3 concentration of somewhat less than 0.3 mol% was reached. However, even at this concentration, which is low by the standards of conventional Nd glass laser material (e.g. phosphate glass), there was observed a fast decay component to the infrared fluorescence, which is probably evidence of clustering of the Nd ions.

In a fibre laser or amplifier the concentration requirement can be greatly relaxed as the pump light is end-launched and therefore can in principle have a very long absorption path available to it. The objections to silica as a laser host are therefore less cogent in a fibre. In fact there are compelling reasons for choosing silica. These arise from the central importance of silica fibre as the preferred medium for optical fibre telecommunication. Add to this the fact that, as a result largely of telecom requirements, there exists a well-developed technology of fibre components and devices compatible with monomode silica fibres, e.g. couplers, polarisation-maintaining fibre etc., then there are reasons for preferring to work with silica fibre as the laser medium, even where the intended use is not in telecom.

The publications which are regarded as the trigger to the now immense amount of activity in active fibre devices are the two comparison papers by Poole et al. [2] and Mears et al. [3]. In [2] a technique was described for introducing rare-earth ions into silica single-mode fibre, based on an extension of the MCVD technique. The technique enabled any of the rare-earth dopants to be introduced into the fibre core, and initial concentrations achieved were in the few hundred parts per million range. This was more than adequate to demonstrate the excellent potential of these active fibres, since fibre lengths of the order of a metre to a few metres were able to provide large pump absorption and large gain. Ainslie et al. [4], also using the MCVD technique, provided a useful early compilation of the absorption and fluorescence spectra of the rare-earths in silica-based fibre. These early fabrication procedures involved limited solubility of the rare-earth ions in the silica host, with clear evidence of clustering and formation of precipitates at greater than a few hundred parts per million of the dopants [5]. Since then, considerable progress has been made in fabrication of fibres with much higher doping concentrations, with values in the region of 1 mol% ($\sim 10{,}000$ ppm), being achievable. This has been made possible by incorporating Al_2O_3 into the core of the fibre, at concentrations of several percent. As noted by Arai et al. [6], working on bulk silica material, the presence of Al_2O_3 and P_2O_5 can significantly increase the solubility of rare-earth dopants in silica. Another important development which has enabled higher doping concentrations to be reached and which has also greatly increased the versatility of fibre fabrication is the adoption of the so-called solution-doping technique [7]. In this technique, based on a procedure first demonstrated by Stone and Burrus [8], the rare-earth is introduced into the fibre core in the convenient form of a solution rather than as a vapour. Besides the greater doping levels that can be achieved in this way [9], the simplification of procedure is very helpful in the fabrication of fibres where two dopants are to be incorporated with good control over their concentrations, as for example in Yb:Er fibre [7].

The relative ease with which low-loss, monomode, doped silica fibre can be fabricated, compared with, say, fluorozirconate fibre, or fibre made from multicomponent glass, has given silica fibre a pre-eminent role as the medium for fibre lasers. More rare-earth dopants have so far been operated successfully as fibre lasers in silica than in other hosts. These now include (see Table 7.1) neodymium (Nd), praseodymium (Pr), [10], samarium (Sm), [11], erbium (Er), [12], holmium (Ho), [13], thulium (Tm), [14] and ytterbium (Yb), [15].

The laser transition at 651 nm is particularly noteworthy, offering a rare example of a visible solid state laser transition in a glass host, and what is more

Table 7.1 Laser transitions in silica fibres

Ion	Transition	Wavelength (μm)
Nd^{3+}	$^4F_{3/2} \rightarrow {}^4I_{11/2}$	1.055–1.14
	$^4F_{3/2} \rightarrow {}^4I_{9/2}$	0.9–0.95
	$^4F_{3/2} \rightarrow {}^4I_{13/2}$	1.36
Er^{3+}	$^4I_{13/2} \rightarrow {}^4I_{15/2}$	1.53–1.60
Pr^{3+}	$^1D_2 \rightarrow {}^3F_4$	1.084
	$^1D_2 \rightarrow {}^3F_2$	0.888
Sm^{3+}	$^4G_{5/2} \rightarrow {}^6H_{9/2}$	0.651
Yb^{3+}	$^2F_{5/2} \rightarrow {}^2F_{7/2}$	1.01–1.162
		0.974
Tm^{3+}	$^3H_4 \rightarrow {}^3H_6$	1.65–2.05
Ho^{3+}	$^5I_7 \rightarrow {}^5I_8$	2.04

a transition for which lasing had not previously been reported in any host. This laser is discussed further in Section 7.8.

A notable feature of a silica host compared with, for example, a fluorozirconate host, is the prominent role played by multiphonon, nonradiative decay in shortening the lifetime of many of the excited states. A

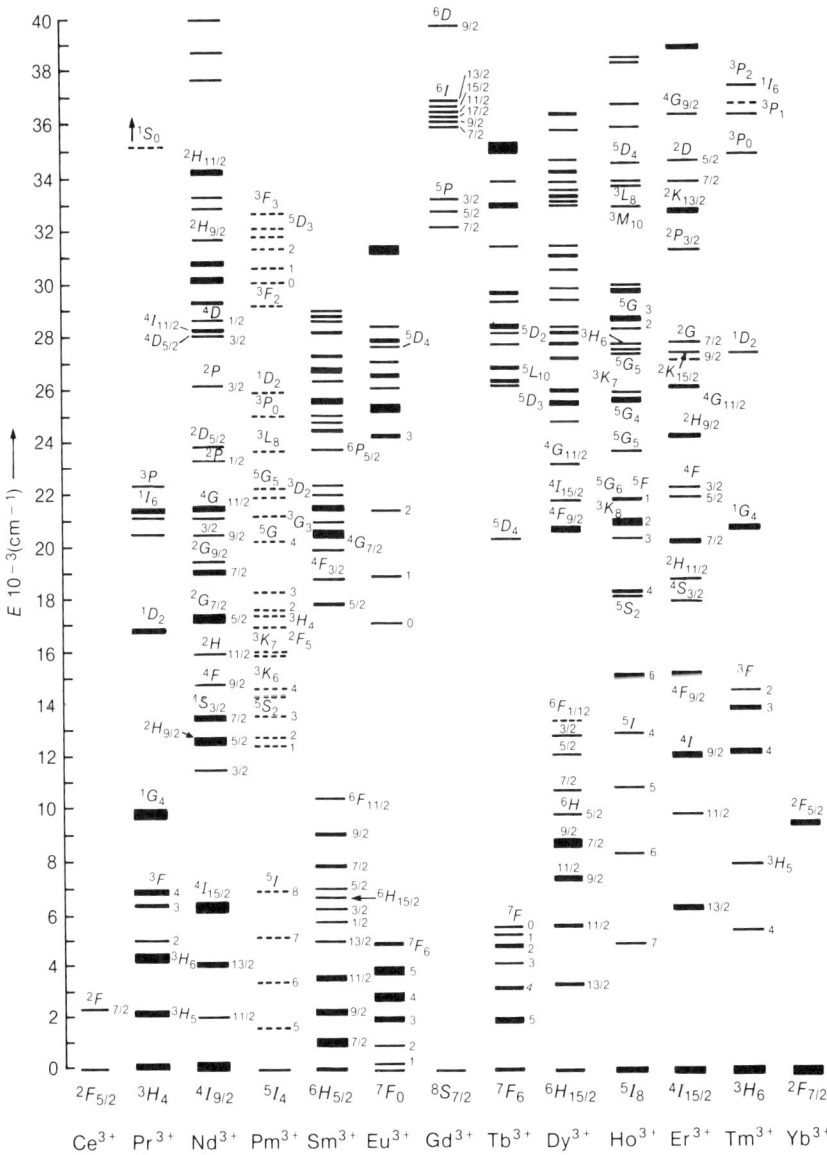

Figure 7.1 Energy levels of tri-valent rare earth ions (after [73]).

cursory perusal of the energy level diagram ('Dieke diagram') in Figure 7.1, of the triply ionised rare earths, RE^{3+}, suggests a veritable multitude of possible laser transitions. However, in the cases of Er^{3+}, Ho^{3+} and Tm^{3+}, only the $^4I_{13/2}$ level, $^5I_7$ level and $^3F_4$ level, respectively, serve as an initial laser level in a silica host, by virtue of being metastable, whereas higher excited levels suffer rapid nonradiative decay. As a rough guide, if the gap between an energy level and the next level below it is less than $\sim 4000\,\text{cm}^{-1}$, nonradiative decay will significantly reduce the fluorescence lifetime in a silica host. By contrast, in a fluorozirconate host, where the phonon energy is less, an energy gap of less than $\sim 3000\,\text{cm}^{-1}$ is needed before significant lifetime shortening occurs. Thus Tm^{3+} in a fluorozirconate host has shown six different laser transitions, to be compared with just the one ($^3H_4 \to {}^3H_6$) in a silica host. Ho^{3+} and Er^{3+} are similarly restricted to one transition each in silica, but many more have operated in fluorozirconate fibre.

Since silica fibre lasers have been more widely available than other fibre types, they have been more widely investigated and subjected to a much wider range of operating conditions. This includes mode-locked and Q-switched operation, discussed in Chapter 9, tuned operation, single frequency operation, and high power operation. The latter three are reviewed in this chapter. Tuneability is one of the operating characteristics that has received considerable attention, exploiting the wide emission bands that are a feature of the silica host. This is discussed in Section 7.2. On the other hand, the broad linewidth would appear to present a significant disadvantage in attempts to achieve single frequency (i.e. single longitudinal mode) operation. Despite this, and despite the close mode-spacing that a long fibre implies, successful single-mode operation has now been seen in fibre lasers using a number of different methods. These results are reviewed in Section 7.3.

If single frequency operation of a fibre laser is seen to be a rather unexpected achievement, then at first sight the idea of high power operation of fibre lasers (i.e. multiwatt powers) might also seem surprising. Nevertheless powers in excess of a watt have been achieved [16]. This is in part due to the favourable conditions for heat removal which apply to fibres generally (small transverse dimension and considerable length), but also due to the excellent power-handling capabilities of silica in particular.

One of the most striking (and most exploitable) consequences of the fibre geometry is that, compared with a bulk laser medium, much higher gain is available for a given pump power. Gains in fibre amplifiers reach many decibels per milliwatt of pump power. One result of this is the quite respectably large gains that can be achieved even on transitions having a low quantum efficiency. Thus, lasers have been successfully operated on transitions with a quantum efficiency as low as $\sim 10\%$, Tm-doped silica being a case in point. This is discussed in Section 7.9, where it is also emphasised that a low quantum efficiency need not imply a low slope efficiency, and the case of Tm-doped silica can be cited, where slope efficiencies of $\sim 36\%$ (photon conversion efficiency

> 80%) have been observed on a transition of much lower quantum efficiency [17]. On the other hand there have been other transitions in doped silica fibre which show much lower efficiency than expected. Some of these cases are discussed in Section 7.9.

The high gain achievable in a fibre amplifier means that it is rather easy to demonstrate superfluorescent emission (more correctly referred to as amplified spontaneous emission, ASE), in which 'mirrorless lasing' occurs [18]. This refers to the situation where the gain is large enough to amplify the spontaneous emission from one end of the fibre to a power level that saturates the gain at the exit end of the fibre. This has been observed in a number of silica fibres, e.g. Nd [19], Yb [20] and results are discussed in Section 7.7.

Two further topics discussed in this chapter are the 1.5 μm Er fibre laser, Section 7.5, and the 1.3 μm Nd fibre laser, Section 7.6. Both of these wavelengths are of particular relevance to telecom applications, where high-gain amplification is the overriding goal. Lasers in these wavelength ranges are also of interest. The 1.3 μm Nd transition ($^4F_{3/2} \rightarrow {}^4I_{13/2}$) provides a model illustration of the unfortunate consequences of excited state absorption (ESA) since in silica the ESA at $\sim$ 1.3 μm [21] rules out an effective amplifier. This finding gave considerable impetus to the investigation of non-silica hosts and modification of the silica composition [22], since a significant variation of such parameters as transition centre wavelength and absorption and emission cross section can occur from one host glass to another. Despite the lack of promise for 1.3 μm amplification, lasing has been demonstrated, illustrating the fact that even if the available gain is low, respectable laser performance is not ruled out.

7.2 Tuneable operation

The most conspicuous spectroscopic attribute of lanthanide ions in a silica-based glass host is the strong broadening which spreads out the optical transitions into bands some 500–1500 cm^{-1} in width. The stimulated emission cross-section at any wavelength is correspondingly reduced, but in the fibre waveguide geometry this loss of gain can be offset by efficient pumping, a small mode area and an extended interaction length. Thus there are a number of transitions which can be operated as tuneable lasers over a significant wavelength range. These systems are potentially interesting as miniature spectroscopic probes, as well as for the range of applications in which wavelength control is desirable. For example the tuning range demonstrated for thulium encompasses absorption bands of water vapour, liquid water, and carbon dioxide, suggesting both medical and atmospheric sensing applications.

The techniques which have been used to tune fibre lasers divide into two groups; those which use an intracavity microscope objective or spherical mirror to collimate the laser mode onto a conventional intracavity wavelength-selecting element, and those which use all-fibre tuning devices such as the variable directional coupler. The chief advantage of the first approach is the speed with which such a cavity can be assembled; all the elements are standard bulk optics, no fusion splicing or other fabrication is required and no specially designed coupler is needed. This is a good way to make a first assessment of the tuning performance of a transition. The disadvantage is the insertion loss associated with the intracavity microscope objective which, if coupled with a low efficiency tuning element will create a rather lossy cavity and limit both the tuning range and the efficiency of the tuned output.

A factor which plays a major role in the performance of this type of cavity is the degree of birefringence of the fibre. All tuning elements to a greater or lesser extent introduce polarisation-dependent losses into the cavity. If the gain medium is a non-polarisation-preserving fibre then light in the cavity will alter its polarisation state in the course of a double pass through the fibre. Since the retardation produced by the fibre is wavelength-dependent, the cavity losses exhibit a complex variation with wavelength which can manifest itself as a pronounced modulation on the tuning curve. The effect can sometimes be eliminated (if the fibre is long enough) using a fibre loop polarisation controller [23, 24]. Each adjustment of the tuning element must then be accompanied by a compensating adjustment of the controller loops.

The insertion loss problem can be alleviated by using an all-fibre tunable ring resonator, such as the device described in [25], in which doped fibre is spliced between an input and an output port of a polished coupler, forming a bidirectional ring resonator. Wavelength tuning of the coupling coefficient and hence the laser output is accomplished by varying the core separation of the coupler.

Of the lanthanide transitions listed in Table 7.1, broad tuning ranges have been observed in Nd [26, 27], Er [27, 28], Yb [24] and Tm [29, 30]. Of particular note is the $\sim 1300\,cm^{-1}$ tuning range of the Yb ion, which is comparable with that of infrared dyes. With the single exception of the lower energy Nd band, these tunable transitions terminate on the ground multiplet of the lanthanide ion, a feature which has a big influence on the tuning behaviour.

The ground multiplets of the Yb, Tm, and Nd ions are split into Stark levels spanning an energy range $> 1000\,cm^{-1}$, several times larger than kT at room temperature ($200\,cm^{-1}$). Consequently at the low energy end of the emission band the terminal levels are not thermally populated and four-level laser action can occur. As the laser is tuned to the higher energy end of the band the effect of ground state reabsorption increases progressively, and the lasing acquires a three-level character. The high energy limit of the tuning range is therefore set by the degree to which the available pump power is able to saturate the transition, which depends in turn on the strength of the pump absorption and the size of the fibre core.

In the case of Er, the Stark splitting of the ground manifold is relatively small, $\sim 400\,\text{cm}^{-1}$, and all of the levels will be thermally populated at room temperature. Er therefore operates essentially as a three-level laser at all wavelengths. This will be discussed further in Section 7.5. By cooling the fibre it is possible to reduce the effects of absorption from the normally populated Stark levels. Cooling to liquid nitrogen temperature, at which $kT \sim 50\,\text{cm}^{-1}$, provides a useful degree of cooling. The fibre geometry lends itself very conveniently to simply being immersed in an open Dewar of liquid nitrogen with the ends left outside to provide access for end-launching. The effect of cooling has been observed, in the case of Yb [24] for example, to give extended tuning to shorter wavelengths.

For the Tm fibre laser in particular the tuning range is sensitively determined by the fibre length. The short-wavelength end of the tuning curve is limited by ground-state reabsorption losses. The pump transition will be saturated over some initial length of the fibre at the input end, producing an inversion on the quasi-three-level laser transition. This length represents an optimum, providing maximum gain for a given available pump power; if the fibre length is longer than this optimum then the unsaturated region at the output end will introduce reabsorption losses into the cavity. At shorter transition wavelengths there is more thermal population in the terminal level and more intense pumping is required to invert the transition. Thus the optimum length decreases with decreasing wavelength. The overall tuning range indicated in Table 7.1 encompasses the tuning ranges achieved with fibres of three different lengths. The resonators were all of the type incorporating conventional wavelength-selecting elements, [29, 30]. In each case the tuning curve had a smooth, flat-topped form which did not follow the profile of the fluorescence spectrum. This may be a consequence of gain saturation associated with the low output coupling used in each of these resonators, as reported recently for a Tm:YAG laser [31].

Although a similar tuning range/fibre length trade-off is found in the Yb fibre laser, the Yb tuning range referred to in Table 7.1 was demonstrated with a single length of fibre. It was, however, found necessary to use two high dispersion intracavity prisms to achieve the required degree of wavelength selection. The large intracavity loss incurred by this configuration necessitated use of highly reflecting mirrors, and output coupling was therefore taken via a Fresnel reflection from a prism surface, thus limiting output power to $\sim 3\,\mu\text{W}$. As with Tm, a smooth and flat-topped tuning curve was obtained.

The results obtained so far from tuned operation of fibre lasers have generally been aimed at demonstration of the impressively wide range that can be covered, with little or no effort put into the achievement of high efficiency. Attention now needs to be paid to the question of devising tunable resonators with extremely low loss, since in this way a wide tuning range should be achievable at high efficiency even with diode laser pumping. This achievement would greatly enhance the prospects for application of tuneable devices.

7.3 Single longitudinal mode operation

Fibre lasers have a number of potential applications, which depend on their operation with a narrow linewidth, for example in the area of sensors. Applications in coherent communications place particularly exacting requirements on linewidth, with stable single longitudinal mode (SLM) operation being essential. A number of different techniques have now successfully produced SLM operation, in Nd and Er doped fibres, despite the fact that mode spacings for typical fibre laser dimensions imply many thousands of modes within the characteristic gain bandwidth of several terahertz. A brief account is given here of the single mode selection techniques used to date.

In Section 7.2 various ways of tuning fibre lasers have been described, in each case involving the use of a line-narrowing element. In principle an element such as a diffraction grating can be used to achieve SLM operation, in much the same way as is routinely done for dye lasers, provided the resonator length can be kept reasonably short. If broad tunability is not required then an integral fibre grating can be used, in which the grating takes the form of a periodic modulation of index imposed into the core, a so-called Bragg reflector. Such a grating can be made, for example by polishing away most of the cladding over a short section of fibre and then imposing a periodic corrugation using ion-beam etching [32]. Using such a grating, narrow line operation has been achieved both in Nd-doped and Er-doped fibre [33, 34], and in the case of Nd-doped fibre, by going to a short enough fibre (51 mm) so that the mode spacing was increased, single longitudinal mode operation was achieved, with a measured linewidth of 1.3 MHz FWHM. The scheme for fabricating the grating can in principle be realised in a much more convenient way, by optically writing a grating into the core by UV light incident transversely into the fibre [35, 36], exploiting the UV-induced index change that occurs in GeO_2 doped silica fibre.

Another approach, demonstrated by Barnsley et al. [37] involves the use of a so-called Fox-Smith resonator, well-known from conventional bulk lasers, in which two incommensurate resonators are coupled by a beam splitter, thus providing a Vernier type of interference between their resonances. This enables a long resonator to have a fine frequency discrimination. In fibre form the beam splitter is simply a fibre coupler.

The schemes described above are standing-wave resonators and rely on the introduction of a strong frequency discrimination to select a single mode. It is, however, well-known that in a homogeneously broadened laser, single frequency operation can be readily achieved with very little frequency discrimination if the laser is made to operate unidirectionally in a ring configuration, i.e. with a travelling wave. The availability of pigtailed polarisation-independent isolators makes the achievement of unidirectional operation straightforward. Fibre couplers and techniques for splicing make the fabrication of a ring resonator also very straightforward. In this way Morkel et al. [38] obtained single frequency operation of an Er-doped travelling wave fibre ring laser with a measured linewidth of 60 KHz.

The basis for single frequency selection via travelling wave operation is that spatial hole burning is avoided and thus the mode with highest net gain will dominate by ensuring that all other modes have their gains driven down by saturation to a net gain of less than unity. Other ways of eliminating spatial hole burning also exist and have been exploited. One of these, effectively equivalent to moving the gain medium longitudinally back and forth to smear out any spatial hole-burning, is reported by Sabert and Ulrich [39]. The effective movement of the fibre is achieved by phase modulators at each end of the fibre acting in a push-pull fashion. A bandwidth of $\sim 10\,\text{kHz}$ has been reported for a Nd-doped fibre laser operated in this way.

Finally a more conventional scheme involving injection locking is reported by Jones and Urquhart [40], in which an external cavity diode laser operating SLM was injected into a conventional standing wave Er fibre laser. Single frequency operation of the Er fibre laser resulted.

The various results described here indicate that fibre lasers lend themselves to single frequency operation in straight-forward, practical schemes with potential for very narrow linewidth operation. While some of the schemes can in principle be applied to fibres other than silica fibres, it is clear that silica fibres currently hold the advantage of availability of the various components required, such as couplers, pigtailed isolators, Bragg reflectors and so on.

7.4 High power operation

The thermal limitations of lamp-pumped laser rods have been discussed extensively in the literature (see Koechner [41] for a review). Generally the situation considered is that of uniform heat deposition throughout the volume of the rod, as appropriate to transverse pumping. For the deposited heat to escape, a radial temperature gradient becomes established. This leads to thermal lensing, via the refractive index dependence on temperature, and birefringence is induced by virtue of the stress caused by differential expansion. Both of these effects can pose problems for rod lasers. In addition, the stress, which is greatest at the surface of the rod, imposes a maximum heat input (and hence laser output), limited by the tensile strength of the material. An analysis [41] leads to the conclusion that there is a certain maximum value of heat deposition per unit length of rod that can be tolerated, but that this value is independent of the diameter of the rod. Owing to its lower thermal conductivity than for crystals, (by typically an order of magnitude), glass will fracture at a much lower thermal input.

The situation for fibre lasers, while formally identical to that of a laser rod differs in practice in a number of important respects. One of the most obvious differences is that a considerable length of fibre can be used so that problems associated with excessive heat deposition per unit length can in principle be avoided simply by using a longer fibre (i.e. of lower doping concentration). Another obvious difference is that the heat deposition is not uniform – it is in fact confined to the core, which occupies only a small fraction of the fibre volume. This situation is formally the same as that of an end-pumped laser rod, where the pump beam occupies a small volume compared to the rod volume. An analysis of such a situation (Hanna *et al.* [21, 42]) reveals that for a given heat deposition per unit length, the maximum stress is greater than for the uniform deposition case by a factor $2 - [R_1/R_2]^2$, where R_1, R_2 are the radius of the core and cladding respectively (or pump beam radius and rod radius, respectively, for an end-pumped rod). Thus the stress can be up to twice as large. However, this disadvantage is more than offset by the real benefit of end-pumping, which is that with a smaller pumped volume, a much higher gain can be produced for a given pump (and hence heat) input. So, the threshold can be reached and efficient lasing achieved, for a low pump power and hence a low heat input. The benefits of end-pumping in bulk laser rods are well illustrated by the demonstration of CW lasing on a three-level transition in an end-pumped Yb:Er glass laser [43]. Fibre lasers give an even more dramatic illustration of the benefit of end pumping since the pumped volume (i.e. the core) is much smaller and only a very low pump input is needed to reach threshold. These low power requirements naturally incline one to think of fibre lasers as low power devices. However, because only a low input power is needed, it does not follow that one is restricted to low input power (and hence output power), and recently attention has begun to be paid to the question of high power operation of fibre lasers [42].

A detailed discussion of the potential limits to high power operation of fibre lasers is beyond the scope of this chapter, so it is restricted to indicating some of the possible limiting factors, and giving a few numerical estimates to provide a feel for the magnitudes involved.

A first, and obvious limitation is imposed by the availability of a high power laser to provide pumping. This needs to be a laser which emits at a wavelength which can be efficiently absorbed by the active ion, and with a beam quality which is good enough to allow it to be efficiently end-launched into the fibre core. It has been noted that Tm doped silica fibre obligingly offers some absorption at 1.06 μm, in the extended high energy wing of the $^3H_6 \rightarrow {}^3H_5$ transition, thus enabling the high power Nd:YAG laser to be exploited as a pump laser [16, 44]. In this way a fibre laser output of 1.3 W has been achieved in the $^3H_4 \rightarrow {}^3H_6$ transition at 2 μm.

Naturally a great deal of interest centres on the question of how to couple power efficiently from high power AlGaAs diode laser arrays into a fibre core, since these operate at wavelengths suitable for pumping a number of rare-earth lasers. The problem, however, is that the beam quality from such an array does not allow direct launching into a monomode core. An elegant solution to this problem has been demonstrated by Snitzer *et al*. [45] via the so-called 'cladding-pumping' technique. In this scheme the monomode core containing the active ion is surrounded by an undoped inner cladding (lower index than the core) and this in turn is surrounded by an outer cladding (of yet lower index). Pump light launched into the inner cladding is confined by total internal reflection at the interface between the claddings, and as it propagates is progressively absorbed into the core. This is a hybrid pumping scheme having the character both of longitudinal and transverse pumping. It has the great merit of coupling all of the power from a low coherence beam into a monomode core, and thus ensuring a monomode output. Cladding-pumping arrangements offer scope for flexible design, as illustrated by Po *et al*. in a report [46] describing the use of an inner cladding of roughly rectangular shape, thus providing a means of overcoming the incompatibility between the geometries of planar array lasers and circular fibre lasers.

Cladding pumping is at its most effective for an ideal four level laser, since, unlike a three-level laser it is not necessary to reach a particular pump intensity in the core to achieve inversion. In fact, however, cladding pumping has also been demonstrated to offer an effective role for the three-level Er^{3+} transition at 1.5 μm, pumped by the green output (0.53 μm) of a frequency doubled Nd:YAG laser [47]. This scheme, which led to $\sim$ 3.9 W of average output power at 1.5 μm, drew on another advantage of cladding pumping, in overcoming the limitations on maximum pump intensity that can be tolerated at the input to the fibre. While silica is a particularly suitable medium for high power operation due to its ability to withstand high optical intensity, optically induced damage does pose an ultimate power limit. To mitigate this limit one can introduce a high pump power into the larger area of the inner cladding, and then extract the fibre laser power as a (monomode, in principle) output from the core.

Another way to overcome optical damage limitations is to use a larger area, multimode core. A disadvantage of this is the fact that the output will in general be multimode, although in practice [16] it is still possible to obtain a monomode output even from a fibre which is capable of supporting higher order modes. Of course, if one moves to progressively larger core diameters, so the problems associated with a high heat input per unit length will progressively assert themselves.

For glass rods/fibres a figure in the region of 1 W/cm of heat dissipation represents a rough guide to the maximum set by consideration of thermally-induced fracture [42]. For a monomode fibre, bearing in mind the typical core area of $\sim 10^{-11}$ m^2 (4 μm diameter), and a typical pump saturation intensity

($I_s \sim 3 \times 10^8$ W/m², for Nd³⁺), one finds a typical saturation power of ~3 mW. If the pump power exceeds the saturation power by a factor (N, say) then the pump will saturate the ions in the core over a length of ~N times the low power extinction length. So, even for the most heavily doped Nd glass materials, where extinction lengths of ~2 mm are possible, it would appear that a 1 W pump would penetrate some ~0.6 m into the fibre, thus leading to only ~1 W/m of heat dissipation. This is two orders of magnitude away from the fracture limit. On the other hand if a multimode core diameter ten times greater is assumed (40 μm), the saturation power is increased by ~100, so a heat dissipation of 1 W/cm is possible and the region of fracture is approached. Naturally, if this proves a problem, a low-doped, longer fibre can provide the means of avoiding the problem. A further point that should be noted is that the estimates above did not consider the situation where the laser is operating many times above threshold. This effectively reduces the pump saturation, thus allowing a greater pump power to be absorbed per unit length as a result.

A final point that should not be overlooked is the fact that if heat is not removed effectively from the fibre, its temperature can rise significantly even though the problems associated with a high temperature gradient (i.e. fracture) are not present. This has been discussed in [42] where an experiment is described which illustrates this point. Conditions were arranged such that a Nd-doped fibre reached melting point well before problems relating to fracture were apparent.

To conclude this section on high power operation of fibre laser, it should be emphasised that this aspect of fibre lasers is in its infancy, but already it has been shown, with output powers in excess of 1 W at 1.5 μm and 2.0 μm, that fibre lasers should no longer be thought of as devices of inherently low average power.

7.5 1.5 μm Er fibre lasers

Of all the optical transitions presented by rare-earth dopants in optical fibre, the $^4I_{13/2} \rightarrow {}^4I_{15/2}$ transition in Er³⁺ at around 1.55 μm has attracted by far the greatest interest as a result of its enormous potential for amplifiers in the third telecom window. The first observation made on this transition in a fibre laser context [12] involved laser oscillation at 1.55 μm, with an argon ion laser at 514.6 nm used as the pump. The absorption spectrum for Er³⁺ reveals a wide choice of possible pump wavelengths (see Figure 7.2 for an energy level diagram). Besides using the Ar ion laser as a pump the frequency doubled output of a Nd:YAG laser, at 532 nm, has proved convenient to use on the same pump transition (the highest average output power of 3.9 W from a Er³⁺ fibre laser was achieved in this way [47]). Other pump wavelengths have included 665 nm from a DCM dye laser (but in principle available from diode

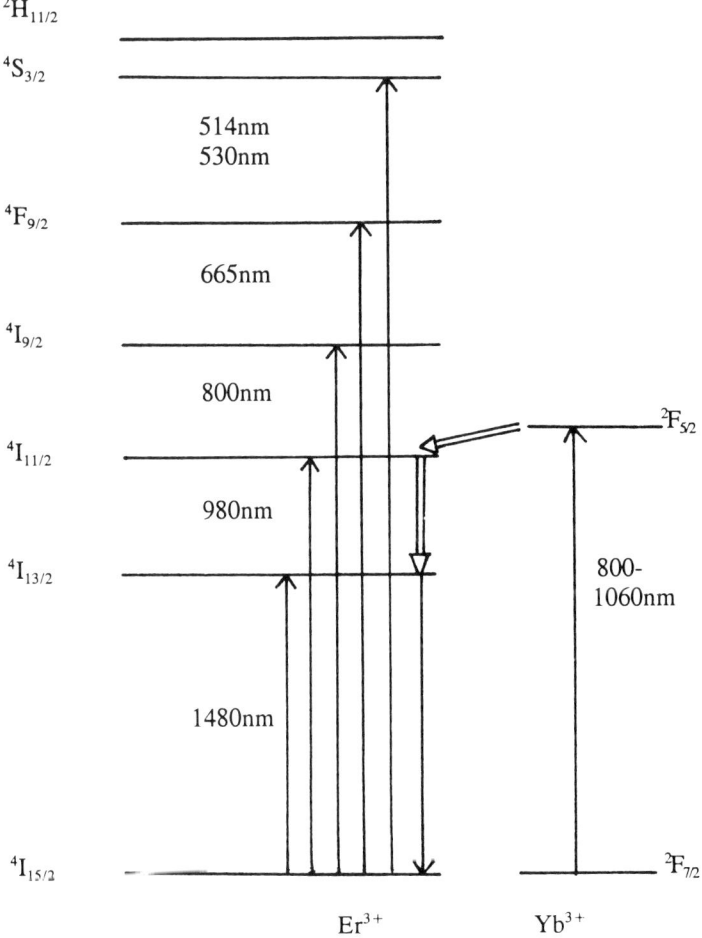

Figure 7.2 Pump transitions for the Er^{3+} laser and the Yb^{3+}-sensitised Er^{3+} laser.

lasers), 807 nm, 980 nm and 1480 nm. The latter three are all available from diode lasers and indeed such lasers have been successfully used for pumping the Er fibre laser. Clearly, for a practical amplifier or laser based on Er doped fibre, diode laser pumping is essential and investigations aimed at identifying the optimum pump wavelength (for amplifiers in particular) have been made [48]. The conclusion drawn was that 980 nm was the ideal pump wavelength giving freedom from pump excited state absorption (ESA) and, therefore, having a considerable advantage over pumping at ~ 800 nm where pump ESA poses a problem. Development of diodes for operation at 980 nm is proceeding rapidly and these are expected to have an important future role as pumps for erbium-doped fibre amplifiers and lasers. Initial attempts to pump in the

800 nm region had led to the disappointing conclusion that pump ESA was a serious obstacle to achieving a high gain at reasonable pump power levels [49]. Nevertheless despite the initial disappointment, efforts were mounted to achieve acceptable performance with 800 nm pumping. First successes were in the form of lasing rather than high gain amplification and using a dye laser rather than a diode [50]. A pump wavelength of ~ 807 nm was found to be best, corresponding to an acceptable compromise between the desired pump absorption and the undesired pump ESA. A low threshold, of 2.5 mW absorbed pump power, was achieved. This was followed by a demonstration of diode laser pumping, at 808 nm, with a similarly low threshold and a rather better, but still low, slope efficiency (3.3%) [51]. By using a diode-array, operating in a self-injection-locked configuration, so that a better quality beam was produced, allowing more efficient launch into the fibre, up to 8 mW output at 1.55 µm was achieved and a slope efficiency of 13% [52]. Thus with suitable care, pumping at ~ 808 nm can produce respectable laser performance. Nevertheless it is clear that pumping with 980 nm is inherently better, with a reported result of 46% slope efficiency to underline this point [53].

With ESA at around 800 nm proving to be problematical, another scheme which can avoid this problem has been investigated. This involves using Er^{3+}-doped fibre which is co-doped (or 'sensitised') with Yb^{3+}. The Yb^{3+} ion shows an absorption extending from ~ 800 nm to beyond 1000 nm (in fact having useful absorption even at the Nd:YAG wavelength of 1.06 µm). By pumping anywhere within that range the Yb^{3+} ion is excited from the $^2F_{7/2}$ level to the $^2F_{5/2}$ level, which, as a result of its close energy match with the Er^{3+} $^4I_{11/2}$ level, can efficiently transfer energy to a neighbouring Er^{3+} ion, and thus lead ultimately to excitation of the $^4I_{13/2}$ level. This energy transfer scheme is a well known one, exploited in bulk glass as long ago as 1966 [54]. More recently it was shown to provide an effective pumping scheme for bulk Er glass lasers, using a Nd:YAG pump laser [43, 55]. Since then, lasing of an Yb:Er fibre at 1.56 µm, using a Nd:YAG pump has been achieved, showing an efficiency of 4% [56], followed by results [57] from a Nd:YLF pump (itself diode-pumped) giving an efficiency of 5%. This co-doped laser offers an extremely wide range of possible pump wavelengths, allowing one to choose a pump wavelength which does not suffer from ESA. The best efficiency reported so far for this system is 17%, with 810 nm pumping [58]. Not surprisingly this is somewhat less than that achievable by 980 nm pumping, but this is in part due to the inefficiency resulting from incomplete transfer from Yb to Er. Nevertheless this result does point to the fact that silica fibre can be doped at sufficient concentration to allow efficient energy transfer, a point to be borne in mind when devising pumping schemes for other laser systems.

The tuning behaviour of Er doped silica fibre pumped at 980 nm has been examined in some detail, in each case using a Ti sapphire laser as the pump [59, 60]. It was found [59], using a grating as the tuning element, that a range from ~ 1.51 µm to 1.58 µm could be covered, with an output power of greater than 250 mW over the range 1.52 μm to 1.57 μm for 540 mW of launched power. The tuning range was in fact limited in this case by lasing occurring on the high gain peak at 1.53 μm. The measurements of Barnes et al. [60] extended the tuning range to longer wavelengths, 1.60 μm being reached. Their measurements clearly demonstrate the dependence of the lasing wavelength on the fibre length and output coupling. This arises since the gain, as a function of wavelength, changes shape with the magnitude of population density in the upper laser level. At low inversion, net gain is only experienced at long wavelengths, where the population in the thermally populated lower level is small. As inversion is increased so the peak gain progressively moves towards shorter wavelengths. The influence of output mirror coupling is: a high output transmission implies the need for a high gain, hence high inversion, and therefore pulls the emission to shorter wavelengths. Equally, by increasing the fibre length, for the same pump power, the peak gain will be shifted to longer wavelengths where reabsorption from the extra fibre length is minimised.

The results on Er-doped fibre lasers indicate that a wide tuning range is possible and that if a particular operating wavelength is specified then appropriate fibre length/doping and output coupling should be chosen to ensure optimum performance at that wavelength. Operation of Er-doped fibre lasers has now covered not only a wide spectral range, but a wide range of operating regimes, including mode-locked operation, Q-switched operation and single frequency operation, all of which are described elsewhere in this book. Each of these different operating regimes is likely to play a significant role in various telecom applications centred on the important 1.5 μm window.

7.6 1.3 μm Nd fibre lasers

The great potential interest of active fibre sources and amplifiers for the 1330 nm telecommunications window prompted work at an early stage on the Nd $^4F_{3/2} \rightarrow {}^4I_{13/2}$ transition, which in a silica glass host gives rise to a broad fluorescence band between 1330 nm and 1430 nm. Laser action could not be demonstrated on this transition, and in a pump and probe experiment using a monolithic Nd:YAG laser probe, pump-induced absorption at 1.32 μm and 1.34 μm was observed [21]. This was attributed to a band of excited state absorption (ESA) on the $^4F_{3/2} \rightarrow {}^4G_{7/2}$ transition, which overlapped the peak of the $^4F_{3/2} \rightarrow {}^4I_{11/2}$ emission band leaving only the long wavelength wing relatively unaffected.

Miniscalco and Andrews have used Judd–Ofelt theory to calculate the linestrengths of the emitting and ESA transitions in a number of different glass host matrices [61]. They found the most favourable ratios in fluoroberyllate glass, then fluorozirconates and fluorophosphates, and then phosphate and phosphorus-doped silica glass. In silica glasses without phosphorus doping the ESA linestrength tended to be greater than laser transition linestrength. These findings have stimulated work on 1.3 μm fluorozirconate fibre lasers and amplifiers, reported elsewhere in this volume, and also work on Nd-doped phosphate and P_2O_5-doped silica glass fibres.

Lasing at 1363 nm has been reported in a monomode fibre fabricated from a commercially available phosphate laser glass by the rod-in-tube method [62]. Nd fluorescence in this host has a maximum intensity at 1325 nm, and gain could be demonstrated at wavelengths longer than 1360 nm, showing that ESA still has a significant effect in this system. The laser was pumped at 815 nm, either by a $Ti:Al_2O_3$ laser or by a diode, and reached threshold at 5 mW of absorbed power, with a slope efficiency above threshold of 10.8% with respect to absorbed power. This is the best result reported to date for this Nd transition in an oxide glass.

It seems at present that although experimenting with host glass composition can undoubtedly achieve some improvement in performance, this Nd transition will always be limited by the excited state absorption and the small branching ratio. Other dopant ions in low phonon energy glass hosts probably offer the most exciting prospects for fibre lasers and amplifiers in this spectral region.

7.7 Superfluorescent operation (ASE operation)

The gains that can be achieved in active monomode fibre are typically in the region of a decibel per milliwatt of absorbed pump power. It is therefore relatively easy to achieve gains of the order of 30–40 dB, even with diode-laser pumps. At these levels of gain strong amplified spontaneous emission (ASE) occurs. This is commonly referred to as 'superfluorescent' emission, although it has been argued by Siegman [63] that such a term should be reserved for a rather different phenomenon. ASE refers to the situation where, in the fibre, spontaneous emission originating from one end of the fibre, is amplified in passing through the fibre (a single pass or double pass situation can be considered) to an intensity which saturates the gain. The output power is therefore comparable to that which would arise if the fibre had been furnished with feedback mirrors to form a laser resonator. This ASE operation is sometimes referred to as mirrorless lasing. An estimate of the required gain for this to occur is derived below.

The attraction of the ASE emission is that it is spatially coherent (assuming the fibre is monomode at the emission wavelength), but has a broad-band emission and thus a very short coherence length. A number of applications call for such a source, in particular the fibre gyro, and this has stimulated efforts to develop and characterise such 'superfluorescent' sources [18].

To derive an expression for the gain required to exhibit ASE, a single pass arrangement is assumed, involving a fibre of core area A, single pass power gain G, and an amplifying transition of saturation intensity I_{SAT}. If the emission bandwidth of the ASE is Δv_e, then the statement that ASE is 'significant', is simply that the noise power in this bandwidth, originating from one end of the fibre, i.e. $hv\Delta v_e$, becomes amplified to a value $hv\Delta v_e G$ at the exit of the fibre, which is sufficient to saturate the gain.

Thus
$$hv\Delta v_e G = AI_{SAT} \qquad (7.1)$$

This can be recast, in a more useful form as
$$G \sim \pi V^2 n^2 \Delta v_{fl}/\Delta v_e (NA)^2 \phi \qquad (7.2)$$

where V is the fibre V value, NA is the numerical aperture, n the refractive index, Δv_{fl} the linewidth of the fluorescent transition and ϕ the radiative quantum efficiency.

Typically G turns out to be 30–40 dB, and therefore for a double-pass (where a mirror is present at one end of the fibre), a value of 15–20 dB is appropriate.

The first observations, by Liu et al. [19], of such superfluorescent emission in a monomode Nd fibre showed onset of ASE for a double pass arrangement for an absorbed pump power of ~ 40 mW, reasonably consistent with the estimate above. Up to 10 mW output was obtained. Besides the need for broadband operation, which the ASE source satisfies, Liu et al. [64] have pointed out that it is important for the source to have very low sensitivity of its emission wavelength to external influence, in particular to temperature. Their measurements indicated that the superfluorescent fibre source should be significantly better in this respect than the other main contender, i.e. superluminescent diodes.

Diode-laser pumping of the superfluorescent source is clearly desirable for a practical system, and Duling et al. [65] have demonstrated the effective combination of cladding-pumping, with an AlGaAs diode array, to produce 80 mW of superfluorescent output from a Nd doped fibre.

ASE is readily observed on other laser transitions, for example on the three-level transition at 974 nm and four-level transition at 1040 nm, in ytterbium-doped fibre [20]. Slope efficiencies of greater than 40% were observed.

It should also be noted that ASE can impose limits on the performance of active fibre devices. Clearly it poses a limit on the maximum gain that can be obtained from a fibre laser, since if in attempts to tune into the low gain wings of a line, the pump power is increased to enhance this gain in the wings, the gain will ultimately be clamped by the onset of ASE at line centre where the gain is very large.

7.8 Visible operation

Although lanthanide ions in crystalline hosts exhibit a number of visible laser transitions, at least two factors tend to limit the performance which can be achieved. The higher energy metastable levels firstly have access to a greater number of pathways for non-radiative decay by cross-relaxation, which can reduce their lifetime significantly, and secondly decay radiatively to many different final levels, so that the branching ratio for any specific transition is likely to be low. In silica glass fibre hosts rapid multiphonon emission rates, and the well-documented tendency for lanthanide dopant ions to cluster, thus enhancing cross-relaxation, have so far prevented lasing on the visible transitions of Pr^{3+}, Er^{3+} and Tm^{3+} ions.

The single example of a visible laser transition in a silica fibre which has been demonstrated to date is a red emission at 651 nm from Sm-doped germanosilicate fibres [66] attributed to the $^4G_{5/2} \to {}^6H_{9/2}$ transition of the trivalent Sm ion. The decay time of the red fluorescence was measured to be ~ 1.5 ms, which is likely to be near to the radiative limit for the transition. The red emission line is strikingly intense and narrow, with a full width at half maximum of 3.3 nm, centred at 651 nm. The peak intensity of this line is more than twice as large as that of any other feature in the spectrum, so that the emission has both a large peak cross-section and a highly favourable branching ratio.

Continuous operation of the Sm laser was achieved by pumping with 488 nm Ar^+ ion laser radiation, which was strongly absorbed by the fibre. The highest reported slope efficiency was 12.7% with respect to absorbed power, obtained with a 40% transmission output coupler. The efficiency is limited by an absorption of ~ 50 dB km^{-1} in the fibre at the laser wavelength, possibly due to the presence of a small proportion of divalent Sm ions. A further problem is that pumping of germanosilicate fibres at 488 nm is known to generate defect centres, increasing the absorption of the fibre across the visible region of the spectrum. Thus pumping at this wavelength aged the fibres, leading to an increase in laser threshold over time. Attempts to operate the Sm laser transition in an aluminosilicate fibre, to circumvent the photochromic damage problem, were unsuccessful [67]. Aluminium co-doping radically changed the fluorescence spectrum, introducing a new feature in the 680–730 nm region with the 651 nm peak correspondingly reduced, and also broadened.

In lanthanide-doped crystals and fluorozirconate fibres, infrared pumped visible lasers have been reported [68, 69], which use energy transfer or stepwise excitation to upconvert the pump laser frequency. The most promising dopant ion for this application in a silica fibre host is Tm, and intense blue fluorescence has been observed from Tm-doped germano- and aluminosilicate fibres under excitation by red pump radiation and also by 1.064 μm infrared radiation from a Nd:YAG laser [70]. Blue laser action has not been achieved in this system, and it appears that rapid multiphonon decay of intermediate metastable levels imposes a severe limit on the efficiency with which the blue-emitting metastable levels can be populated. By contrast the heavy metal fluoride glasses, in which multiphonon emission is suppressed, are very promising hosts for upconversion laser operation.

7.9 Laser transitions with low quantum efficiency

Since the optical fibre geometry offers the possibility of very high gain for low pump power, it follows that transitions with a low quantum efficiency may also be induced to lase, where in a bulk medium this could be precluded by an excessive pump power requirement. This proves indeed to be the case. Before discussing this further it is important, however, to first clarify the meaning of the term quantum efficiency, ϕ. In fact ϕ represents the fraction of absorbed pump photons which lead to an ion subsequently undergoing a radiative decay on the radiative channel of interest. It can be separated into two contributions; the pumping quantum efficiency, ϕ_p, which is the fraction of pump photons which lead to excitation of ions to the upper laser level of relevance; and the radiative quantum efficiency, ϕ_r, which is the fraction of ions so excited that subsequently radiate via the relevant radiative transition. The quantum efficiency ϕ is therefore given by $\phi = \phi_r \phi_p$. It is important to note that the threshold power is $\alpha \phi^{-1}$, but the slope efficiency is $\alpha \phi_p$ and is limited to ϕ_p as a maximum. Thus, although a low radiative quantum efficiency implies a high threshold, it does not imply a low slope efficiency. The Tm $^3H_4 \rightarrow ^3H_6$ transition provides a good illustration of this [17], with a low ϕ_r (perhaps 10–20% typically) but a slope efficiency of greater than 80% (measured in photons rather than energy).

While the above comments point to an encouraging aspect of fibre lasers, i.e. that good laser performance is possible from transitions of low radiative quantum efficiency, it also has to be added that some laser transitions in doped silica fibres have yielded rather low efficiencies, for reasons that are not yet conclusively identified. These are now briefly discussed. The 2.04 μm transition reported in a Ho-doped germanosilicate fibre [13] appears at first sight spectroscopically similar to the 1.8 μm Tm:silica transition. The lifetimes of the upper laser levels in Tm and Ho were measured to be 500 μs and 600 μs respectively. The radiative lifetime of the Tm transition is estimated from absorption data to be ~ 3.4 ms, corresponding to a radiative quantum efficiency of $\sim 15\%$. The concentration of the Ho-doped fibre was known only approximately so that the radiative lifetime could not be reliably determined, but it is presumably in the range 5–15 ms, corresponding to a radiative quantum efficiency between 12 and 4%. The major contrast between these two systems is in the threshold for laser operation. Whereas the Tm system has lased with ~ 4 mW of absorbed pump power in a resonator with 10% output coupling [30], the Ho system was found to require ~ 45 mW of absorbed pump power to reach threshold in a resonator with $\sim 2\%$ output coupling [13].

The explanation for the poor efficiency of the Ho system must involve the pump transition, a broad, intense absorption band at wavelengths < 500 nm due to several overlapping multiplets including $^5F_2$, $^3K_8$ and $^5G_6$. However, it is not the case that population created in these levels is transferred inefficiently to the $^5I_7$ metastable level since in a silica host there is no radiative emission from any of the intermediate levels which relax by multiphonon emission. Moreover the strong pump absorption should make possible saturation of the ground state population, facilitating operation on this quasi-three-level transition. It seems therefore most likely that the high threshold reflects the poor transmission characteristics of germanosilicate fibres at the 457.9 nm wavelength of the Ar^+ ion pump laser beam, which is known to induce photochromic damage [71]. The possibility cannot be ruled out that a secondary problem also exists, in the form of excited state absorption of pump or laser radiation.

Pr-doped germanosilicate fibre is another system in which poor efficiency appears to have so far ruled out practical application of the observed laser transitions. The attraction of Pr is that in a silica host it exhibits four fluorescence bands at wavelengths from the red to the near infrared, each broadened by $\sim 1000\,cm^{-1}$ and therefore potentially capable of wide tuning. Laser action has been demonstrated on two of these, $^1D_2 \rightarrow {}^3F_{3,4}$, $^1D_2 \rightarrow {}^3H_6$ $^3F_2$ [10], but with too low a gain to permit useful tunable operation. Since both are four-level terminating on a multiplet which is rapidly depopulated by multiphonon emission, it is possible to deduce in each case from the observed threshold a value for the quantum efficiency ϕ. The values of ϕ calculated for these transitions $^1D_2 \rightarrow {}^3F_{3,4}$, $^1D_2 \rightarrow {}^3H_6$, $^3F_2$ are respectively 0.02 and 0.007 [10].

Since no Judd–Ofelt analysis has so far been performed on the Pr:silica system, the best interpretation of the values that can be made at present must rely on published data relating to a Pr:fluorozirconate glass system [72]. Assuming, therefore, that the calculated radiative lifetime of $\sim 600\,\mu s$ for fluorozirconate is roughly correct also for germanosilicate, then the observed $\sim 120\,\mu s$ fluorescence lifetime in the germanosilicate fibre (a value typical of the $^1D_2$ multiplet in silicate glass) would imply an overall radiative quantum efficiency of $\sim 20\%$ for this metastable level. The branching ratios derived from [72] are $\sim 6\%$ for the $^1D_2 \rightarrow {}^3F_{3,4}$ transition (1080 nm) and $\sim 40\%$ for the $^1D_2 \rightarrow {}^3H_6$, $^3F_2$ transition (888 nm). Thus taking into consideration purely the relative probabilities of the various radiative and nonradiative decay paths from the $^1D_2$ multiplet, a value of 1–2% is predicted for the 1080 nm transition, quite consistent with the measured value. The corresponding calculation for the 888 nm transition, however, predicts $\sim 8\%$, an order of magnitude larger than the measured value. This discrepancy may indicate that the 888 nm laser photons suffer excited state absorption, possibly on the $^1G_4 \rightarrow {}^3P_2$ transition.

Discrepancies such as those referred to above in Ho and Pr doped silica fibre serve as a salutary reminder that each rare-earth-doped fibre is a system that needs extensive investigation before a full picture can emerge of all the factors that influence and limit its behaviour. So far most of the effort has been directed at Er-doped fibre for obvious reasons of application potential, but it is likely that with a relatively small effort directed at other dopants other laser systems will prove to have much improved performance characteristics and prospects for application.

References

1. H. Namikawa, K. Arai, K. Kumata, Y. Ishi and H. Tanaka, *Jpn. J. Appl. Phys.* **21** (1982) L360–362.
2. S.B. Poole, D.N. Payne and M.E. Fermann, *Electron. Lett.* **21** (1985) 737–738.
3. R.J. Mears, L. Reekie, S.B. Poole and D.N. Payne, *Electron. Lett.* **21** (1985) 738–740.
4. B.J. Ainslie, S.P. Craig and S.T. Davey, *J. Lightwave Technol.* **6** (1988) 287–293.
5. B.J. Ainslie, S.P. Craig and S.T. Davey, *J. Mat. Sci. Lett.* **6** (1987) 1361–1363.
6. K. Arai, H. Namikawa, K. Kumata, T. Honda, Y. Ishii and T. Handa, *J. Appl. Phys.* **59** (1986) 3430–3436.
7. J.E. Townsend, S.B. Poole and D.N. Payne, *Electron. Lett.* **23** (1988) 329–331.
8. J. Stone and C.A. Burrus, *Appl. Phys. Lett.* **23** (1973) 388–389.
9. B.J. Ainslie, S.P. Craig, S.T. Davey and B. Wakefield, *Mater. Lett.* **6** (1988) 139–144.
10. R.M. Percival, M.W. Phillips, D.C. Hanna and A.C. Tropper, *IEEE J. Quant. Electron.* **25** (1989) 2119–2123.
11. M.C. Farries, P.R. Morkel and J.E. Townsend, *Electron. Lett.* **24** (1988) 709–710.
12. R.J. Mears, L. Reekie, S.B. Poole and D.N. Payne, *Electron. Lett.* **22** (1986) 159–160.
13. D.C. Hanna, R.M. Percival, R.G. Smart, J.E. Townsend and A.C. Tropper, *Electron. Lett.* **25** (1989) 593–594.
14. D.C. Hanna, I.M. Jauncey, R.M. Percival, I.R. Perry, R.G. Smart, P.J. Suni, J.E. Townsend and A.C. Tropper, *Electron. Lett.* **24** (1988) 1222–1223.
15. D.C. Hanna, I.M. Jauncey, R.M. Percival, I.R. Perry, R.G. Smart, P.J. Suni, J.E. Townsend and A.C. Tropper, *Electron. Lett.* **24** (1988) 1111–1113.
16. D.C. Hanna, I.R. Perry, J.R. Lincoln and J.E. Townsend, *Opt. Lett.* (in press).
17. D.C. Hanna, R.M. Percival, R.G. Smart, and A.C. Tropper, *Opt. Comm.* **75** (1990) 283–286.
18. M.J.F. Digonnet, *J. Lightwave Technol.* **LT-4** (1986) 1631–1639.
19. K. Liu, M. Digonnet and H.J. Shaw, *Electron. Lett.* **23** (1987) 1320–1321.
20. D.C. Hanna, I.R. Perry, R.G. Smart, P.J. Suni, J.E. Townsend and A.C. Tropper, *Opt. Comm.* **72** (1989) 230–234.
21. I.P. Alcock, A.I. Ferguson, D.C. Hanna and A.C. Tropper, *Opt. Comm.* **58** (1986) 405–408.
22. P.R. Morkel, M.C. Farries and S.B. Poole, *Opt. Comm.* **67** (1988) 349–352.
23. H.C. Lefevre, *Electron. Lett.* **16** (1980) 778.
24. D.C. Hanna, R.M. Percival, I.R. Perry, R.G. Smart, P.J. Suni, and A.C. Tropper, *J. Mod. Opt.* **37** (1990) 517–525.
25. C. Yue, J. Peng and B. Zhou, *Electron. Lett.* **25** (1989) 101–102.
26. I.P. Alcock, A.I. Ferguson, D.C. Hanna and A.C. Tropper, *Opt. Lett.* **11** (1986) 709–711.
27. L. Reekie, R.J. Mears, S.B. Poole and D.N. Payne, *J. Lightwave Technol.* **LT-4** (1986) 956–960.
28. R. Wyatt, *Electron. Lett.* **25** (1989) 1498–1499.
29. D.C. Hanna, R.M. Percival, R.G. Smart, and A.C. Tropper, *Opt. Comm.* **75** (1990) 283–286.
30. W.L. Barnes and J.E. Townsend, *Electron. Lett.* **26** (1990) 746–747.
31. R.C. Stoneman and L. Esterowitz, *Opt. Lett.* **15** (1990) 486–488.
32. I.M. Jauncey, L. Reekie, R.J. Mears, D.N. Payne, C.J. Rowe, D.C.J. Reid, I. Bennion and C. Edge, *Electron. Lett.* **22** (1986) 987–988.
33. I.M. Jauncey, L. Reekie, R.J. Mears and C.J. Rowe, *Opt. Lett.* **12** (1987) 164–165.
34. I.M. Jauncey, L. Reekie, J.E. Townsend, D.N. Payne and C.J. Rowe, *Electron. Lett.* **24** (1988) 24–26.
35. R. Kashyap, J.R. Armitage, R. Wyatt, S.T. Davey and D.L. Williams, *Electron. Lett.* **14** (1989) 823–825.
36. G.A. Ball, W.W. Morey and J.P. Waters, *Electron. Lett.* **26** (1990) 1829–1830.
37. P. Barnsley, P. Urquhart, C. Millar and M. Brierley, *J. Opt. Soc. Am. A.* **5** (1988) 1339–1346.
38. P. Morkel, G.J. Cowle and D.N. Payne, *Electron. Lett.* **26** (1990) 632–634.
39. H. Sabert, R. Ulrich, *OFS 1990*, Sydney, Australia, Paper MOA2.3.

40. J.D.C. Jones and P. Urquhart, *Opt. Comm.* **76** (1990) 42–46.
41. W. Koechner, *Solid State Laser Engineering*, Vol. 1 Springer Series in Optical Sciences, Springer-Verlag, New York (1976).
42. D.C. Hanna, M.J. McCarthy, P.J. Suni, *SPIE Proc 1171*, Fiber Laser Sources and Amplifiers (1989) 160–166.
43. D. Espie, D.C. Hanna, A. Kazer and D.P. Shepherd, *Opt. Comm.* **69** (1988) 153–155.
44. D.C. Hanna, M.J. McCarthy, I.R. Perry, P.J. Suni, *Electron. Lett.* **25** (1989) 1365–1366.
45. E. Snitzer, H. Po, F. Hakimi, R. Tumminelli and B.C. McCollum, in *Digest of Conference on Optical Fiber Sensors*, Optical Society of America, Washington D.C. (1988), Paper PD5.
46. H. Po, E. Snitzer, R. Tumminelli, L. Zenteno, F. Hakimi, N.M. Chu and T. Haw, *Optical Fiber Communication Conference* (1989) Paper PD7.
47. V.P. Gapontsev, P.I. Sadovsky and I.E. Samartsev, *Conference on Lasers and Electro-optics*, Anaheim (1990) Paper CPDP-38.
48. R. Laming, M.C. Farries, P.R. Morkel, L. Reekie and D.N. Payne, *Electron. Lett.* **25** (1989) 12–14.
49. R. Laming, S.B. Poole and E.J. Tarbox, *Opt. Lett.* **13** (1988) 1084–1086.
50. C.A. Millar, I.D. Miller, B.J. Ainslie, S.P. Craig and J.R. Armitage, *Electron. Lett.* **23** (1987) 865–866.
51. L. Reekie, I.M. Jauncey, S.B. Poole and D.N. Payne, *Electron. Lett.* **23** (1987) 1076–1077.
52. R. Wyatt, B.J. Ainslie and S.P. Craig, *Electron. Lett.* **24** (1988) 1362–1363.
53. W.L. Barnes, P.R. Morkel, L. Reekie and D.N. Payne, *Opt. Lett.* **14** (1989) 1002–1004.
54. E. Snitzer and R. Woodcock. *Appl. Phys. Lett.* **6** (1966) 45–46.
55. D.C. Hanna, A. Kazer and D.P. Shepherd. *Opt. Comm.* **63** (1987) 417–420.
56. M.E. Fermann, D.C. Hanna, D.P. Shepherd, P.J. Suni and J.E. Townsend, *Electron. Lett.* **24** (1988) 1135–1136.
57. G.T. Maker and A.I. Ferguson, *Electron. Lett.* **24** (1988) 1160–1162.
58. W.L. Barnes, S.B. Poole, J.E. Townsend, L. Reekie, D.J. Taylor and D.N. Payne, *J. Lightwave Technol.* **7** (1989) 1461–1465.
59. R. Wyatt, *Electron. Lett.* **25** (1989) 1498–1499.
60. W.L. Barnes, P.R. Morkel, L. Reekie, D.N. Payne, *Opt. Lett.* **14** (1989) 1002–1004.
61. W.J. Miniscalco and L.J. Andrews, *Materials Science Forum* **32–33** (1988) 501–510.
62. S.G. Grubb, W.L. Barnes, E.R. Taylor and D.N. Payne, *Electron. Lett.* **26** (1990) 121–122.
63. A.E. Siegman, *Lasers*, Oxford University Press, (1986) 551.
64. K. Liu, M. Digonnet, K. Fesler, B.Y. Kim and H.J. Shaw, *Electron. Letts.* **24** (1988) 838–840.
65. I.N. Duling, W.K. Burns and L. Goldberg, *Opt. Lett.* **15** (1990) 33–35.
66. M.C. Farries, P.R. Morkel and J.E. Townsend, *Electron. Lett.* **11** (1988) 709–710.
67. M.C. Farries, P.R. Morkel and J.E. Townsend, *IEE Proceedings 137* (1990) 318–322.
68. F. Tong, W.P. Risk, R.M. Macfarlane and W. Lenth, *Electron. Lett.* **25** (1989) 1389–1390.
69. J.Y. Allain, M. Monerie and H. Poignant, *Electron. Lett.* **26** (1990) 261–262.
70. D.C. Hanna, R.M. Percival, I.R. Perry, R.G. Smart, J.E. Townsend and A.C. Tropper, *Opt. Comm.* **78** (1990) 187–194.
71. C.A. Millar, S.R. Mallinson, B.J. Ainslie and S.P. Craig, *Electron. Lett.* **24** (1988) 590–591.
72. J.L. Adam and W.A. Sibley, *J. Non-Cryst. Solids* **76** (1985) 267–279.
73. G.H. Diecke and H.M. Crosswhite, *Appl. Opt.* **2** (1963) 675.

8 Fluoride fibre lasers and amplifiers

P.W. FRANCE and M.C. BRIERLEY

8.1 Introduction

Fluoride glasses now provide an interesting alternative to silica-based systems for rare-earth (lanthanide) doped fibres. In earlier chapters it has been shown how silica fibres can be used effectively as amplifiers at 1.53 μm, but with less efficiency at most other wavelengths. The reasons for this are fairly straightforward and, as shall be seen later, result from high non-radiative transition rates. Fluoride glasses on the other hand can provide more efficient laser host materials with high radiative efficiencies and correspondingly low non-radiative rates. As will be shown in this chapter, this leads to a wealth of radiative transitions to explore.

Fluoride glasses based on zirconium fluoride (fluorozirconates) were initially discovered at the University of Rennes, in France, in 1974 [1]. Other fluoride systems had been known but were either highly toxic (BeF_2) or very unstable (AlF_3), and this was the first fluoride system that was essentially practical to use. Their accidental discovery followed attempts to fabricate large pieces of ZrF_4 crystals doped with Nd ($NdZrF_7$) for laser applications. One of the samples turned out to be vitreous rather than crystalline and the fluorozirconate glasses had been discovered. Initial interest in the materials was for improved glasses in high-energy fusion lasers, and in 1978 Weber *et al.* concluded that these glasses should be considered as candidates for laser materials [2]. However, a second potential use was proposed when the infrared transmitting properties were investigated. The intrinsic losses of these glasses are significantly below those in silica and they are now being investigated for use in ultra-low-loss fibres [3]. These two interests were finally amalgamated in 1987 when the first fluoride fibre laser was demonstrated at 1.05 μm using Nd as a dopant [4]. Following this work many new dopants have been investigated and many different lasing lines have been reported. Moreover, fibre amplifiers have been demonstrated at 1.53, 1.34 and 2.7 μm, the latter two wavelengths being especially significant since they are as yet unavailable in silica-based fibre. Fluoride glasses are therefore proving themselves to be an extremely important alternative glass host to silica and will allow a useful extension of working devices.

8.2 Fluoride glasses as laser host materials

Although many different fluorozirconate compositions have been investigated, the most stable discovered to date and the one preferred for fibre drawing is ZrF_4–BaF_2–LaF_3–AlF_3–NaF or ZBLAN for short. Most of the work described in this chapter will therefore be restricted to this composition. An index difference between the core and cladding glasses can, among other ways, be achieved by adding PbF_2 to the core or HfF_4 to the cladding [3], and again many of the fibres described here will contain Pb. The index of the glass is about 1.5, and melting temperatures are in the range 500–700°C [3].

One interesting feature of this composition is that the glass already contains about 4 mol% LaF_3 as an integral component, and it can easily be substituted by an alternative rare-earth dopant without significantly effecting the properties of the glass. The structure of fluoride glasses is not well understood, but if the classical concepts of glass networks as developed by Zachariasen [5] are followed then in silicate systems Si (and Ge) acts as a network former, and Na and Ba break up the lattice and act as network modifiers. Poulain [6] has extended the network model to fluorides and a two-dimensional schematic representation of a ZBLA glass is shown in Figure 8.1. Here Zr and Al are thought to be network formers, La to be an intermediate (in this case playing

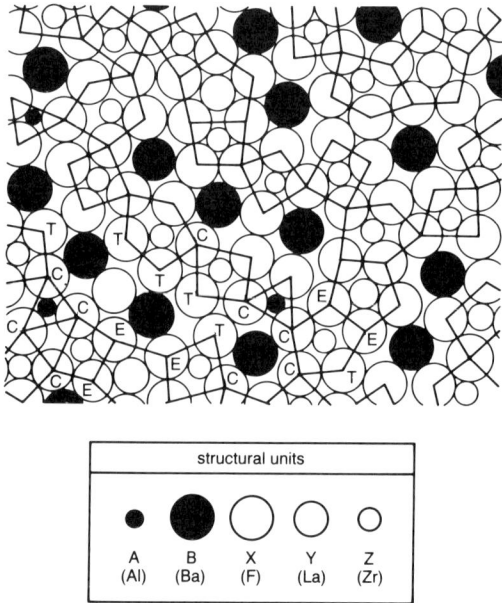

Figure 8.1 Schematic representation of the structure of a fluorozirconate glass (after [6]). There are three types of fluorine ions: terminal (T), edge (E) and corner (C).

the role of a former) and Ba and Na to be network modifiers. The network of connecting fluorine ions is outlined in the diagram, and shows that La^{3+} sits in the matrix in much the same way as Zr^{4+}. This is in sharp contrast with rare earths in a silicate glass where La plays much the same role as Na, acting as a modifier and breaking up the network. The net result of this is that rare earths are compatible with ZBLAN, and fairly high dopant concentrations can be tolerated without modification to the glass. Moreover, since the site symmetry of the lanthanide ion is likely to be more uniform in ZBLAN, inhomogeneous broadening of the fluorescence linewidths is likely to be narrower.

Another important feature of fluoride glasses is their infrared transmitting property. In general these glasses consist of heavy elements and, together with the fact that the fluoride ion is only singly charged this results in the fundamental phonon energies lying at longer wavelengths ($525\,cm^{-1}$) than silicates. Besides leading to low intrinsic losses, Reisfeld first suggested that this property should also lead to lower non-radiative decay rates in fluoride glasses [7]. Since the probability for non-radiative decay depends on the number of phonons that have to be generated in order to lose energy to the lower level, then in general fluoride glasses should exhibit lower probabilities. The non-radiative decay rate W_{nr} has been found to obey the empirical formula:

$$W_{nr} = W_0 \exp(-\alpha \Delta E) \qquad (8.1)$$

where W_0 and α are material constants and ΔE is the energy gap to the next lowest level. These have been measured empirically in several glass hosts, and are illustrated in Figure 8.2, taken from [8]. In general the lower the fundamental phonon energies in the material the lower the non-radiative decay rates. In Chapter 2, the quantum efficiency η in a material is defined as:

$$\eta = \frac{\sum A_r}{\sum A_r + W_{nr}} \qquad (8.2)$$

where A_r is the radiative spontaneous decay rate for a particular transition and W_{nr} is the non-radiative decay rate. In general, therefore, fluoride glasses have higher quantum efficiencies than many other glass systems. Moreover, the minimum energy gap ΔE_{min} required for fluorescence, corresponding to non-radiative decay rates in excess of $10^3\,s^{-1}$, is about $3100\,cm^{-1}$ which compares with a value of about $4600\,cm^{-1}$ in silica. (For ΔE below these values the systems will essentially lose energy non-radiatively, thereby quenching the fluorescence.) This implies that transitions greater than about $3.2\,\mu m$ in ZBLAN and $2.2\,\mu m$ in silicates would not be expected. One further point is that fluoride glasses will exhibit many more fluorescing transitions than other hosts, and a good example is the fluorescence spectrum of Ho^{3+} in ZBLAN which exhibits several bands that are not observed in silicates.

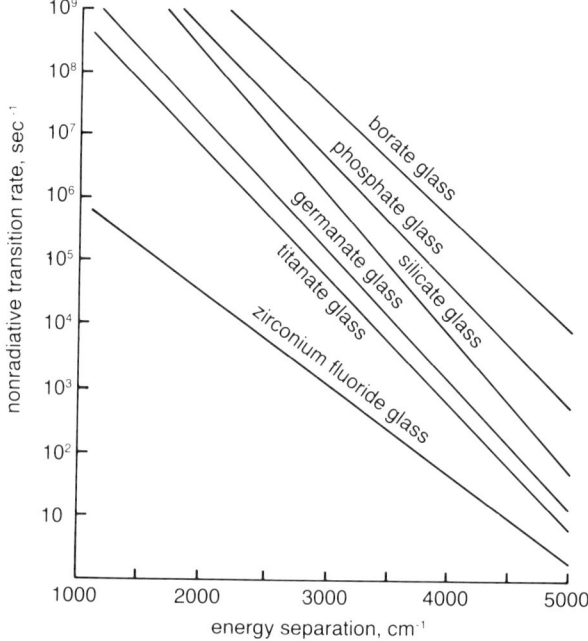

Figure 8.2 Non-radiative decay rates for several glasses (after [8]).

Also in Chapter 2 the overall lifetime τ of a state is given by:

$$\tau = \frac{1}{\sum A_r + W_{nr}}. \tag{8.3}$$

Hence for many levels the lifetimes in fluoride glasses will be longer than in other glass hosts. Besides the possibility of lower thresholds for lasing, this also offers the possibility for up-conversion lasers. The longer lifetimes allow a greater probability for two-photon (or more) absorption, so that IR to UV lasers become more feasible. Indeed the first up-conversion lasers have now been demonstrated in ZBLAN and will be described in Section 8.6.

Another term that could be used for up-conversion is excited-state absorption (ESA). This was again described in Chapter 2, and there it was noted that although ESA is required for up-conversion lasers, it can be a detrimental property in the case of some fibre amplifier systems. For example, the emission cross-section of Nd close to 1.3 μm is reduced by signal excited-state absorption. However, different glass hosts will have different values of ESA cross-sections and, as shown by Andrews et al. [19], the cross-section for this transition is by chance lower in ZBLAN than in silica. Fibre amplifiers have now been demonstrated close to 1.3 μm in ZBLAN, and will be described in Section 8.7.

8.3 Spectroscopy of rare earths in fluoride glasses

8.3.1 *Absorption and emission spectra*

A pre-requisite for lasing is, among other features, that there must be a strong fluorescence at the required wavelength when pumped with light from a suitable source. Absorption and fluorescence spectra of rare-earth ions in the glass host are therefore very useful firstly to identify possible pump bands and secondly in identifying possible lasing wavelengths. Several authors have published spectra concerning individuals ions, but a comprehensive set of both absorption and emission spectra of lanthanides in a ZBLANP host has been given by Davey and France [10]. These spectra are discussed below.

Absorption and emission spectra for {Pr, Nd, Sm}, {Eu, Tb, Dy}, and {Ho, Er, Tm} are given in Figures 8.3 to 8.5. The absorption spectra have been presented on a restricted wavelength scale covering 0.2–1.8 μm since useful

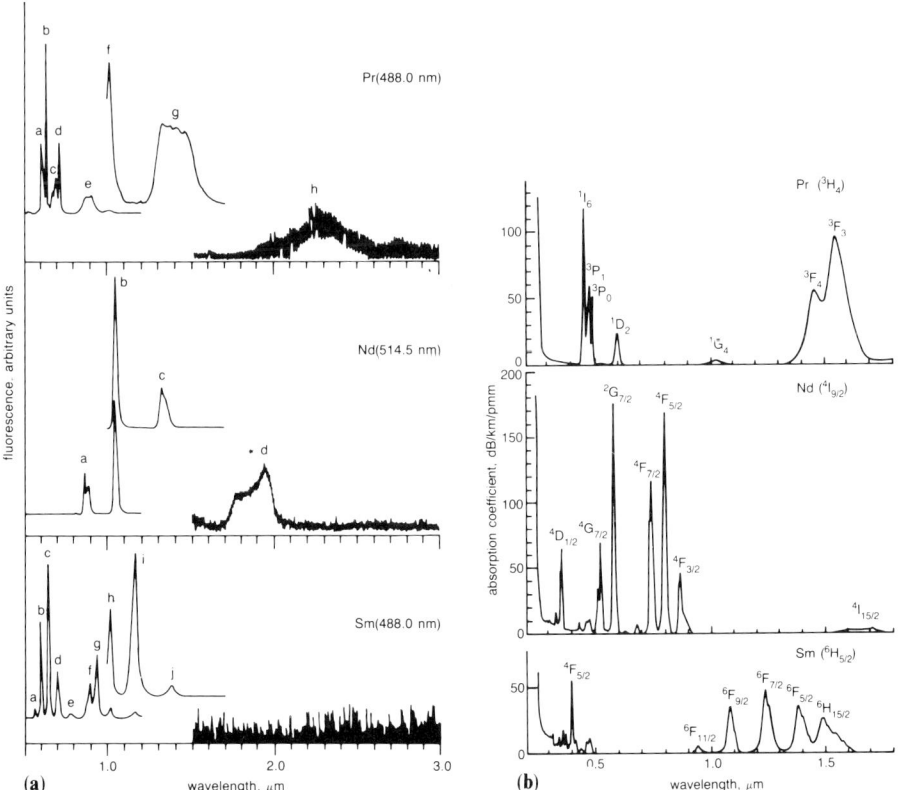

Figure 8.3 (a) Fluorescence spectra of Pr^{3+}, Nd^{3+} and Sm^{3+} in ZBLANP glass (after [10]); (b) absorption spectra of Pr^{3+}, Nd^{3+} and Sm^{3+} in ZBLANP glass (after [10]).

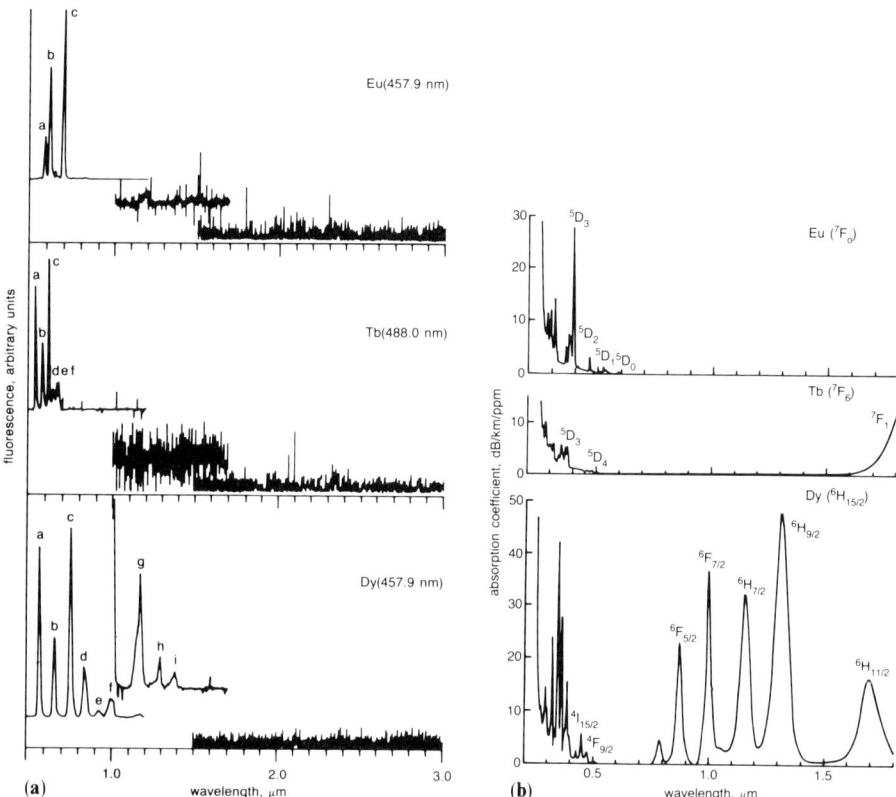

Figure 8.4 (a) Fluorescence spectra of Eu^{3+}, Tb^{3+} and Dy^{3+} in ZBLANP glass (after [10]); (b) absorption spectra of Eu^{3+}, Tb^{3+} and Dy^{3+} in ZBLANP glass (after [10]).

pump sources will be restricted to this range. The emission spectra, on the other hand, have been presented over a wider range covering 0.5–3.0 μm, in order to determine all potential emission wavelengths. Excitation wavelengths are given in parentheses. Measurements were actually made out to 4.0 μm although no emissions were observed at room temperature between 3 and 4 μm, which is consistent with the statement made above that for an emission to be observed ΔE_{min} should be greater than 3100 cm^{-1} (3.2 μm). Each emission spectrum was split into three distinct ranges: 0.5–1.2 μm; 1.0–1.7 μm; and 1.5–3.0 μm, corresponding to different detectors and gratings. Fluorescence was not observed from Lu, Ce or Gd (as expected). Yb, however, was found to have a single fluorescence band at 0.98 μm when excited using a beam of 30 keV electrons from a scanning electron microscope. Similarly, the absorption spectrum of Yb consisted of a single band centred at 0.974 μm with a peak height of 70 dB/km/ppm.

Chapter 5 mentioned that useful pump wavelengths would be 800, 980 and

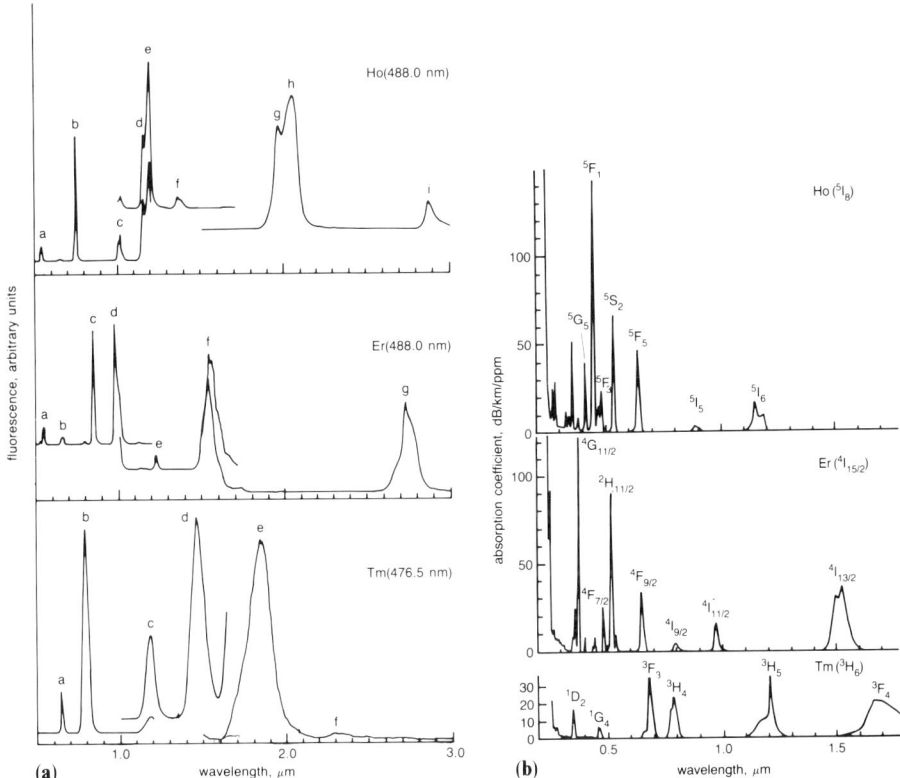

Figure 8.5 (a) Fluorescence spectra of Ho^{3+}, Er^{3+} and Tm^{3+} in ZBLANP glass (after [10]); (b) absorption spectra of Ho^{3+}, Er^{3+} and Tm^{3+} in ZBLANP glass (after [10]).

1480 nm from state of the art laser-diode pumps. However, it was also shown in Chapter 2 that other useful wavelengths that may be available in the future would be 503, 650, 1060 and 1300 nm. Hence the suitability of a particular pump wavelength can be assessed by examining the absorption spectra.

A list of peak emission wavelengths and assigned transitions is given in Table 8.1, made by reference to the rare-earth ion energy diagram given in Figure 2.3. As expected, the emission spectra are rich in lines confirming that the ZBLAN host has a relatively high radiative efficiency. Emissions are observed over the whole range 0.5–3.0 μm, offering the possibility of an extensive set of lasing emissions. For example, the spectrum of Ho^{3+} contains nine significant lines from 0.543 μm to 2.848 μm. On the other hand, published data on the spectra of Ho^{3+} in silica based glasses exhibit only one significant transition close to 2.1 μm. More precise energy levels diagrams for Ho, and also including Nd, Er and Tm, have been constructed from the absorption spectra and these are presented in Figure 8.6. The assigned emissions are labelled in order to give a clearer picture for each dopant.

Table 8.1 Fluorescence transitions of rare earths in ZBLANP. Transitions and wavelengths (in μm) for the peaks labelled in Figures 8.3–8.6. Wavelengths given in brackets are for subsidiary peaks and shoulders near the main peak.

Peak	Pr	Nd	Sm	Eu	Tb
a	$^3P_0 \rightarrow {}^3H_6$ 0.603	$^4F_{3/2} \rightarrow {}^4I_{9/2}$ 0.867 (0.891)	$^4G_{5/2} \rightarrow {}^6H_{5/2}$ 0.560	$^5D_0 \rightarrow {}^7F_1$ 0.591	$^5D_4 \rightarrow {}^7F_5$ 0.543
b	$^3P_0 \rightarrow {}^3F_2$ 0.635	$^4F_{3/2} \rightarrow {}^4I_{11/2}$ 1.048	$^4G_{5/2} \rightarrow {}^6H_{7/2}$ 0.595	$^5D_0 \rightarrow {}^7F_2$ 0.615	$^5D_4 \rightarrow {}^7F_4$ 0.585
c	$^1D_2 \rightarrow {}^3H_5$ 0.695 (0.674)	$^4F_{3/2} \rightarrow {}^4I_{13/2}$ 1.318	$^4G_{5/2} \rightarrow {}^6H_{9/2}$ 0.641	$^5D_0 \rightarrow {}^7F_4$ 0.698	$^5D_4 \rightarrow {}^7F_3$ 0.621
d	$^3P_0 \rightarrow {}^3F_4$ 0.717	$^4F_{3/2} \rightarrow {}^4I_{13/2}$ 1.945	$^4G_{5/2} \rightarrow {}^6H_{11/2}$ 0.704		$^5D_4 \rightarrow {}^7F_2$ 0.649
e	$^3P_0 \rightarrow {}^1G_4$ 0.908 (0.883)		$^4G_{5/2} \rightarrow {}^6H_{13/2}$ 0.782		$^5D_4 \rightarrow {}^7F_1$ 0.668
f	$^1D_2 \rightarrow {}^3F_4$ 1.014		$^4G_{5/2} \rightarrow {}^6H_{15/2}$ 0.896		$^5D_4 \rightarrow {}^7F_0$ 0.681
g	$^1G_4 \rightarrow {}^3H_5$ $^1D_2 \rightarrow {}^1G_4$ 1.326 (1.364) (1.406)(1.460)		$^4G_{5/2} \rightarrow {}^6F_{5/2}$ 0.936		
h	$^1G_4 \rightarrow {}^3F_2$ $^3H_6 \rightarrow {}^3H_4$ 2.300		$^4G_{5/2} \rightarrow {}^6F_{7/2}$ 1.022		
i			$^4G_{5/2} \rightarrow {}^6F_{9/2}$ 1.162		
j			$^4G_{5/2} \rightarrow {}^6F_{11/2}$ 1.376		

Peak	Dy	Ho	Er	Tm	Yb
a	$^4F_{9/2} \to {}^6H_{13/2}$ 0.575	$^5S_2 \to {}^5I_8$ 0.543	$^4S_{3/2} \to {}^4I_{15/2}$ 0.543 (0.550)	$^1G_4 \to {}^3F_4$ 0.649	$^2F_{5/2} \to {}^2F_{7/2}$ 1.0
b	$^4F_{9/2} \to {}^6H_{11/2}$ 0.661	$^5S_2 \to {}^5I_7$ 0.750	$^4F_{9/2} \to {}^4I_{15/2}$ 0.655 (0.666)	$^1G_4 \to {}^3H_5$ $^3H_4 \to {}^3H_6$ 0.799	
c	$^4F_{9/2} \to {}^6H_{9/2}$ $^4F_{9/2} \to {}^6F_{11/2}$ 0.750	$^5S_2 \to {}^5I_6$ 1.015	$^4S_{3/2} \to {}^4I_{13/2}$ 0.847	$^1G_4 \to {}^3H_4$ $^3F_3 \to {}^3F_4$ $^3H_5 \to {}^3H_6$ 1.184	
d	$^4F_{9/2} \to {}^6H_{7/2}$ $^4F_{9/2} \to {}^6F_{9/2}$ 0.834	$^5I_6 \to {}^5I_8$ 1.154	$^4I_{11/2} \to {}^4I_{15/2}$ 0.977	$^1G_4 \to {}^3F_3$ $^3H_4 \to {}^3F_4$ 1.464	
e	$^4F_{9/2} \to {}^6H_{5/2}$ 0.924	$^5I_6 \to {}^5I_8$ 1.191	$^4S_{3/2} \to {}^4I_{11/2}$ 1.219	$^3F_4 \to {}^3H_6$ 1.847	
f	$^4F_{9/2} \to {}^6F_{5/2}$ 1.000	$^5S_2 \to {}^5I_5$ 1.381	$^4I_{13/2} \to {}^4I_{15/2}$ 1.538	$^3H_4 \to {}^3H_5$ 2.307	
g	$^4F_{9/2} \to {}^6F_{5/2}$ 1.170	$^5S_2 \to {}^5I_4$ 1.953	$^4I_{11/2} \to {}^4I_{13/2}$ 2.719		
h	$^4F_{9/2} \to {}^6F_{3/2}$ 1.294	$^5I_7 \to {}^5I_8$ 2.039			
i	$^4F_{9/2} \to {}^6F_{1/2}$ 1.384	$^5I_6 \to {}^5I_7$ 2.848			

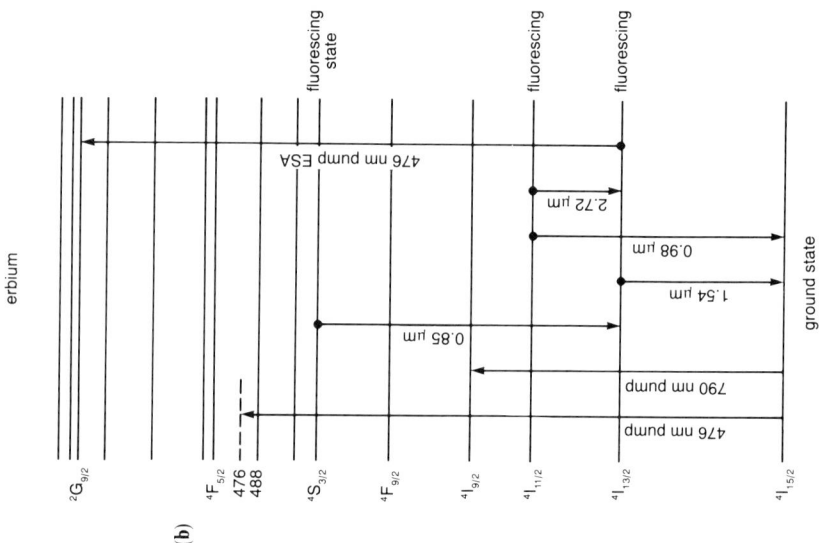

(a)

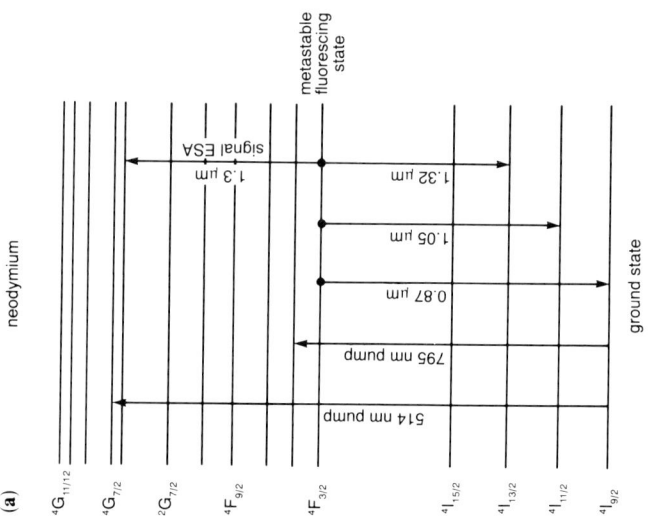

(b)

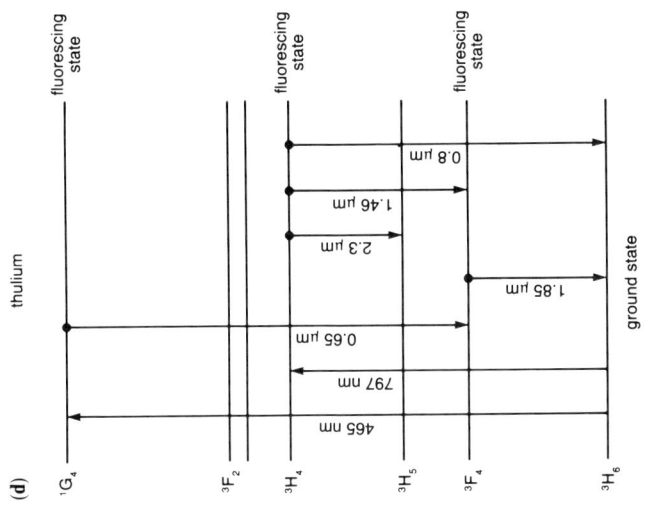

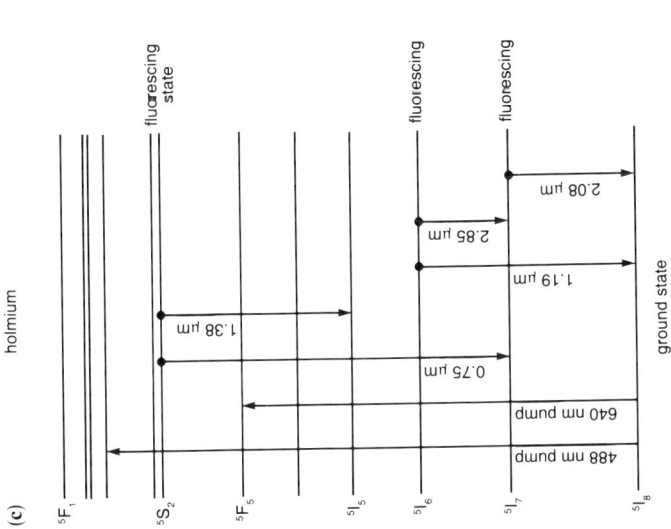

Figure 8.6 Energy level diagrams for Nd^{3+}, Er^{3+}, Ho^{3+} and Tm^{3+} in ZBLANP glass.

From these data it becomes more evident why Ho has a much richer spectrum in ZBLAN than in silicate glasses. For example, if we consider the $^5S_2$ state, ΔE to the next lowest level ($^5F_5$) is 3042 cm^{-1}. Since we decided that ΔE_{min} was about 3100 cm^{-1} in ZBLAN this implies that $^5S_2$ just about satisfies this condition and is therefore a weak metastable level from which fluorescence can occur, giving transitions at 0.543, 0.750, 1.015, 1.381 and 1.953 μm. Unfortunately, no fluorescence has been seen at 3.29 μm from this level. Moreover, $^5I_6$ also satisfies the fluorescence condition and gives emissions at 1.19 and 2.85 μm. Since the energy level diagrams are largely independent of the host glass, the energy differences will be much the same in silicates. However, since ΔE_{min} for silica is 4600 cm^{-1}, in this case the $^5S_2$ and $^5I_6$ levels are not fluorescing states. Indeed the only state to fulfil the condition in silicates is $^5I_7$, which is responsible for the only observed line.

Tm is another interesting system also shown in Figure 8.6. In particular the $^3H_4$ level gives a transition at 2.3 μm which can be accessed by direct pumping at 800 nm. This level also has a transition at 1.46 μm which may prove useful. Two conventions have arisen for the labelling of the energy levels of the trivalent thulium ion and, as a consequence, both conventions have been employed by authors on the subject of thulium-doped fluoride fibre lasers. To avoid further confusion in this section, we shall use the convention employed in Figure 8.6 – that is, to label the first level above the ground state as $^3F_4$ and the third level above the ground state as $^3H_4$. The effect of this will be that the transition assignments of references [24] and [25] will be as published, but these two levels will be interchanged from those given in the other published references in this and following sections of this chapter where thulium is the active ion.

8.3.2 Detailed spectroscopy of Nd

It is well known that Nd^{3+} has an emission close to 1.3 μm and this $^4F_{3/2} \rightarrow {}^4I_{13/2}$ transition is also seen in our ZBLAN host at 1.32 μm. Of course this wavelength is particularly interesting for telecommunications since it coincides with the second transmission window where silica-based fibres have low losses and low dispersion, and for these reasons we shall consider the spectroscopy of Nd in more detail. An expanded version of the emission spectrum for the transitions from $^4F_{3/2}$ at 0.87, 1.05 and 1.32 μm is given in Figure 8.7. Emission cross-sections have also been calculated from the absorption and emission spectra using the equations

$$\sigma_p = \frac{\lambda_p^2 A_r}{8\pi n^2 \Delta v}$$

$$= \frac{g_i \int k(v) dv}{g_j \Delta v N_0} \quad (8.4)$$

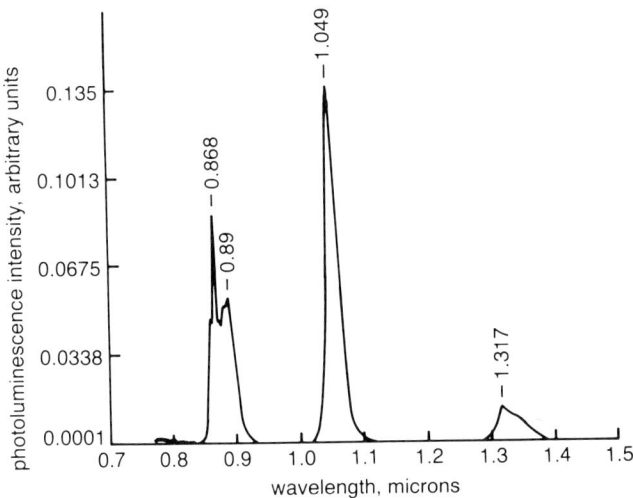

Figure 8.7 Expanded normalised fluorescence spectrum of Nd^{3+} in ZBLANP glass.

where λ_p is the peak emission wavelength, A_r is the radiative transition rate for the given transition, Δv is the effective emission linewidth ($\int \sigma \, dv/\sigma_p$), g_i and g_j are the initial and final state degeneracies, $k(v)$ is the absorption coefficient and N_0 is the concentration of dopant ions.

Hence σ_p can be determined for the $^4F_{3/2} \rightarrow {}^4I_{9/2}$ transition at 0.87 μm, and is related to $\sigma_p(1.05)$ and $\sigma_p(1.32)$ using:

$$A_{ij} = \frac{A_{0.87} h v(0.87) \int_{ij} \sigma(v) dv}{h v(ij) \int_{0.87} \sigma(v) dv} \quad (8.5)$$

(note $ij \equiv 1.05$ or 1.32). Calculated values of σ_p, τ_{rad} (where $\tau_{rad} = 1/\sum A_r$) and τ_{meas} are given in Table 8.2, where we see confirmation that lifetimes and efficiencies in fluoride glasses are somewhat higher than in silicates.

A significant point to emerge from these data is that although the spontaneous transition rate A_r is significantly higher at 1.05 than at 1.32 μm,

Table 8.2 Nd^{3+} in several glass hosts.

| Host class | Ref. | $\sigma_p \times 10^{20}$ cm² | | | τ_{rad} (μs) | τ_{meas} (μs) |
		0.87 μm	1.05 μm	1.32 μm		
ZBLAN	This work	1.23	3.29	0.60	463	480
NS918 (SiO₂)	[11]	—	1.7	—	—	450
Na–Ca–SiO₂	[12]	—	1.65	1.35	820	400
LHG-8 (phosphate)	[11]	—	4.2	—	—	315

because of the wavelength difference the cross-sections are not dissimilar. This implies that high gains are also possible close to 1.3 μm, and that quantum efficiencies close to 1 can be obtained for both the 1.05 and 1.32 μm transitions. Indeed, as we shall see in Section 8.5, recent work confirms that high efficiencies can be obtained at both wavelengths. The high spontaneous emission rate at 1.05 μm, however, may present problems in other ways. One of the major objectives has been to obtain high amplifier gains close to 1.32 μm and we can see here that as the pumping level is increased in order to increase gain at 1.32 μm, the system may be limited by lasing from amplified spontaneous emission (ASE) at 1.05 μm. This point is also discussed further in Section 8.7.

Another point mentioned earlier was that signal ESA at 1.32 μm may also limit the gain that is available. ESA exists to a certain extent in most glass hosts and generally coincides with the peak of the emission spectrum at 1.32 μm. In these cases the overall cross-section for gain takes the form

$$\sigma_{gain} = \sigma_e - \sigma_{ESA} \tag{8.6}$$

where σ_e and σ_{ESA} are the stimulated emission and excited-state absorption cross-sections respectively. Hence gain can only be obtained in regions where $\sigma_e > \sigma_{ESA}$. This often occurs at the longer wavelength end of the emission and hence lasing has been restricted to 1.36 μm in P-doped silica fibre [13] and 1.346 μm in ZBLAN fibre. ESA can be estimated by Judd–Ofelt analysis of the spectroscopic data as outlined in Chapter 2. This has been undertaken by Miniscalco et al. for Nd [9] and they have estimated approximately half the values in ZBLAN than in silica. For these reasons overall gain values close to 1.32 μm are expected to be significantly higher in ZBLAN and, as discussed in Section 8.7, gain has now been measured in this host over the range 1.32–1.36 μm.

8.3.3 Detailed spectroscopy of Er

The $^4I_{13/2} \rightarrow {}^4I_{15/2}$ transition at 1.54 μm in Er^{3+} is the most interesting for telecommunication applications. Earlier chapters have had detailed information on this transition in silica-based fibres, but it also exists as expected in ZBLAN.

The emission and absorption cross-sections for this transition are given in Figure 8.8, determined using Equation (8.4). The peak value of σ_e is 4.7×10^{-21} cm², which compares reasonably well with other published values [8, 14]. It also compares favourably with AlP-doped silica fibre measured by Miniscalco et al. [14]. These values for σ_p and also τ_{rad} and τ_{meas} are given in Table 8.3.

Because of the discrepancy in the emission and absorption spectra in the short-wavelength region of the transition, once again there is the possibility of exciting the emission by directly pumping into the band. In this case the

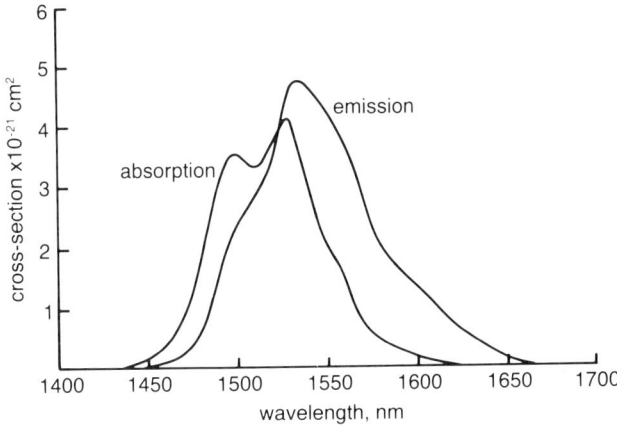

Figure 8.8 Emission and absorption cross-sections of Er^{3+} in ZBLANP glass.

Table 8.3 Er^{3+} in several host glasses

Host glass	Ref.	$\sigma_p \times 10^{20}$ cm^2	τ_{rad} (ms)	τ_{meas} (ms)
ZBLAN	This work	0.47	11	11
ZBLAN	[14]	0.63	–	–
Al–P–SiO$_2$	[14]	0.55	10	10

quantum efficiency that is likely to be achieved will be close to 100% since the non-radiative rates are extremely low. However, high efficiencies are also possible in silica when pumping into this level although probably closer to 95%. In ZBLAN the optimum pump wavelength is likely to be 1495 nm rather than 1480 nm as in the case of silica. Recent results, presented later in the chapter, show that pumping at 1480 nm into ZBLAN has been successful.

Paradoxically when pumping via higher levels at 800 or 980 nm, the overall efficiency of silicate glasses at 1.54 μm is likely to be higher. This is because high non-radiative decay in the case of silicate will result in a high population of the $^4I_{13/2}$ state. In the case of ZBLAN, however, ions may skip this level by higher energy radiative transitions at 980 or 800 nm, down to the ground state. Therefore, 1.54 μm amplifiers in ZBLAN will most likely be feasible only with 1495 nm pumping.

8.4 Fabrication techniques

Although several fibre-drawing techniques have been proposed for fluoride glasses certainly the most successful has been the preform process, whereby glass melts are converted into preforms which are finally drawn down into

Table 8.4 Glass compositions

	Zr	Hf	Ba	La	Al	Na	Pb	Index
ZBLANP 2	51.5	–	19.5	5.3	3.2	18.0	2.5	1.5056
ZBLAN 2	53.0	–	20.0	4.0	3.0	20.0	–	1.4985
ZHBLAN 1	39.7	13.3	18.0	4.0	3.0	22.0	–	1.4925

optical fibre. The exact composition of the glass used to make the fibres determines their resulting quality and properties. For example, the most stable HMF glasses so far discovered are based on ZBLAN and, consequently, this glass is the most favoured for fibre making.

A list of some popular compositions is given in Table 8.4. As usual acronyms are used based on the first letter of each component, and constituents are given in mol%. The effect on refractive index (measured at sodium D line of 589 nm) has been measured for two popular modifiers, PbF_2 and HfF_4 [3].

8.4.1 Glass melting and preform fabrication

The starting materials are in general anhydrous fluorides and all materials are stored in glove boxes to prevent hydrolysis. The melting crucibles are either vitreous carbon or Pt/Au. A typical melting schedule includes an initial hold at 400°C to allow fluorination, followed by a rise in temperature to 850°C to allow fusion and melting. Oxygen can be introduced in order to oxidise the melt. The melt is then cooled to 670°C prior to casting in a brass mould.

The casting technique, suggested by Tran *et al.* [15], is based on an old-established method used in other systems to form tubes by rotational casting. The technique is illustrated in Figure 8.9. In this case the mould is only partially filled with molten cladding glass, and then swung to a horizontal position and spun until the glass cools to form a tube. In this way controlled

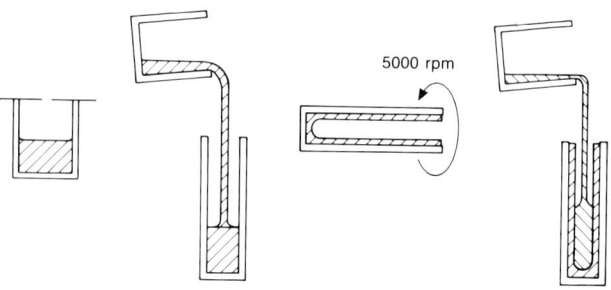

(1) melt   (2) pour cladding glass   (3) spin cladding tube   (4) pour core glass

Figure 8.9 Rotational casting technique for multimode preforms.

cladding tubes can be fabricated and the core glass is then cast into this tube. A small lathe is used for spinning the mould and a typical spin speed is 5000 rpm. The moulds are often fabricated in brass, although some authors have suggested gold coating, and are split along the centre-line for easy removal of the preform. After casting the mould is transferred into a separate annealing oven situated below the glove box, again without coming into contact with ambient. Since the outer surface of the molten glass is brought into contact with a potentially contaminating metal during the casting process, several techniques have been developed to remove the outer layer from the preform. They include mechanical polishing and chemical etching and the techniques are able to improve both the fibre strength and the loss.

8.4.2 Fibre drawing

The techniques used to draw preforms into fibre are based on standard methods developed for silica fibre. The main difference is the draw furnace since much lower temperatures are required for fluoride fibres. A feeder is used to lower the preform into the furnace at a constant rate. The preform is mounted inside a silica liner which is sealed at the top end around the stainless steel rod so that dry nitrogen can be flushed through the chamber. In this way surface crystallisation during reheating of the preform to draw fibre can be kept to a minimum. A diameter monitor is used to measure the fibre diameter and the output from the monitor is then servo-linked to the winding drum speed controller. In this way the diameter is controlled to $\pm 2\,\mu\text{m}$. Polymer coatings cured by UV light can be applied in the usual way. Essentially, epoxy acrylates can be applied in the liquid monomer form using a coating cup applicator. The coating is then cured by passing the fibre through a UV lamp, flushed with N_2. A concentricity monitor is used to ensure even coating thickness around the fibre (usually about $50\,\mu\text{m}$).

8.4.3 Monomode fibre

The rotational casting process is really only suitable for multimode fibre because of the difficulties of casting glass into the restricted core volume required for monomode. However, a few adaptations of the casting technique are now being developed and one such example is illustrated in Figure 8.10. Here a normal multimode preform is made as the first part of the process. The second stage is to stretch out this preform to a final diameter of about 1 mm from an initial diameter of, say, 10 mm. The third stage is to sleeve the stretched preform with another cladding glass tube and to collapse the tube onto the preform. In this way the resultant new preform may have an outer diameter of 10 mm but a core diameter of less than 1 mm and would be suitable for drawing into monomode fibre. Care must be taken at all stages of the process to ensure that surface crystallisation does not occur.

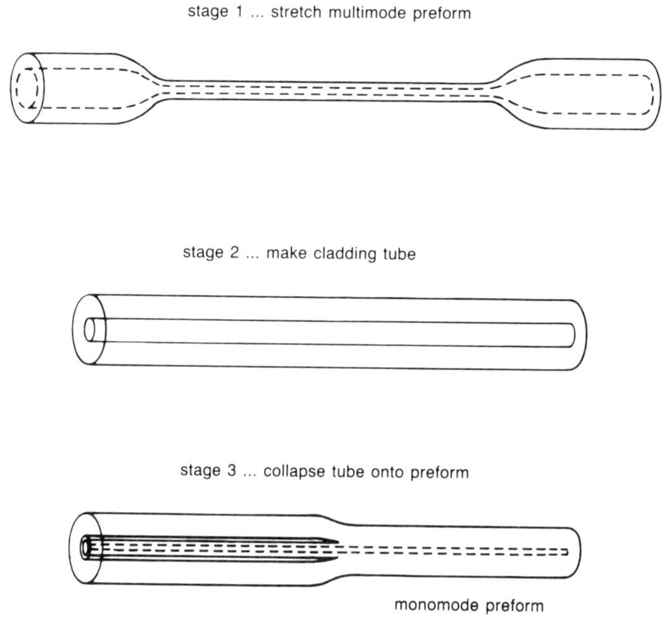

stage 1 ... stretch multimode preform

stage 2 ... make cladding tube

stage 3 ... collapse tube onto preform

monomode preform

Figure 8.10 'Stretch and overclad' process for monomode preforms.

8.5 Fluoride fibre lasers

As we have seen in Sections 8.2 and 8.3, fluoride glasses offer considerable potential as a laser host material, but it was not until 1987 that the first fluoride fibre laser was demonstrated [4]. Indeed that was the first laser of any kind to be demonstrated in a fluoride glass, and even now the vast majority of laser reports in fluoride glass hosts have been in fibre form and we shall concentrate on those in this section. To date laser operation has been demonstrated on 20 different transitions in neodymium, holmium, thulium and erbium, ranging from $0.455\,\mu$m in the blue to $2.7\,\mu$m in the mid-IR, almost all of them continuous wave (CW), and almost all at room temperature. There are many other possibilities for lasing in these and other rare-earth ions, so we expect to see the number of reported laser transitions increase. We shall study each active ion separately, and it will help the reader to make frequent use of the energy level diagrams in Figure 8.6. We shall begin with neodymium because that is where the development of fluoride fibre lasers began.

8.5.1 *Neodymium-doped fluoride fibre lasers*

There are three obvious candidates for laser transitions in neodymium-doped fluoride fibres, $^4F_{3/2} \rightarrow {}^4I_{9/2}$ around $0.9\,\mu$m, $^4F_{3/2} \rightarrow {}^4I_{11/2}$ around $1.05\,\mu$m, and $^4F_{3/2} \to {}^4I_{13/2}$ around 1.32 μm. Only the last two have been demonstrated so far, the first has not been addressed. The first papers on this subject [4, 16, 17] employed argon ion lasers at 0.514 μm as pump sources, which inefficiently energise the $^2G_{9/2}$ level to populate the upper laser level by radiative and non-radiative decay. This process is also inefficient because of direct radiative decay to lower levels and to the ground state. However, improvements in threshold and slope efficiency from 390 mW and 0.03% to 33 mW and 16.8% were achieved on the 1.05 μm transition, and from 670 mW and 2.6% to 84 mW and 3.2% around 1.34 μm. However, much greater efficiency can be achieved by exciting the $^4F_{5/2}$ level by a semiconductor laser at 0.795 μm [18]. This scheme efficiently populates the $^4F_{3/2}$ level by direct non-radiative decay due to the small energy gap between the levels (Sections 8.2 and 8.3 above). This allows low-threshold operation (3 and 20 mW) and close to 100% quantum efficiency operation on both of these transitions. Figure 8.11 shows the laser characteristics of both transitions, using optimised output mirrors, taken from [18]. The slope efficiencies obtained were 70% at 1.05 μm and 57% at 1.345 μm, relating to quantum efficiencies of 92.5 and 97%. As mentioned earlier in this chapter, the $^4F_{3/2} \to {}^4I_{13/2}$ transition suffers from ESA, and the effect of this is to move the maximum gain wavelength to 1.345 μm, where lasing occurs in a free-running cavity, from the fluorescence maximum at 1.32 μm. More detail of this will be given in Section 8.7 when amplification is considered.

All of the above work was done on multimoded fibres, and to date there have been no specific reports of lasing in neodymium using single-mode fluoride fibres. However, close study of the modal structure of the output of these multimode fibre lasers shows that, with some control of the cavity, single transverse mode operation can be achieved [19]. In the course of amplification studies, lasing at 1.05 μm from small reflections or amplified spontaneous

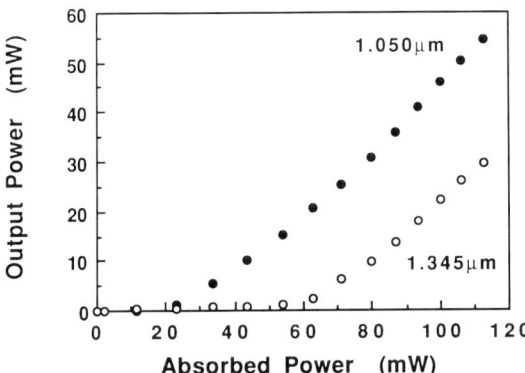

Figure 8.11 Lasing characteristics of optimised neodymium-doped lasers. The upper curve is the 1.05 μm characteristic, and the lower is 1.345 μm.

emission causes limitations to gain on the 1.32 μm transition [20, 21], so it is apparent that efficient operation in single-mode fibres should also be achieved.

8.5.2 Holmium-doped fluoride fibre lasers

One obvious advantage of fluoride glass fibres is their transparency from the UV to the mid-IR, and with the now proven suitability of these glasses as laser hosts for rare-earth ions, an obvious development of this field was to examine the possibility of producing mid-IR fibre lasers. The first attempt to achieve this was to examine the $^5I_7 \rightarrow {}^5I_8$ three-level transition in holmium [22]. This is a well-known laser transition in crystal hosts and occurs around 2 μm wavelength. In this host, lasing occurs at 2.08 μm. Pumping was CW at 0.488 μm into the $^5K_3$ energy level. The output was not CW but consisted of a series of spikes, similar to Q-switched output, with a repetition rate of 3 kHz. Since this was a non-optimised three-level laser it is likely that improvements in fibre quality and cavity length optimisation would lead to CW lasing. Observing the absorption and fluorescence spectra for holmium (Figure 8.5) above shows that holmium also exhibits a four-level transition, which is attributed to $({}^5S_2 + {}^5F_4) \rightarrow {}^5I_5$, around 1.35 μm. Reference [22] also reports lasing on that transition, also pumped at 0.488 μm. Here the threshold was 1.12 W of launched power with a slope efficiency of 0.28%. Lasing occurred at 1.38 μm, which is longer than the fluorescence peak, suggesting that yet again signal ESA is dominant at shorter wavelengths. Very recently lasing has been achieved on the $^5I_6 \rightarrow {}^5I_7$ transition between 2.83 and 2.95 μm [23].

Lasing in holmium also occurs around 0.54 and 0.75 μm, but these transitions employ up-conversion processes and will be dealt with in Section 8.6.

8.5.3 Thulium-doped fluoride fibre lasers

So far, despite the abundance of radiative transitions in erbium, thulium has the greatest number of demonstrated lasing transitions – seven in all, three of them via up-conversion processes. The four conventionally pumped transitions will be examined here, and the remainder in Section 8.6. The four are the $^3H_4 \rightarrow {}^3H_6$ three-level transition at 0.82 μm, the $^3H_4 \rightarrow {}^3F_4$ normally self-terminating transition at 1.48 μm, the $^3F_4 \rightarrow {}^3H_6$ three-level transition at 1.88 μm, and the classic four-level $^3H_4 \rightarrow {}^3H_5$ transition at 2.3 μm. Early work concentrated on the 2.3 μm transition, pumped first by a pulsed Alexandrite laser at 0.786 μm [24] and then a CW semiconductor diode at 0.79 μm [25] pumping directly into the upper laser level. These both used a small core fibre to increase pump power density, and the latter produced CW lasing with a 3.5 mW absorbed power threshold and 10% slope efficiency, which is a quantum efficiency of approximately 29%. Some evidence of output power saturation was seen in this last laser, and this has been attributed to saturation

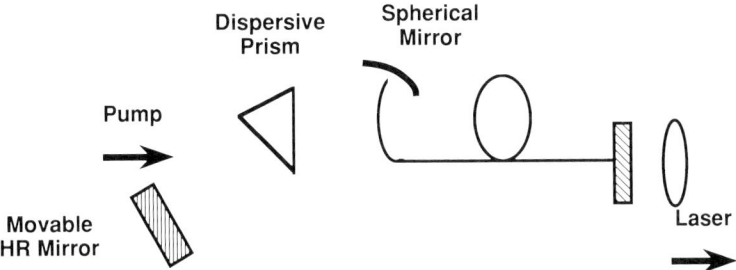

Figure 8.12 Laser cavity used by Allain et al. for CW, tuneable, and Q-switched operation.

of the long-lived $^3F_4$ energy level below the lower lasing level, which would reduce the depopulation rate of the lower laser level, and hence the population inversion. Tuneable CW lasing has been achieved on all four transitions [26]. The pump source was the 0.6764 μm line from a krypton–ion laser, pumping into the $^3F_3$ level for this work, and, as in other papers by the same group, an unusual cavity design was used (Figure 8.12) which allows both tuning and Q-switching via a combination of a movable mirror and a dispersive prism, at the expense of course of higher cavity losses and hence higher threshold powers. The tuning range achieved for the 2.3 μm transition was 2.27–2.4 μm, and on the 1.88 μm transition 1.84–1.94 μm. Threshold powers were below 50 mW, and slope efficiencies around 3.5%. On the 1.48 μm transition – which is normally self-terminating since the upper state lifetime is ~ 1 ms, the lower state lifetime is ~ 10 ms, and other parameters such as branching ratios are unfavourable – CW lasing was achieved by allowing simultaneous lasing at 1.9 μm, reducing the effective lifetime of the $^3F_4$ level. Threshold power was 63 mW and slope efficiency 1.6%. Tuning was achieved between 1.46 and 1.51 μm for pulsed pumping at twice threshold. Here, as in the case of the 1.3 μm transition in neodymium, the gain and fluorescence spectra are different, with the peak of the gain occurring at longer wavelengths. This is due to ESA of shorter wavelength signal photons from $^3F_3$ to $^1G_4$. On the 0.82 μm transition tuning was obtained from 0.815 to 0.825 μm with 1.14 times the threshold pump power of 45 mW. All of this work was carried out with the same fibre length and pump wavelength so some improvement would be expected with optimisation of these parameters for each transition, and greater pump power.

Finally, diode-pumped operation, using a 0.795 μm laser diode, has also been reported on the $^3F_4 \rightarrow {}^3H_6$ 1.9 μm transition [27]. In this case the fibre was multimoded. CW oscillation was observed with a threshold of 40 mW launched power (20 mW absorbed) and slope efficiency 0.3%. Interestingly, lasing occurred at 1.97 μm, which is much longer than the peak of the fluorescence, and also outside of the tuning range of [26] above. This is possibly due to the fibre being too long for optimum three-level operation, making the cavity effectively lossy at the emission wavelength, thus giving higher probability to the quasi four-level operation at longer wavelengths. The authors suggest that one way to improve the pumping efficiency for this scheme would be to force simultaneous operation of the 2.3 μm transition to favourably change branching ratios.

8.5.4 Erbium-doped fluoride fibre lasers

Erbium in fluoride glass fibre displays over 20 fluorescent transitions from 0.255 μm in the UV to 2.7 μm in the mid-IR [28]. One of these transitions occurs at a wavelength directly applicable to telecommunications, around 1.5 μm, and another is close to the loss minimum of the glass host itself. CW lasing has been observed on six of these transitions, $^4S_{3/2} \to {}^4I_{13/2}$, at 0.85 μm, $^4I_{11/2} \to {}^4I_{15/2}$, around 0.98 μm, $^4I_{13/2} \to {}^4I_{15/2}$ at 1.563 μm, $^2H_{11/2} \to {}^4I_{9/2}$ at 1.66 μm, $^4S_{3/2} \to {}^4I_{9/2}$ at 1.72 μm, and the $^4I_{11/2} \to {}^4I_{13/2}$ transition at 2.7 μm. The first, and last three, of these transitions are normally self-terminating, but CW operation has been achieved at 0.85 and 2.7 μm in multimode fibre by efficiently depopulating the $^4I_{13/2}$ level by ESA of the pump wavelength of 0.4765 μm [28–30]. This approach was suggested as possible by Kaminskii [31], though more recent work in single-mode fibre [32] suggests that for diode pumping at an appropriate wavelength, 0.82 μm, CW operation at 2.7 μm may be possible without ESA because branching ratio can be favourable. However employing ESA at a pump wavelength around 0.79 μm does lead to more efficient operation.

Surprisingly, lasing on the 1.5 μm transition has received little attention, but has been observed at 1.563 μm [33] and 1.55 μm [34]. This transition in fluoride glass fibres will probably find application as an amplifier for telecommunications systems, as it already does in silica-based fibres. This will be discussed in Section 8.7. However, lasing on the transition around 0.98 μm appears to be exclusive to this host and a number of papers have appeared. CW [28, 34, 35], tuneable, and Q-switched operation [36] have all been demonstrated. Tuning has been observed from 0.981 to 1.004 μm when pumped at 0.5145 μm and using the cavity arrangement of Figure 8.12, and the same setup gives Q-switched pulses of 120 ns duration, 10 W peak power, at a repetition rate of 50 Hz. Recently self-terminated operation has also been observed on both the $^4S_{3/2} \to {}^4I_{9/2}$ and $^2H_{11/2} \to {}^4I_{9/2}$ transitions at 1.72 and 1.66 μm respectively [37] with thresholds of 90 mW absorbed power at 0.5145 μm.

8.5.5 Systems experiments

Despite the number of reports of fibre lasers, both fluoride and silica based, operating within telecommunications windows at 1.3 and 1.5 μm, there have been no reports of the use of these lasers in telecommunications systems. This is largely due to the ready availability of semiconductor devices, developed for systems in these windows, which can be directly modulated at high speeds. Near the loss minimum of fluoride fibres however, semiconductor devices are not yet readily available, so it is possible that fibre lasers may have a role to play in that area. As shown in Section 8.5.4, fluoride fibre lasers have been made at 2.3 and 2.7 μm, which are spaced either side of that loss minimum, and one of these lasers, an erbium-doped 2.7 μm laser, has been used in a system demonstration [38]. The laser was modulated by means of a lithium niobate external modulator operating at 34 Mbit/s and the whole system easily met the target error rate figure of 1 in 10^9. A much more probable use of doped fibre technology in telecommunications is in the role of amplification, and the contribution of fluoride fibres in this field will be discussed in Section 8.7.

8.6 Up-conversion lasers

As shown in Section 8.5, ESA of signal photons can modify the spectral gain characteristics of some transitions, and ESA of pump photons can facilitate CW lasing on otherwise self-terminating transitions. The nett effect of either of these phenomena is to energise ions to higher level, often to much higher energy levels than the original pump level. This allows the possibility of 'up-conversion lasers', which generally means lasers operating at higher energies, shorter wavelengths, than the pump source, although the term could equally well apply to longer wavelength lasers operating on higher level transitions which require up-conversion mechanisms to populate the upper laser level. It is an attractive possibility that semiconductor diode pumped visible and UV lasers could be made, and these would undoubtedly find application in optical data storage and compact disc equipment. Up-conversion can occur through energy transfer between ions in the same excited state, a process which generally requires dopant concentrations of 1% or greater, or through ESA of either pump or signal wavelengths, or indeed of additional pump wavelengths. At the concentrations common in fluoride fibre work it is likely that ESA is the dominant mechanism. High-energy up-conversion emissions occur in all of the dopants discussed above to a greater or lesser extent (erbium-doped fibre glows bright green when pumped near 0.8 μm), but so far there have been only two successful reports, both from the same group, of lasing on any of these transitions in doped fluoride fibres. The first [39] pumped a thulium-doped fibre, cooled to 77 K, simultaneously at 0.6471 μm and 0.6764 μm from a krypton ion laser. Chopping the pump produced lasing at 0.455 μm on the $^1D_2 \rightarrow {}^3F_4$ self-terminating transition (labelling as described in Section 8.3.1 above) and on the $^1G_4 \rightarrow {}^3H_6$ three-level transition at 0.48 μm. In another fibre CW lasing was observed at 1.51 μm on the $^1D_2 \rightarrow {}^1G_4$ transition, a high-level transition which required up-conversion processes to populate the upper state. The second [40] pumped holmium-doped fibre at room temperature at 0.6471 μm to produce tuneable CW lasing between 0.54 and 0.553 μm on the three-level $^5S_2 \to {}^5I_8$ transition, and CW lasing at 0.753 µm on the $^5S_2 \to {}^5I_7$ transition, which again required up-conversion to populate the upper level.

8.7 Fluoride fibre amplifiers

In telecommunications, probably the main application of doped fibre technology will be in providing simple amplifiers for trunk and under-sea systems, local networks, and for broadband distribution systems. In the 1.5 µm window erbium-doped silica fibre amplifiers have demonstrated that systems longer than 900 km can be possible without the use of electronic regenerators [41], and that the use of one amplifier can extend local networks to serve thousands of customers [42]. The development of amplifier technology in fluoride fibres is currently a long way behind that of silica-based fibre amplifiers, but fluorides do have a contribution to make. The first report of gain measurements in an erbium-doped fluoride multimoded fibre [43] on the $^4I_{13/2} \to {}^4I_{15/2}$ transition demonstrated that the bandwidth would be greater than that of erbium-doped germanium/silica amplifiers, and recent work [44, 45] has shown that in fact under similar conditions, an erbium/fluoride amplifier almost exactly matches the gain characteristics of the best erbium/silica amplifier, but exhibits less gain ripple, 1 dB compared to 3 dB

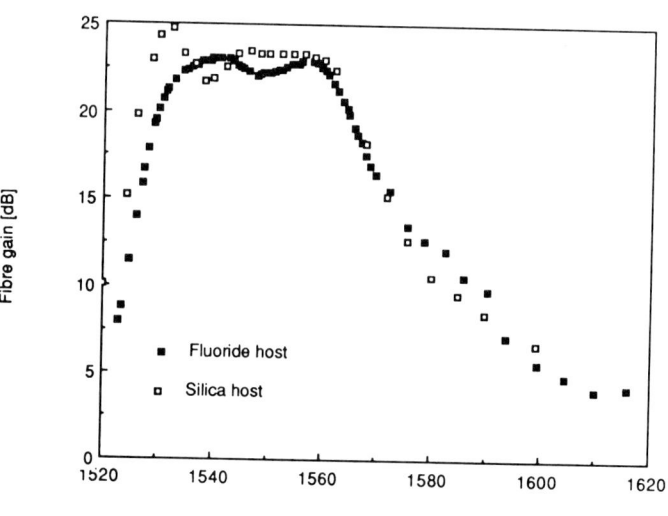

Figure 8.13 Gain spectra of erbium in single-mode fluoride and silica fibres pumped under nearly identical conditions near 1.48 µm.

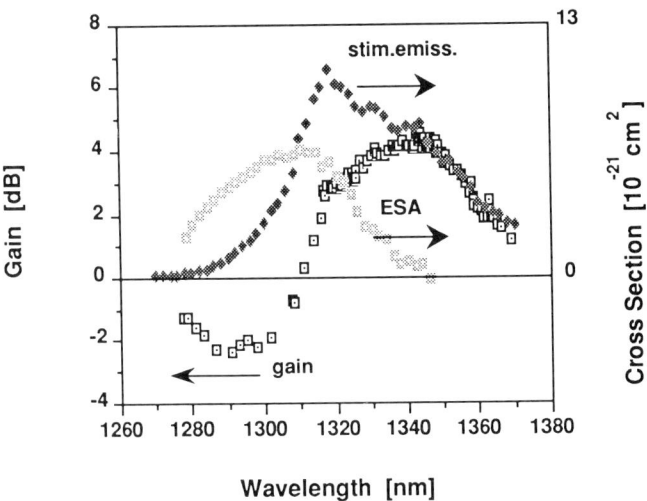

Figure 8.14 Gain spectrum, signal ESA and emission cross-sections of a single-mode neodymium-doped fibre pumped at 0.795 μm.

(Figure 8.13). Gain of 23 dB has been demonstrated for a 28 mW pump power at 1.482 μm, directly pumping the upper level.

For the 1.3 μm window, however, no silica-based amplifier has been demonstrated because the most likely candidate, the $^4F_{3/2} \rightarrow {}^4I_{13/2}$ transition in neodymium, suffers from dominant signal ESA at wavelengths shorter than 1.36 μm even in the best silica-based fibre [13]. In fluoride fibres the ESA transition is both at shorter wavelengths and less strong [21]. This has led to the recent development of amplifiers for the 1.3 μm window using neodymium-doped fluoride fibres [21, 46, 47]. Figure 8.14 illustrates the gain spectrum, ESA and emission cross-sections associated with these amplifiers, although the gain figures in this diagram are well below the best now achieved (10 dB has been reported). Interestingly, because this is a four-level system, the gain spectrum can be scaled as gain increases, unlike the three-level erbium amplifier whose spectrum changes with gain. However, there is a further limiting factor to gain on this transition, and that is competition from the $^4F_{3/2} \rightarrow {}^4I_{11/2}$ and $^4F_{3/2} \rightarrow {}^4I_{9/2}$ transitions at 1.05 and 0.9 μm. The 1.05 μm transition is dominant here, and when sufficient ASE has built up, which occurs where the single-pass gain on that transition is around 40 dB, the onset of superfluorescent emission clamps the population of the $^4F_{3/2}$ level, and hence clamps the gain. Hence the maximum gain that can be achieved in a single fibre is governed by the branching ratio of the 1.3 and 1.05 μm transitions, and the single-pass gain required to reach superfluorescence. An indication of the power handling capabilities of these amplifiers can be obtained from the small-signal saturation characteristics [47], which show input signal saturation at around +10 dBm (Figure 8.15). The system

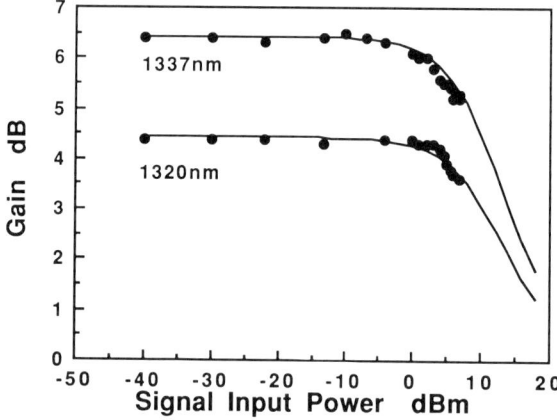

Figure 8.15 Signal saturation characteristics of a single-mode neodymium-doped fibre amplifier. The upper curve is near to the peak gain wavelength (see Figure 8.14) where ESA is small, and the lower is at the wavelength of zero dispersion in standard silica fibres, where ESA is significant.

performance of these amplifiers has also been characterised at 2.4 Gbit/s [48] where no degradation in receiver sensitivity was found when the device was used as a power amplifier, and only 1.5 dB degradation when used as a pre-amplifier. The wavelength used for these experiments was 1.338 μm, which is near to the peak of the gain curve and where the ESA cross-section is small. If more gain were available the noise penalty would be greater, up to 3 dB, even for operation at shorter wavelengths, due to the ESA of signal photons. Gain of 18.3 dB has also been observed at 2.707 μm in an erbium-doped fibre on the $^4I_{11/2} \rightarrow {}^4I_{13/2}$ transition, pumped at 0.647 μm [49].

8.8 Other systems

Rare earths are of course not the only dopants that can be added to glasses in order to induce lasing effects. Other well-known systems include the 3d transition metals and the actinides. These materials will be examined here as potential dopants in fluoride glasses, and further uses of the rare earths in co-doped systems will also be looked at.

The transition metals formed the basis of some of the first successful lasing materials ever produced and include, of course, Cr^{3+} which is the active component of ruby. In general, however, transition metals are less efficient as lasing materials than the rare earths. A comparison of their absorption spectra [3] shows that the transition metals have much broader bands and this, in turn, implies that the effective separation between energy levels will be reduced. Since non-radiative decay depends on this separation, the overall emission efficiencies of the transition metals will be reduced. Also, the ionic energy levels of rare earths are largely independent of the host, due to shielding of the outer 5s and 5p electrons, whereas in transition metals they are very host dependent. For example, the absorption spectra of transition metals in fluoride glasses have some notable differences to those in oxide glasses [3]. In particular, the lower ligand field strength in fluorides leads to lower energy separations, which in turn leads to higher non-radiative decay rates. Moreover, the preferred coordination symmetry of the transition metals in fluorides is octahedral, as opposed to tetrahedral in many oxides. This in turn implies lower absorption and hence lower emission cross-sections in the fluoride hosts.

In summary, transition metals are not expected to have efficient emissions in fluoride glasses and, indeed, preliminary investigations indicate that they do not give any detectable emissions in ZBLAN glasses over the wavelength range 0.5–3.0 μm. At present, therefore, the transition metals are less interesting than the rare earths, although they may find applications as sensitisers used to improve efficiencies in co-doped glasses.

Another group of elements that has been found to lead to lasing properties is the actinides. This series, of course, starts with Th and includes U. Again these elements should be compatible with fluorides since ThF_4 has been used as a component of fluorozirconate glasses. To date the only dopant from this series that has been investigated is U. U^{3+} has been shown to give lasing close to 2.5 μm in a CaF_2 crystalline host, and since this wavelength is close to the predicted minimum loss wavelength that may occur in fluoride glasses it is also of interest for fibre lasers and amplifiers. The major problem is that U prefers to take the U^{4+} state as opposed to the more useful U^{3+} form. However, work at Sheffield University [50] has shown that it is possible to induce the preferred tripositive state in fluoride glasses and a fluorescence spectrum covering the 2.5 μm band taken from this work is shown in Figure 8.16. Further work will hopefully lead to a useful fibre laser device at this wavelength.

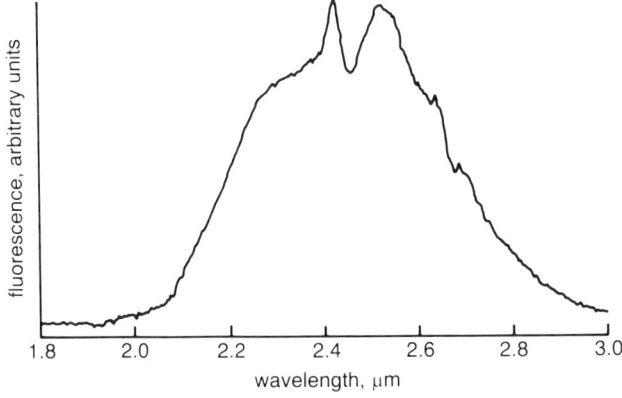

Figure 8.16 Fluorescence spectrum of U^{3+} in ZBLANP glass (after [50]).

The final category of 'other systems' that is discussed in this section is that of rare-earth co-doped glasses. Co-doping is a well-known technique that has been used to improve the performance of some laser systems. So far there have been few reports of co-doping in fluoride glasses and no reports of working devices. However, Pafchek et al. [51] have demonstrated that the addition of Ho to a Tm-doped fluoride glass can help deplete the lower lasing level $^3F_4$ in Tm and hence help to induce lasing at 1.47 μm. They claim that results in this system are superior to those in a Tb–Tm co-doped glass reported by Rosenblatt et al. [52].

References

1. M. Poulain, M. Poulain and J. Lucas, *Mat. Res. Bull.* **10** (1975) 243.
2. J. Lucas, M. Chanthanasinh, M. Poulain, M. Poulain and M.J. Weber, *J. Non-Cryst. Solids* **27** (1978) 273.
3. P.W. France, M.G. Drexhage, J.M. Parker, M.W. Moore, S.F. Carter and J.V. Wright, *Fluoride Glass Optical Fibres*, Blackie/CRC Press (1990).
4. M.C. Brierley and P.W. France, *Electron. Lett.* **23** (1987) 815.
5. W.H. Zachariasen, *J. Amer. Chem. Soc.* **545** (1932) 3840.
6. M. Poulain, Chap. 2, *Fluoride Glasses* (ed. A.E. Comyns), Wiley (1989).
7. R. Reisfeld, *Proc. Int. Conf. on Lasers, New Orleans* (15–19 December 1980), p. 349.
8. Zheng Haixing and Gan Fuxi, *Chinese Phys.* **6** (1986) 978.
9. L.J. Andrews, W.J. Miniscalco and T. Wei, *Proc. First Int. School on Excited States of Transition Elements, Poland* (20–25 June 1988), p. 9.
10. S.T. Davey and P.W. France, *Brit. Telecom Technol. J.* **7** (1989) 58.
11. T. Yamashita, *Proc. SPIE* **1171** (1989) 291.
12. P.H. Sarkies, Cross-section measurements of neodymium laser glasses, PhD Thesis, Sheffield University (1971).
13. F. Hakimi, H. Po, R. Tumminelli, B.C. McCollum and E. Snitzer, *Opt. Lett.* **14** (1989) 1060.
14. W.J. Miniscalco, L.J. Andrews, B.A. Thompson, T. Wei and B.T. Hall, *Proc. SPIE* **1171** (1989) 93.
15. D.C. Tran, C.F. Fisher and G.H. Sigel, *Electron. Lett.* **18** (1982) 657.
16. W.J. Miniscalco, L.J. Andrews, B.A. Thompson, R.S. Quimby, L.J.B. Vacha, and M.G. Drexage, *Electron. Lett.* **24** (1988) 28.
17. M.C. Brierley and C.A. Millar, *Electron. Lett.* **24** (1988) 438.
18. M.C. Brierley and M.H. Hunt; *Proc. SPIE* **1171** (1989) 157.
19. C.A. Millar, S.C. Fleming, M.C. Brierley, M.H. Hunt, *IEEE Photon. Technol. Lett.* **2** (1990) 415.
20. J.E. Pedersen, M.C. Brierley, S.F. Carter and P.W. France, *Electron. Lett.* **26** (1990) 329.
21. J.E. Pedersen and M.C. Brierley, *Electron. Lett.* **6** (1990) 819.
22. M.C. Brierley, P.W. France and C.A. Millar, *Electron. Lett.* **24** (1988) 539.
23. L. WetenKamp, *Electron. Lett.* **26** (1990), 883.
24. L. Esterowitz, R. Allen and I. Aggarwal, *Electron. Lett.* **24** (1988) 1104.
25. R. Allen and L. Esterowitz, *Appl. Phys. Lett.* **55** (1989) 721.
26. J.Y. Allain, M. Monerie and H. Poignant, *Electron. Lett.* **25** (1989) 1660.
27. J.N. Carter, R.G. Smart, D.C. Hanna and A.C. Tropper, *Electron. Lett.* **26** (1990) 600.
28. M.C. Brierley, C.A. Millar and P.W. France, *Digest of CLEO* (1989), Paper TUJ22.
29. M.C. Brierley and P.W. France, *Electron. Lett.* **24** (1988) 935.
30. J.Y. Allain, M. Monerie and H. Poignant, *Electron. Lett.* **25** (1989) 28.
31. A.A. Kaminskii, *Laser Crystals* (2nd edn), Springer Series in Optical Sciences, Springer-Verlag (1990), p. 52.
32. R. Allen and L. Esterowitz, *Appl. Phys. Lett.* **56** (1990) 1635.

33. M.C. Brierley, P.W. France, M.W. Moore and S.T. Davey, *Digest of CLEO* (1988), Paper TUM29.
34. M. Monerie, J.Y. Allain, H. Poignant and F. Auzel, *Digest of 15th ECOC* (1989), Paper TuB12-6 232.
35. J.Y. Allain, M. Monerie and H. Poignant, *Electron. Lett.* **25** (1989) 318.
36. J.Y. Allain, M. Monerie and H. Poignant, *Electron. Lett.* **25** (1989) 1082.
37. R.G. Smart, J.N. Carter, D.C. Hanna and A.C. Tropper, *Electron. Lett.* **26** (1990) 649.
38. M.C. Brierley, P.W. France, R.A. Garnham, C.A. Millar and W.A. Stallard, *Digest of OFC* (1989), Postdeadline paper PD14.
39. J.Y. Allain, M. Monerie and H. Poignant, *Electron. Lett.* **26** (1990) 166.
40. J.Y. Allain, M. Monerie and H. Poignant, *Electron. Lett.* **26** (1990) 261.
41. N. Edagawa, Y. Yoshida, H. Taga, S. Yamamoto, K. Mochizuki and H. Wakabayashi, *Digest of 15th ECOC* (1989), Postdeadline paper PDA-8.
42. A.M. Hill, D.B. Payne, K.J. Blyth, D.S. Forrester, J.W. Arkwright, R. Wyatt, J.F. Massicott, R.A. Lobbett, P. Smith and T.G. Hodgkinson, *Electron. Lett.* **26** (1990) 605.
43. C.A. Millar, M.C. Brierley and P.W. France, *Digest of 14th ECOC*, IEE Conference Publication Number 292 – Part 1 (1988), p. 66.
44. C.A. Millar and P.W. France, *Electron. Lett.* **26** (1990) 634.
45. D.M. Spirit, G.R. Walker, P.W. France, S.F. Carter and D. Szebesta, *Electron. Lett.* **26** (1990) 1218.
46. Y. Miyajima, T. Komoukai and T. Sugawa, *Electron. Lett.* **26** (1990) 194.
47. Y. Miyajima, T. Komoukai, T. Sugawa and Y. Katsuyama, *Digest of OFC* (1990), Postdeadline paper PD16.
48. J.E. Pedersen, M.C. Brierley and R.A. Lobbett, *Photon. Technol. Lett.* **2** (1990) 750.
49. D. Ronarch, J.Y. Allain, M. Guibert, M. Monerie and H. Poignant, *Electron. Lett.* **26** (1990) 903.
50. A.G. Clare, J.M. Parker, D. Furniss, E.A. Harris and T.M. Searle, *J. Phys. Condens. Matter* **1** (1989) 8753.
51. R. Pafchek, J. Aniano, E. Snitzer and G.H. Sigel, *Materials Research Society Symp. Proc.* **172** (1990) 347.
52. G.H. Rosenblatt, R.J. Ginther, R.C. Stone and L. Esterowitz, Paper presented at the *Tunable Solid State Laser Conf., Cape Cod, Mass.* (1989).

9 Q-switched and mode-locked fibre lasers

N. LANGFORD and A.I. FERGUSON

9.1 Introduction

Fibre lasers are important systems for the generation of short high-power laser pulses. Mode-locking, Q-switching and combined mode-locking and Q-switching were among the first modes of operation to be studied in fibre lasers. Mode-locking of a fibre laser is interesting both as a practical source for short-pulse generation which is compatible with communications and fibre sensor systems and for fundamental studies of mode-locking. The relatively large bandwidth of most fibre laser transitions means that they are capable of supporting very short pulses with high peak powers. A unique feature of mode-locked fibre lasers is that the long interaction length (several metres) in a very small core diameter means non-linear effects such as self-phase modulation, which are negligible or are just a small perturbation to the performance of a conventional bulk laser, become dominant in the fibre laser and dictate many of the operational features. A new range of aspects of mode-locking behaviour may be studied using the unique geometry of the fibre, and new insights into the fundamentals of mode-locking of lasers may be revealed by a careful study and modelling of the laser behaviour.

The peak power available from fibre lasers may be increased by several orders of magnitude over the continuous-wave power by Q-switching the system. The relatively long upper state lifetimes of the common rare-earth dopants means the Q-switching works particularly well with fibre lasers. Further increases in the peak power may be obtained by a combination of mode-locking and Q-switching. In this case the peak powers attained may lead to a wide variety of non-linear optical processes.

In this chapter some methods of mode-locking and Q-switching of fibre lasers are described and the importance of non-linear processes particularly in the case of mode-locking indicated. Some experimental results for a variety of systems are given. A detailed understanding of the interplay between optical non-linearities and the mode-locking dynamics in a fibre laser is one of the most interesting challenges in laser transient phenomena. This deeper understanding, together with the convenient and efficient geometry of fibres, suggests that short-pulse generation in fibres will have many important practical applications.

9.2 Mode-locking of fibre lasers

9.2.1 Introduction

The simplest fibre laser consists of an amplifying medium located between two mirrors forming the resonator. If the optical length between the two mirrors is l, a series of longitudinal modes may be supported which have a frequency separation Δf, given by

$$\Delta f = \frac{c}{2l} \tag{9.1}$$

with c the speed of light.

If a coherent phase relationship exists between the longitudinal modes they can constructively interfere, leading to the production of a series of mode-locked ultrashort pulses. To attain this mode-locked state the modes must communicate with each other, and this can be achieved in a variety of ways [1].

The long upper state lifetimes and small gain cross-sections associated with the rare-earth ions used as the active media in fibre lasers result in a high saturation flux. As a result of this the techniques applied primarily to mode-lock dye and colour-centre lasers, such as passive and synchronous mode-locking [1], cannot be used with fibre lasers. Mode-locking, however, can be induced with schemes commonly applied to Nd:YAG lasers, such as amplitude and phase modulation [2]. The geometry and waveguiding nature of the optical fibre host may enhance a variety of linear and non-linear processes present in bulk lasers, such as Nd:YAG lasers, which can affect the mode-locking mechanisms.

A typical single-mode fibre has a core diameter of $\sim 5\,\mu$m, and high power densities can be generated in the core even when modest average power signals are injected into the core. The waveguiding nature of the fibre, combined with the low losses associated with present-day fibres, enables this power density to be maintained over long lengths of fibre. This allows the development of a high optical gain in the fibre together with a low lasing threshold and enables the introduction of bulk elements into the cavity, such as microscope objectives or modulators, without a major degradation in performance of the laser. A disadvantage associated with this high-gain nature is that laser action can be observed with the cleaved facets of the fibre acting as the cavity mirrors. Thus care has to be taken to prevent reflections from intracavity elements re-entering the fibre as this can lead to the formation of subcavities inside the main cavity which can affect the mode-locking process.

The shape and flexibility of an optical fibre, enabling it to be stored on a drum, is a bonus as long cavity lengths (tens to hundreds of metres) can be established without the need for large physical separations between the mirrors facilitating the generation of mode-locked pulses at repetition frequencies not accessible with bulk lasers. This, too, introduces its own set of problems. In a typical mode-locked Nd:YAG laser with a cavity length of 2 m the gain medium is normally several tens of millimetres long and so can be regarded as a discrete element in the cavity. On the other hand, for a fibre laser the active rare-earth ion is distributed throughout the core of the optical fibre and as the majority of the cavity is formed from the fibre, the gain medium can no longer be considered as a discrete element. Thus, dispersion arising from either the optical fibre host or transitions associated with the active ion may become important factors in pulse formation. Similarly, the ability to maintain a high power density along the fibre, combined with the long fibre length, means that there is a long interaction length for non-linear processes, such as self-phase modulation [3], to build up which may assist in pulse production. The various dispersive and non-linear processes that can contribute to mode-locking will be discussed in later sections.

In order to exploit both the non-linear and dispersive properties of the optical fibre, pulse formation has to be initiated in the fibre laser and, as already mentioned, this can be achieved through either amplitude or phase modulation. The next two sections give a brief outline of the physical concepts involved in the mode-locking process.

9.2.2 Loss modulation

The basic concept of loss modulation can be understood with either time-domain or frequency-domain arguments. If a mode exists with its frequency v_0 at the peak of the gain curve and it propagates through a loss modulator operating at a frequency f_m, sidebands are generated at a frequency given by $v_0 \pm f_m$. If f_m matches the mode spacing of the cavity, the modes adjacent to the lasing mode are coupled together with a well-defined phase and amplitude relationship.

Alternatively, in the time domain, the process can be viewed as follows. When the modulation period is matched to the cavity round-trip time, light which is incident at one particular point in the modulation cycle will be incident at the same point after one round trip of the cavity. Thus any light that suffered a loss after one round trip will always experience a loss. In fact, all the light in the cavity will experience a loss except that which passes through the modulator at the zero loss point.

Following an analysis made by Kuizenga and Siegman for loss modulation in a homogeneously broadened gain medium [2], the pulse duration in a steady state, $t_p(AM)$, with no frequency chirp, is given by

$$t_p(AM) = \frac{(2\ln 2)^{1/2}}{\pi} \left(\frac{1}{f_m \delta_{am} \Delta v}\right)^{1/2} (g_{fo}L)^{1/4} \qquad (9.2)$$

with f_m the modulation frequency, Δv the gain bandwidth, g_{fo} the saturated gain coefficient, L the active length of the gain medium and δ_{am} the amplitude modulation index which is related to the single-pass amplitude transmission of the modulator $m(t)$ through

$$m(t) = \cos(\delta_{am} \sin 2\pi f_m t). \tag{9.3}$$

9.2.3 Phase modulation

An alternative way to induce the necessary communication between the modes is to use an electro-optic phase modulator. When light passes through a phase modulator it experiences a frequency shift which is proportional to the rate of change of phase with respect to time, unless it passes through the modulator at an extremum of the phase modulation cycle. At either of these two points no frequency shift is induced upon the light. Successive transits through the modulator result in a shifting of the light outside of the gain bandwidth of the active medium thus inducing a loss for that light. However, the presence of two points in the modulation cycle where the light experiences no frequency shift, leads to the production of two sets of pulses from the laser, one of which can be selected by optimising the cavity alignment.

For a Gaussian pulse operating in a steady-state regime Kuizenga and Siegman [2] have shown that the resultant pulse duration, $t_p(\text{FM})$, satisfies the relationship

$$t_p(\text{FM}) = \frac{(2\ln 2)^{1/2}}{\pi} \left(\frac{g_{fo}L}{\delta_{fm}}\right)^{1/4} \left(\frac{1}{f_m \Delta v}\right)^{1/2} \tag{9.4}$$

where g_{fo} is the saturated gain coefficient, L the length of the gain medium, Δv the gain bandwidth and δ_{fm} the single-pass phase retardation.

The driving signal applied to the phase modulator is often a sine wave and when the pulse arrives at the modulator the phase change appears quadratic in nature leading to the development of a frequency chirp, β_{fm}, across the pulse given by

$$\beta_{fm} = \pi^2 \Delta v f_m \sqrt{\frac{\delta_{fm}}{4g_{fo}L}}. \tag{9.5}$$

Both loss and phase modulation techniques have been used to mode-lock fibre lasers and pulses of the order of 10 ps have been generated. The average output powers associated with these pulses are of the order of 10 mW, thus peak powers of watts can be realised in the core of the fibre and these powers are sufficiently intense to induce non-linear effects in the optical fibre.

9.2.4 Optical non-linearities in fibres

When an electric field, $E(\mathbf{z}, t)$, of the form

$$E(\mathbf{z}, t) = \tfrac{1}{2}\hat{z}[\tilde{E}(\mathbf{z}, t)\exp -i\omega_0 t + \text{c.c.}] \tag{9.6}$$

(with c.c. the complex conjugate) is applied across a transparent material, the refractive index n can be written in terms of a linear component n_0 and a non-linear field-dependent term n_{2e} such that

$$n = n_0 + n_{2e}E^2(\mathbf{z}, t). \tag{9.7}$$

This field dependence can also be realised when an optical pulse traverses the material. In this case the refractive index is expressed as

$$n = n_0 + n_2 I(t) \tag{9.8}$$

with $I(t)$ the optical field intensity.

If the pulse propagates down an optical fibre of effective length l, a phase change will occur, with the non-linear component, $\Delta\phi$, given by

$$\Delta\phi = -kn_2 I(t) l \tag{9.9}$$

with k the propagation constant for the pulse.

As the intensity is a function of time, the pulse develops a frequency shift $\Delta\omega$ from the carrier frequency ω_0 such that

$$\Delta\omega = -kn_2 l \frac{dI(t)}{dt}. \tag{9.10}$$

The frequency components in the leading part of the pulse are down-shifted while those in the trailing portion are up-shifted and over the central portion of the pulse a linear frequency chirp evolves. Although the temporal profile of the pulse remains unchanged after propagation through the medium the spectral profile is broadened. This broadening and chirping of the pulse spectrum is referred to as self-phase modulation and the power density I_0, at which the spectrum is broadened by $\sqrt{2}$, is given by

$$I_0 = \frac{\lambda}{2\pi n_2 l}. \tag{9.11}$$

For light with a wavelength of 1 μm propagating along a 1 m length of fibre with a non-linear refractive index coefficient, n_2, of 3×10^{-16} cm^2/W [4], the power density, I_0, is 5.3×10^8 W/cm^2. Such power densities are attainable in optical fibres with modest average powers and quite broad pulses; for example, a pulse train of 10 ps pulses with a repetition frequency of 100 MHz and average power of 25 mW propagating down a 10 m length of fibre of core diameter 5 μm has an intrafibre power density of 1.2×10^9 W/cm^2, which is in excess of the onset power for self-phase modulation.

The refractive index of a material also exhibits a wavelength dependence which leads to the development of a group velocity dispersion. The propagation constant $\beta(\omega)$ for a pulse in a dispersive medium is given by

$$\beta(\omega) = n(\omega)k \tag{9.12}$$

where k is the free space propagation constant and $n(\omega)$ is the frequency-dependent refractive index. The dispersive effects can be obtained by expanding the propagation constant β as a Taylor series about ω_0, the centre frequency of the pulse, such that

$$\beta(\omega) = \beta_0 + \beta'(\omega - \omega_0) + \tfrac{1}{2}\beta''(\omega - \omega_0)^2 \tag{9.13}$$

where the primes refer to the first and second derivatives of β with respect to frequency. β' gives the group velocity of the pulse while β'' gives the group velocity dispersion of the material, which is related to the material dispersion $\partial^2 n/\partial \lambda^2$, by

$$\beta'' = \frac{\partial^2 \beta}{\partial \omega^2} \approx -\frac{\lambda^3}{2\pi c^2} \frac{\partial^2 n}{\partial \lambda^2} \tag{9.14}$$

where λ is the central wavelength of the pulse and c the speed of light; β'' can also be expressed in terms of the dispersion parameter of the fibre D, such that

$$D = \frac{\partial^2 \beta}{\partial \omega^2} \frac{2\pi c}{\lambda^2} \equiv -\frac{\lambda}{c} \frac{\partial^2 n}{\partial \lambda^2}. \tag{9.15}$$

It is well known that there is a wavelength (see, for example, ref. [3]), the zero dispersion wavelength, where the material dispersion $\partial^2 n/\partial \lambda^2$, changes sign and this has interesting repercussions on pulse propagation down an optical fibre. When a pulse propagates in a dispersive medium, such as an optical fibre, it can be shown [5] that the pulse experiences both a phase and a temporal change, the magnitudes of which are related to $\partial^2 \beta/\partial \omega^2$. For positive values of $\partial^2 \beta/\partial \omega^2$, the pulse acquires an up-chirp while for negative values a down-chirp is introduced across the pulse spectrum.

The combined effects of self-phase modulation and group velocity dispersion on pulse propagation in the fibre results in two different regimes, depending on the sign of the group velocity dispersion. If a pulse has a wavelength shorter than the dispersion minimum wavelength and a peak power sufficient to induce a strong self-phase modulation, the up-chirp arising from the group velocity dispersion combines with the up-chirp from the self-phase modulation to linearise the frequency up-chirp across the pulse. The temporal profile of the pulse is also altered: the pulse broadens and the peak of the pulse becomes square [6]. If the pulse is then sent through a negatively dispersive delay-line pulse compression can be achieved where the up-chirp arising from the optical fibre is balanced by the down-chirp from the dispersive delay line [7].

Conversely, if the central wavelength of the pulse is longer than the zero dispersion wavelength the chirp induced by group velocity dispersion is opposite to that generated by self-phase modulation. Thus the fibre itself can act as its own compensating medium, removing the need for an external delay line [8]. If the up-chirp induced by self-phase modulation is compensated for by the down-chirp arising from group velocity dispersion, a soliton can be generated which will then propagate down the fibre without change in its profile or phase. The existence of solitons in an optical fibre was first pointed out by Hasegawa and Tappert [9]. In order to generate a soliton from a pulse of duration t_p and wavelength λ, there is both a critical power level P_s and an optimum fibre length, Z_0, known as the soliton period. These two quantities are defined as

$$Z_0 = 0.322 \frac{\pi^2 c t_p^2}{\lambda^2 D} \qquad (9.16)$$

$$P_s = 0.776 \frac{\lambda^3 D A}{\pi^2 c n_2 t_p^2} \qquad (9.17)$$

with D, the dispersion parameter of the fibre, c the speed of light, n_2 the non-linear refractive index of the material and A the effective core area of the fibre.

If the input power is an integer multiple of the fundamental soliton power then higher order solitons may exist. If a higher order soliton passes down a fibre, the pulse shape changes depending on the distance propagated along the fibre, but the pulse always returns to its original profile at an integer multiple of the soliton period Z_0.

The frequency dependent nature of the refractive index of the optical fibre core is not the only source of dispersion in the fibre. In fact the electronic transitions associated with the laser cycle of the active ion may lead to the introduction of an atomic dispersion. If the laser transitions occur off resonance then a frequency chirp may be imposed on the pulse spectrum as it propagates through the fibre gain medium. This frequency chirp arises from the non-linear variation of the real part of the electric susceptibility with frequency which is related to the absorption and emission line shapes of the gain medium through a Kramers–Kronig relationship. The sign and magnitude of the frequency chirp depends on the detuning of the transition frequency from the resonance frequency of the atomic line shape, and the presence of this atomic dispersion can substantially modify the overall dispersive properties of the fibre and thus affect the pulse-forming processes occurring in the fibre.

9.2.5 *Mode-locking of doped fibres with bulk modulators*

The high gain presented by a fibre laser allows the introduction of bulk components into the cavity without affecting overall performance of the laser. This has enabled the mode-locking of a variety of doped fibres with both bulk acousto-optic and phase modulators [10–13]. The cavity configuration used typically is illustrated in Figure 9.1. The pump light is coupled through the mirror M_1 by the microscope objective MO_1 into the fibre which is butted up to the mirror. The laser radiation emitted from the output end of the fibre is collected by the microscope objective MO_2 and focused onto the output

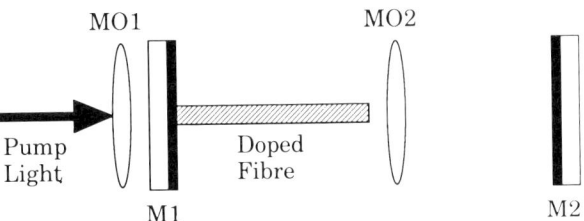

Figure 9.1 A generalised fibre laser cavity.

coupler M_2. The modulator, tuning elements and polarisation-preserving devices can then be inserted in the air space between the microscope objective, MO_2, and the output coupler M_2. With a cavity similar to that just described with a bulk acousto-optic modulator and a Nd^{3+}-doped fibre as the gain medium, Alcock *et al.* [10] have generated pulses as short as 1 ns (detector limited) with pulse energies of 17 pJ at a repetition frequency of 20.73 MHz.

One major drawback with this form of cavity is the presence of subcavities. The high optical gain presented by a fibre laser, which allows the introduction of bulk components into the laser cavity, does present problems. If the pump power is sufficiently high laser action can be observed from the cleaved facets of the fibre, or from light which is fed back into the fibre from cavity optics, such as reflections off the microscope objectives. These oscillations can be prevented by either depositing the dielectric mirror coating onto the cleaved ends of the fibre, wedging the ends of the fibre or by immersing the cleaved end of the fibre in index-matching fluid. With this latter technique Phillips and coworkers have generated pulses of 50 ps duration from an acousto-optically mode-locked Nd^{3+}-doped fibre laser [11].

The low pump power thresholds observed for fibre lasers combined with the fact that both Nd and Er ions exhibit absorptions around 800 nm results in fibre lasers being ideal candidates for diode laser pumping. This allows the configuration of compact laser systems with high electrical conversion efficiencies. Duling *et al.* [12] have demonstrated the acousto-optic mode-locking of a diode-pumped Nd^{3+} fibre laser. This laser generated pulses as short as 120 ps (detector limited) with 5 mW of average power.

As outlined in Section 9.2.3, active mode-locking can also be achieved using a phase modulator to engender the necessary modal communication. Phillips *et al.* have used a bulk $LiNbO_3$ phase modulator to generate bandwidth limited pulses as short as 20 ps from a Nd^{3+} fibre [11], as illustrated in Figure 9.2. The main advantage of using a phase modulator comes from the interaction of the pulse spectrum with the gain medium. The authors found that the pulses obtained from the laser were shorter than the value predicted by the Kuizenga and Siegman analysis (see Equation (9.4)) and optimum pulse formation occurred when there was a frequency mismatch between the cavity mode-spacing (see Equation (9.1)) and the modulation frequency. These

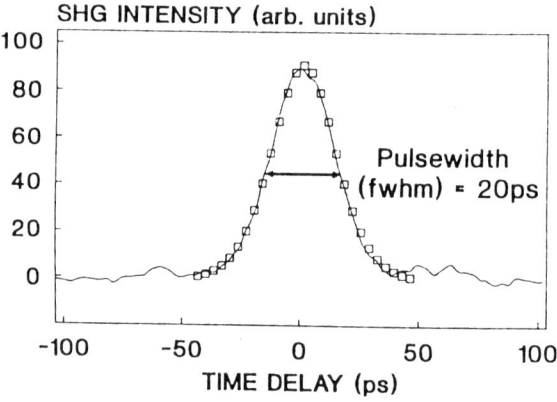

Figure 9.2 An auto-correlation of a 20 ps pulse from an FM mode-locked Nd^{3+}-doped fibre laser (taken from ref. [11]).

discrepancies were interpreted as follows. When the cavity length is detuned from the matched condition the pulse spectrum is frequency shifted away from the line centre of the emission band (see Section 9.2.3). This results in the production of an atomic dispersion [14] which is both greater than and opposite in sign to the material dispersion. This induces a down-chirp across the pulse, which when combined with the up-chirp arising from both self-phase modulation and the group velocity dispersion of the fibre, allows solitonic processes to assist in the mode-locking dynamics. The frequency shift necessary to generate the atomic dispersion does not occur when an acousto-optic modulator is used to mode-lock a fibre laser, and as a result solitonic pulse shaping does not occur.

As discussed in Section 9.2.4, the group velocity dispersion of an optical fibre changes sign at a wavelength of 1.32 μm, and solitons can be generated beyond this wavelength [15]. Erbium-doped fibres have been identified as viable sources of laser radiation at 1.55 μm [16] and it should be possible to use the solitonic pulse-shaping processes associated with the fibre to enhance the mode-locking kinetics of the actively mode-locked system. To this end Hanna et al. have mode-locked a $Yb^{3+}:Er^{3+}$ fibre laser with a bulk phase modulator to produce pulses as short as 37 ps at a wavelength of 1.55 μm [13]. These pulses were not quite bandwidth limited, indicating that solitonic shaping was not dominating the pulse-formation process and it is thought that the complex nature of the Er ion lasing transition may introduce dispersion features which can change sign depending on the operating conditions [17].

The mode-locking of an Er^{3+} fibre laser with a bulk phase modulator has recently been demonstrated [18]. When the fibre laser was mode-locked and the lasing bandwidth was restricted using a birefringent filter, pulses of ~35 ps duration with 100 mW output power were recorded. Removal of the tuning element combined with a reduction in the pump power to the fibre laser resulted in the generation of pulses as short as 2 ps with a time-bandwidth product close to that associated for sech^2 pulse profiles, again indicating that solitonic kinetics are assisting in the mode-locking process.

9.2.6 Mode-locking of doped fibre lasers with integrated devices

Although mode-locking has been demonstrated from a variety of fibre lasers using bulk modulators, the use of integrated optical devices for certain applications, such as telecommunications, to produce the necessary phase coupling between the longitudinal modes is advantageous. It has been demonstrated that $LiNbO_3$ is a suitable material for the construction of both bulk and integrated modulators. Several doped fibre lasers have been mode-locked using integrated $LiNbO_3$ devices [19–23]. For example, Geister and Ulrich have demonstrated the FM mode-locking of a Nd^{3+}-doped fibre laser with a $Ti:LiNbO_3$ phase modulator with the production of pulses ranging from 25 to 50 ps [20].

Pulse production at 1.55 μm from ytterbium:erbium-doped optical fibres has been demonstrated with bulk phase modulators [16, 18]. As doped fibres are compatible with standard telecommunications fibre it should be possible to configure a pulse source, based on an integrated modulator, which is compact and operates at the minimum loss wavelength of the telecommunications fibre. Kafka and coworkers [21] used a fibre ring laser consisting of a 70 m length of Er^{3+}-doped fibre spliced to a length of telecommunications fibre. The ring was completed by the inclusion of an electro-optic amplitude modulator and dichroic coupler as illustrated in Figure 9.3. Once the pump power coupled into the fibre laser and the length of the telecommunications fibre were optimised, pulses as short as 4 ps were generated from the laser with 1 mW average output power, as illustrated in

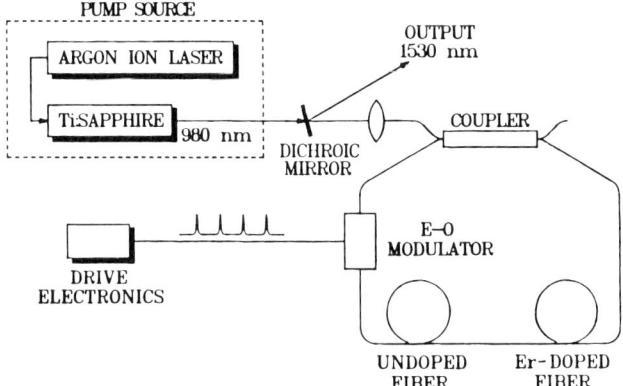

Figure 9.3 The cavity configuration used to AM mode-lock an Er^{3+}-doped fibre laser with an integrated electro-optic modulator (taken from ref. [21]).

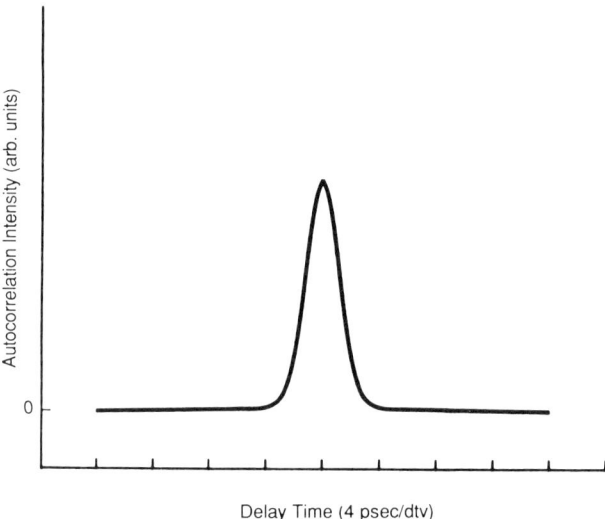

Figure 9.4 An auto-correlation of a 4 ps pulse produced by the laser illustrated in Figure 9.3 (taken from ref. [21]).

Figure 9.4. Again the authors believe that solitonic pulse-shaping processes are assisting in the mode-locking kinetics of the laser as the output power and pulse duration are close to those predicted by Equations (9.16) and (9.17). Mode-locking was also demonstrated at a repetition rate of 50 kHz, with pulses with durations of 500 ps and peak powers in excess of 100 W being generated.

Similar studies on amplitude modulation of Er^{3+} fibres with integrated modulators have been undertaken by Schlager and colleagues [22] who generated 18 ps pulses at 40 MHz from a ring fibre laser and then compressed the pulses to 10 ps by propagating them along a length of standard telecommunications fibre.

The integrated electro-optic amplitude modulators described so far have been operated at modulation frequencies ranging from 50 to 100 MHz. For telecommunication purposes higher operating frequencies would be more favourable. The low-gain cross-sections associated with optical fibres means that long lengths of fibre have to be used to provide significant gain. For example, typical fibre lengths are of the order of 1 m, which corresponds to a fundamental cavity mode frequency spacing of 100 MHz. In order to operate at higher modulation frequencies harmonic mode-locking has to be used. To this end Takuda and Miyazawa [23] have mode-locked an Er^{3+}-doped fibre at a frequency of 30 GHz with an integrated $Ti:LiNbO_3$ amplitude modulator and generated pulses as short as 8 ps.

A severe limitation placed on the use of integrated optic devices fabricated out of $LiNbO_3$ arises from the refractive index mismatch between the fibre core and the $LiNbO_3$ [19, 21]. This is large and, as a consequence, high insertion losses are encountered when the device is butted up to or coupled to the optical fibre. This tends to reduce the overall performance of the laser and so minimises the advantages obtained from miniaturisation. An alternative approach is to use a fibre phase modulator [24] which modulates the phase of the intracavity flux while it is still in the fibre. Using such a device Phillips et al. [25] have been able to generate pulses with duration of ~80 ps (detector limited) at a frequency of 417 MHz from a Nd^{3+}-doped fibre laser. Although this device has allowed the mode-locking of a fibre laser several problems are encountered with its operation. Firstly, the modulator has to be placed close to the output mirror of the fibre laser to ensure that effective mode-coupling can occur. Thus the output mirror has to be butted up to the end of the fibre. This means that no cavity length control can be included in this laser, and to match the cavity length to the modulation frequency the fibre had to be cut to the correct length with 0.5 mm accuracy [25]. A second problem encountered with the fibre phase modulator scheme is that there is no bandwidth restricting element in the cavity. As a result the time-bandwidth product measured for this laser was well in excess of that expected for bandwidth limited operation. This problem can be overcome by using an external grating to restrict the bandwidth, and with this technique Phillips and coworkers have generated pulses of 80 ps duration (detector limited) with time-bandwidth products of 0.8 [26].

Recently the mode-locking of an Er^{3+} fibre laser with a modulated diode laser acting as the mode-locker has been demonstrated [27]. This system produced pulses that were 4 ps in duration, but in order to prevent damage to the laser diode the average power of the Er^{3+} fibre laser has to be restricted to ~4 mW.

9.3 Q-switching of optical fibre lasers

9.3.1 *Introduction*

When operated in a continuous-wave fashion, fibre lasers are capable of generating output powers of the order of tens to hundreds of milliwatts. For some applications pulses with much higher peak powers and energies are desirable, and although mode-locking may produce pulses with high peak powers these pulses are not very energetic. If the fibre is Q-switched it is possible to generate pulses with both high peak powers and energies.

The operation of a Q-switched laser can be understood as follows. The Q of an optical cavity is defined as the ratio of energy stored in the cavity to that lost per round trip. By varying the Q of a cavity in a well-defined manner, pulses may be generated. Pulse formation occurs in the following way. Initially the losses of the cavity are set to an abnormally high value, reducing the Q of the cavity, and the gain medium is excited. As the losses are high the laser is unable to reach threshold. This allows the population inversion to build up to a value well in excess of the value that would be achieved under normal operating conditions. The Q of the cavity is then switched to a value where the round trip gain exceeds the losses of the cavity. The light in the cavity, present due to spontaneous emission, then builds up into a rapidly rising burst which quickly saturates the gain. The gain saturation is so strong that it drives the inverted population well below the threshold population level, thus inhibiting laser action again. It is this rapid rise and fall of the intracavity flux which forms the Q-switched pulse.

In general, the output pulse duration is much shorter than the pumping interval over which the population inversion was created and, as a result, the peak power of the Q-switched pulse is much greater than that attained from a continuous-wave laser operating under the same pumping conditions.

From simple rate equation analysis [28] the energy, E, and duration, t_p, of a Q-switched pulse can be found from

$$E = \frac{Vh\nu(n_i - n_f)}{\psi} \cdot \frac{\ln(1/R)}{\ln(1/R) + L} \qquad (9.18)$$

and

$$t_p = t_c \left\{ \frac{n_i - n_f}{n_i - n_t[1 + \ln(n_i/n_t)]} \right\}. \qquad (9.19)$$

where n_i is the inverted population prior to opening of the Q-switch, n_t is the threshold population, n_f is the inverted population after Q-switch pulse has formed, R is the reflectivity of the output coupler of the laser, L the intrinsic losses of the fibre, V the volume of the gain medium, ν the lasing frequency, h Planck's constant, t_c the photon lifetime in the cavity, and ψ the ratio of the absorption and emission line shapes.

It can be seen that both the duration and energy of the pulse are determined by the population n_i at the opening time of the Q-switch device. Thus, in order to obtain an energetic pulse from the system the difference between n_i and n_t must be large. In this respect rare-earth doped optical fibres are ideal because they have small-gain cross-sections together with long-lived upper state lifetimes. This means the excited-state population does not decay via spontaneous emission before the Q of the cavity is altered, resulting in large values of n_i. The energy of the pulse is related to the intrinsic losses in the cavity and it has been demonstrated that these losses are extremely small [29]. Hence, it should be possible to exploit these features to produce energetic pulses from the laser.

Also the pulse duration depends on the photon lifetime in the cavity t_c, which in turn is governed by the length of the optical fibre. As fibre lasers can be constructed from long lengths of fibre the possibilities exist for the generation of energetic broad pulses with high peak powers per pulse.

9.3.2 Q-switched and Q-switched mode-locked operation of fibre lasers with bulk devices

The standard cavity used for the Q-switching of a fibre laser is similar to that illustrated in Figure 9.1 except the modulator is replaced by a Q-switching device. Using an acousto-optic Q-switch Alcock *et al.* [30] were able to generate pulses of 200 ns duration with peak powers of 8 W at a frequency of 100 Hz from a Nd^{3+} fibre laser, as depicted in Figure 9.5. Using a mechanical chopper to vary the Q of the cavity they were also able to generate pulses that had durations of 450 ns with peak powers per pulse of 4 W. A bulk acousto-optic Q-switch has also been used to generate pulses from an Er^{3+}-doped fibre laser [31]. When Q-switched at 200 Hz, pulse durations ranging from 60 to 100 ns with peak powers close to 2 W were produced.

The ability to match the emission wavelength of low-power diode lasers to the absorption band of Nd^{3+} ions in a silica host has allowed the Q-switching of a diode-pumped Nd^{3+} doped fibre laser [32]. In this case pulses with durations of 300 ns with peak powers of 300 mW at frequencies up to 4 kHz were routinely obtained, indicating that efficient pulsed operation from these devices is possible.

Simultaneous Q-switching and mode-locking has also been demonstrated in Nd^{3+}-based fibre systems. Alcock *et al.* [10] have shown that by inserting a mechanical chopper into the cavity of a mode-locked fibre laser, a mode-locked and Q-switched pulse train may be obtained. The Q-switched envelope had a duration of 690 ns and the pulses in the envelope had durations of < 3 ns (detector limit). The energy associated with each pulse at the peak of the Q-switched envelope was estimated to be 20 nJ. The authors believed that the pulses may not have been much shorter than the value recorded on the

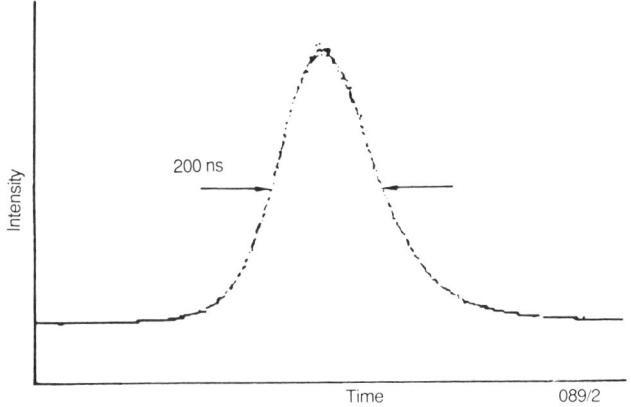

Figure 9.5 An example of the Q-switched pulses generated by a Q-switched Nd^{3+}-doped fibre laser (taken from ref. [30]).

photodiode because the open time of the Q switch was not long enough to allow sufficient transits of the cavity for effective pulse formation to occur.

9.3.3 Q-switched and Q-switched mode-locked fibre lasers with integrated optic devices

The development of integrated optical devices (see Chapter 5) has allowed the production of compact amplitude and phase modulators. Recently, Geister and Ulrich [19] have shown that an integrated Q-switch device can be used to generate pulses from a fibre laser. A Ti:$LiNbO_3$ chip was fabricated which had both a Q switch and a phase modulator on it, to allow simultaneous Q-switching and mode-locking to take place. Again the $LiNbO_3$ chip was butt-coupled to the output end of the fibre introducing high insertion losses, as outlined in Section 9.2.6, and to minimise the possibility of étalon effects occurring at the butt, the ends of both the fibre and the chip were wedged. When operated in a Q-switched mode, pulses with 100 ns duration were observed with pulse energies of 0.16 µJ. Pulse repetition frequencies ranging from 0.1 Hz to 9 Hz were used. Simultaneous mode-locked Q-switched operation resulted in the production of pulses with durations of ∼ 50 ps under a Q-switched envelope of 300 ns. The energy per pulse was estimated at 10 nJ corresponding to a peak power of 200 W [20].

The above schemes used for both bulk and integrated optic Q-switching are all active in that an external field is applied to the Q switch to induce the necessary pulse formation. Recently, however, Zenteno et al. [33] have demonstrated that Q-switching of a double cladding Nd^{3+} fibre laser [34] can be accomplished using a passive Q switch. In this case the Q switch was a saturable dye embedded in an acetate sheet ∼ 125 µm thick. The duration of Q-switched pulses originating from this laser was between 600 and 800 ns with an estimated energy of 0.8 µJ per pulse. This corresponds to a peak power of 8.75 W per pulse. The authors observed that the Q-switched pulse was structured, and this was attributed to the mode-locking of the fibre due to the fast recovery time of the Q-switch dye. Mode-locked pulses with durations of ∼ 5 ns were observed. The duration of these pulses was limited by the lifetime of the dye, which was ∼ 3 ns. By lengthening the fibre to 23 m, the mode-locked pulses disappeared. This was thought to be a consequence of amplified spontaneous emission which competes for, and saturates, the gain available for short-pulse formation. Thus by using a passive Q-switch device the authors have demonstrated that both Q-switched and mode-locked operation from a fibre laser may be attained.

9.4 Conclusions

The results outlined in this chapter indicate that fibre lasers are viable optical pulse sources. Little attention, however, has been paid to the physical kinetics involved in the pulse-formation processes. For example, the analysis presented by Kuizenga and Siegman [2], which predicts the pulse durations for mode-locked bulk lasers, may not be a suitable model for fibre lasers in which optical non-linearities, such as self-phase modulation, play an important role. In fact a clearer insight into the pulse-formation kinetics may only be elucidated when a rigorous solution to the non-linear Shrödinger equation is obtained.

Although mode-locking has been accomplished with both amplitude and phase modulation techniques little attention has been paid to coupled-cavity mode-locking. In a coupled-cavity mode-locked laser pulse formation can be enhanced or initiated with the introduction of a linear or non-linear element into a branch cavity [35–42]. The observation of self-starting laser action from a coupled-cavity laser has led to a re-evaluation of the conditions necessary to observe pulse formation from a coupled-cavity laser [43]. It is now thought that gain media which exhibit long upper state gain lifetimes combined with small gain cross-sections and broadband tuneability will be ideal candidates for coupled-cavity mode-locking. As outlined in Chapters 2 and 7, fibre lasers satisfy these requirements and so it may be possible to configure fibre lasers, using either bulk optical or integrated optical components, to exploit the enhanced mode-locking features offered by coupled-cavity techniques.

References

1. C.V. Shank, *Ultrashort Laser Pulses and Applications* (W. Kaiser, ed.), Chap. 2, Springer Verlag, Berlin (1988).
2. D.J. Kuizenga and A.E. Siegman, *IEEE J. Quant. Electron.* **QE-6** (1970) 694.
3. G.P. Agrawal, *Nonlinear Fibre Optics*, Academic Press (1989).
4. K. Tai, A. Hasegawa and N. Bekki, *Opt. Lett.* **13** (1988) 392.
5. S. De Silvestra, P. Laporta and O. Svelto, *IEEE J. Quant. Electron.* **QE-20** (1989) 533.
6. W.J. Tomlinson, R.H. Stolen and C.V. Shank, *J. Opt. Soc. Amer.* **B1** (1984) 139.
7. C.V. Shank, R.L. Fork, R. Yen, R.H. Stolen and W.J. Tomlinson, *Appl. Phys. Lett.* **42** (1982) 761.
8. L.F. Mollenauer, R.H. Stolen and J.P. Gordon, *Phys. Rev. Lett.* **45** (1980) 1095.
9. A. Hasegawa and F. Tappert, *Appl. Phys. Lett.* **23** (1973) 142.
10. I.P. Alcock, A.I. Ferguson, D.C. Hanna and A.C. Tropper, *Electron. Lett.* **22** (1986) 268.
11. M.W. Phillips, A.I. Ferguson and D.C. Hanna, *Opt. Lett.* **14** (1989) 21.
12. I.N. Duling III, L. Goldberg and J.F. Weller, *Electron. Lett.* **24** (1988) 1333.
13. D.C. Hanna, A. Kazer, M.W. Phillips, D.P. Shepherd and P.J. Suni, *Electron. Lett.* **25** (1989) 96.
14. A.E. Siegman, *Laser*, Chap. 27, University Science Books (1986).
15. L.F. Mollenauer, R.H. Stolen, J.P. Gordon and W.J. Tomlinson, *Opt. Lett.* **8** (1983) 289.
16. R.S. Mears, L. Reekie, S.B. Poole and D.N. Payne, *Electron. Lett.* **22** (1986) 159.
17. E. Desurvivre and J.R. Simpson, *J. Light. Technol.* **7** (1989) 835.
18. K. Smith, J.R. Armitage, R. Wyatt, N.J. Doran and S.M.J. Kelly, *Electron. Lett.* **26** (1990) 1149.
19. G. Geister and R. Ulrich, *Opt. Commun.* **68** (1988) 187.
20. G. Geister and R. Ulrich, *Appl. Phys. Lett.* **56** (1990) 509.
21. J.D. Kafka, T. Baer and D.W. Hall, *Opt. Lett.* **14** (1989) 1269.
22. J.B. Schlager, Y. Yamubayashi, D.L. Franzen and R.I. Juneau, *IEEE Photonics Technol. Lett.* **1** (1989) 264.
23. A. Takuda and H. Miyazawa, *Electron. Lett.* **26** (1990) 216.
24. D.B. Patterson, A.A. Godil, G.S. Kino and B.T. Khuri-Yakub, *Opt. Lett.* **14** (1989) 248.

25. M.W. Phillips, A.I. Ferguson, G.S. Kino and D.B. Patterson, *Opt. Lett.* **14** (1989) 680.
26. M.W. Phillips, A.I. Ferguson and D.B. Patterson, *Opt. Commun.* **75** (1990) 33.
27. D. Burns, W. Sibbett and R.A. Baker, *Proc. Conf. Lasers and Electro-Optics Anaheim, California* (1990), Paper CFR5.
28. W. Koechner *Solid State Laser Engineering*, Chap. 8, Springer Verlag, Berlin (1988).
29. D.C. Hanna, R.G. Smart, P.J. Suni, A.I. Ferguson and M.W. Phillips, *Opt. Commun.* **68** (1988) 128.
30. I.P. Alcock, A.C. Tropper, A.I. Ferguson and D.C. Hanna, *Electron. Lett.* **22** (1986) 84.
31. R.J. Mears, L. Reekie, S.B. Poole and D.N. Payne, *Electron Lett.* **22** (1986) 159.
32. I.M. Jauncey, J.T. Lin, L. Reekie and R.J. Mears, *Electron. Lett.* **22** (1986) 199.
33. L.A. Zenteno, H. Po and N.M. Cho, *Opt. Lett.* **15** (1990) 115.
34. H. Po, E. Snitzer, R. Rumminelli, L.A. Zenteno, F. Hakimi and N.M. Cho, *Digest of Optical Fibre Conference* (1989), Paper PD7.
35. L.F. Mollenauer and R.H. Stolen, *Opt. Lett.* **9** (1984) 13.
36. F.M. Mitschke and L.F. Mollenauer, *IEEE J. Quant. Electron.* **22** (1986) 2242.
37. K.J. Blow and D. Wood, *J. Opt. Soc. Amer.* **B5** (1988) 629.
38. K.J. Blow and B.P. Nelson, *Opt. Lett.* **13** (1988) 1026.
39. P.N. Kean, X. Zhu, D.W. Crust, R.S. Grant, N. Langford and W. Sibbett, *Opt. Lett.* **14** (1989) 39.
40. J. Mark, L.Y. Liu, K.L. Hall, H.A. Haus and E.P. Ippen, *Opt. Lett.* **14** (1989) 48.
41. J. Goodberlet, J. Wang, J.G. Fujimoto and P.A. Schulz, *Opt. Lett.* **14** (1989) 1125.
42. P.M.W. French, S.M.J. Kelly and J.R. Taylor, *Opt. Lett.* **15** (1990) 378.
43. E.P. Ippen, L.Y. Lui and H.A. Haus, *Opt. Lett.* **15** (1990) 183.

10 Future directions

C.A. MILLAR

10.1 Introduction

The research and development effort into fibre lasers and amplifiers described in the preceding chapters has been driven largely by the interest in optical telecommunications applications. Historically, advances in optical fibre communications systems have been associated with the need and the desire to overcome the attenuation in the signal path between the transmitter and the receiver, as well as the search to find means to reduce or circumvent pulse dispersion. These pursuits have gone hand-in-hand with progress towards the realisation of laser sources that have large output powers and optimum spectral properties, while receivers have become more responsive to faster data rates. It is clear that the introduction of fibre lasers and amplifiers could be seen as an inevitable occurrence along the route of this historical progression. However, to get a feel for their influence on the likely future direction of optical communications, it may be beneficial to answer the question 'Why the interest and activity in this field *now*?'

While recognising the outstanding contributions of the pioneering workers pre-1985 [1-6], it is nonetheless apparent that the technology did not attract the great level of interest it currently enjoys until Poole and coworkers [7] incorporated rare-earth ions into telecommunications-grade single-mode silica fibre in 1985. Rather than deriving from the imperatives of the laser physicist, this breakthrough came from fibre-makers in the Optical Fibre Group at Southampton University in the UK, who, with remarkable foresight, realised that the fundamental loss limits of the silica fibre transmission medium had been achieved and would probably not be bettered in any other fibre material in the 1300 or 1500 nm telecommunications 'windows'. The search for a fibre-based amplifier or laser source had begun in earnest, because it was realised that 'passive' fibre technology had reached a fundamental performance limit that could be surpassed using optical amplification directly in the fibre. Coincidentally, with semiconductor laser sources providing launched power levels limited at around 1 mW, and with opto-electronic receiver sensitivities improving only relatively slowly, it was recognised by the Southampton Group that optical system performance as a whole would have advanced at a relatively slow rate – or indeed may even have stagnated.

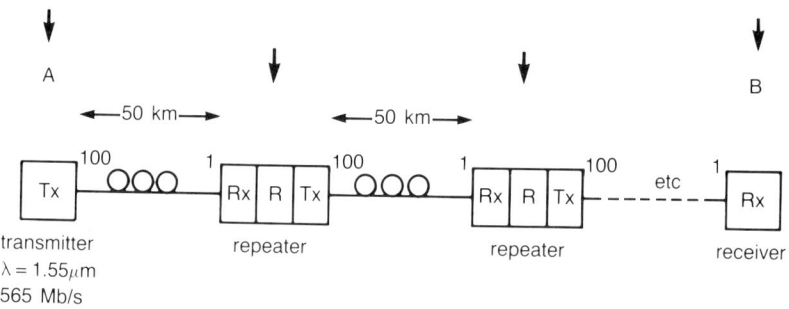

Figure 10.1 Typical state-of-art 1550 nm window optical AM system. R denotes electronic regeneration circuit. ↓ denotes electrical power feed. Numbers denote typical signal power levels in microwatts.

A purpose of this chapter is to demonstrate that the impact of active fibre technology will indeed have a profound effect on our local and global optical communication capabilities. As a benchmark against which to judge the influence of the new devices described in this book, consider that a state-of-the-art installed point-to-point direct detection (AM) communication system in the latter part of the 1980s would typically (perhaps even optimistically) have the performance attributes shown in Figure 10.1. Typical laser-launched powers at 1550 nm would be around 0.1 mW, giving, at the receiver sensitivities at a bit-rate of 565 Mbit/s, a maximum system span between optical–electrical–optical regeneration of about 50 km. Electrical power feeding for the regenerators would be required, either locally or remotely, if the latter was a submarine cable. Electronic signal reprocessing at these data-rates would be demanding and costly, as the equipment would have to be extremely reliable in such strategic and demanding environments. The technologically viable alternative to active fibre devices – semiconductor laser travelling-wave optical amplifiers – although under intense development since the beginning of the 1980s, would not have seen field service until the mid-1990s. It was therefore considered that long-haul optical communications systems would have advanced only slowly beyond this performance plateau, and that breaking the fundamental constraint of a stagnating power margin (that is, the margin in the difference between the launched power and the receiver sensitivity, taking into account all the transmission path losses) was necessary to take the subject to another realm of capability. This realisation spread to other laboratories soon after the early Southampton University publications, and activity mushroomed in the research establishments of major telecommunications operators in 1987, particularly after Mears [8] provided the first indication of the superior performance of an erbium silica fibre amplifier at 1536 nm.

From a communications viewpoint, the impact of the work on fibre lasers and amplifiers since 1985 should not be understated. Breaking the loss barriers

FUTURE DIRECTIONS 231

has been the historical reality of fibre communications research – for example, the 20 dB/km fibre loss barrier, SELFOC fibre, CVD single-mode fibre, coherent detection techniques, etc. – each stage having brought about dramatic increases in the ability to communicate further or with larger information capacity than before. Optical amplification in fibres is part of this successful trend, but with one additional important factor. We are on the verge of a new revolution in communications, taking the first steps towards an *all-optical* network concept, with electronics being reduced or eliminated in the transmission path. This may radically alter communication network architectures, and have a knock-on effect in the way in which the switching and processing nodes of such networks function. The active fibre devices at the heart of the technology discussed between the covers of this book will be the key elements in this process.

Although fibre travelling-wave amplifiers will have this important influence on the future development of optical telecommunications, the unique properties of fibre lasers should also find application in advanced fibre systems. For example, generating bandwidth transform-limited pulses from mode-locked fibre lasers could provide high-power ultrafast sources (possibly diode-laser pumped) for soliton transmission systems. Advantageous thermal properties, coupled with homogeneously broadened absorption and emission bands across a wide spectral range, contribute to new laser possibilities in spectroscopy, medicine and sensors. Indeed, in the sensor field, the importance of active fibre sensors with gain or enhanced backscatter properties has hardly been addressed. The injection of research interest inspired by the telecommunications imperative has promoted interest in new materials for the fibre laser host glass, leading to entirely new laser systems (for example, ZBLAN in Chapter 8). These lasers operate from the blue region via up-conversion processes, through the normal communications windows between 1 and 2 μm, and into the mid-IR for sensors and communications between 2 and 2.7 μm, medical lasers at 2.8 μm, and free-space communications at 3 μm.

The impact of the work on fibre lasers will therefore ripple outwards from the central communications application to affect many branches of optical science and technology. Predicting the future in such a wide and rapidly developing field is hazardous and ultimately will be overtaken by time and hindsight. Nevertheless, the reader is presented with a view of the impact that fibre lasers might have in several application areas, the most assured being telecommunications.

10.2 Fibre amplifiers – the impact on optical communications

As described in Chapter 6, the fibre amplifier is characterised by excellent figures of merit for gain and noise, high saturation output power, linear behaviour in the small-signal regime, cross-talk immunity in multichannel operation, low polarisation sensitivity, wide optical bandwidth and inherent compatibility with the fibre of the transmission system. The linearity of fibre amplifiers in the small-signal regime allows their use in digital or analogue systems, the latter being very important in video distribution systems using microwave multiplexing techniques based on direct modulation of laser sources [9]. The glass active medium is robust and immune to optical damage and shows no signs of ageing. The reliability of the amplifier as an entire device equates largely to the reliability of the semiconductor laser pump source, which in the case of GaAlAs or InGaAsP technology is acceptably high. There seem few technical restrictions or disadvantages in the implementation of fibre amplifiers. With this impressive performance specification, where are they likely to be used in optical communications networks? There are four fibre amplifier configurations that offer distinctive benefits to a systems designer. Each will now be discussed in turn, and the impact of each option on the development of new opportunities will be emphasised.

10.2.1 Fibre power amplifiers

A significant difference between a fibre amplifier and a semiconductor laser amplifier is that the pump and signal propagate over long distances through the inverted medium, the background propagation loss for both fluxes being negligible. In the distinctive fibre case, amplifying material can be added *ad infinitum* to the device, which, if inverted, converts the pump energy into signal energy even although the medium is heavily saturated. Accumulation of gain ceases to be exponential in growth with distance, and the signal output power becomes linearly dependent on the pump power available to invert the ions. The usual definition of the 'saturation' signal intensity in an inverted medium is [10]

$$I_{sat} = \frac{h\nu_s}{\sigma_{21}\tau} \tag{10.1}$$

For the erbium ion in silica, the emission cross-section, σ_{21}, is approximately 0.5 picometers squared, and the fluorescence lifetime τ, is about 11 ms, giving a value of P_{sat} of 2 mW in a fibre core of effective core area of 80 square microns.

This term provides a fundamental material constant for the analysis of the propagating fields in the fibre amplifier, but should not be regarded as a term placing upper bounds on the saturation performance of the device. It is relatively straightforward to construct a pair of coupled differential equations describing the propagation of the signal and pump fields in a lossless fibre amplifier when contradirectionally pumped [11], as follows:

$$\frac{dQ}{dZ} = -\frac{\sigma_{13}}{\sigma_{21}} \cdot SQ \bigg/ \left[\left(1 + \frac{\sigma_{12}}{\sigma_{21}}\right)S + Q\right] \tag{10.2}$$

$$\frac{dS}{dZ} = -SQ \Big/ \left[\left(1 + \frac{\sigma_{12}}{\sigma_{21}}\right)S + Q\right] \quad (10.3)$$

where

$$S = \frac{P_{signal}\sigma_{21}\tau}{hv_s A_s} \quad (10.4)$$

$$Q = \frac{P_{pump}\sigma_{13}\tau}{hv_p A_p} \quad (10.5)$$

and

σ_{13} = pump absorption cross-section
σ_{12} = signal absorption cross-section.
σ_{21} = signal emission cross-section.

Equations 10.2 and 10.3 only hold in the limit where the signal flux is large, i.e.

$$S \gg 1.$$

The equations have a common denominator, so taking ratios, we find the relationship between the pump and signal in this regime is

$$\frac{dS}{dQ} = -\frac{\sigma_{21}}{\sigma_{13}}. \quad (10.6)$$

For clarity, let us assume that the pump and lasing fields are identical (a good approximation for amplifiers pumped at 1480 nm wavelength), giving, from Equations (10.6), (10.4) and (10.5),

$$P_{signal} = \text{constant.} \frac{\lambda_p}{\lambda_s} \cdot P_{pump}(\text{absorbed}). \quad (10.7)$$

It can be seen from Equation (10.7) that the signal power is linearly proportional to the absorbed pump power, and that the constant of proportionality (the extraction efficiency) is the ratio of the signal and pump photon energies. In practice, the fibre background loss, the internal inefficiency of the lasing process and imperfect field overlap terms, all tend to diminish the measured extraction efficiency.

A fibre amplifier operating in a heavily saturated regime is characterised by an extraction efficiency from the pump power to the signal power, which can be as large as 80% in practice [12]. Given that we have the potential to provide hundreds of decibels of single-pass small-signal gain using a fibre amplifier with a large number of inverting ions, *and* that we have sufficient pump power to invert them, the output power of a fibre amplifier is determined only by the availability of copious amounts of pump power and an efficient laser system. Massicott [13] has shown that an erbium–alumina–silica fibre power amplifier can give 30 dB gain for a 0.5 mW input signal level, giving a 500 mW signal output power level. The launched pump power required for this experiment was 1000 mW at a wavelength of 980 nm, derived from a

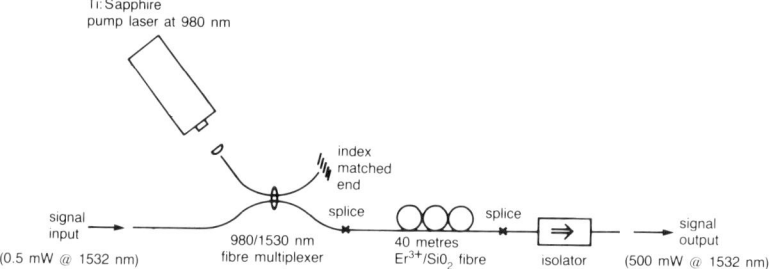

Figure 10.2 Layout of an experimental fibre power amplifier (after [13]).

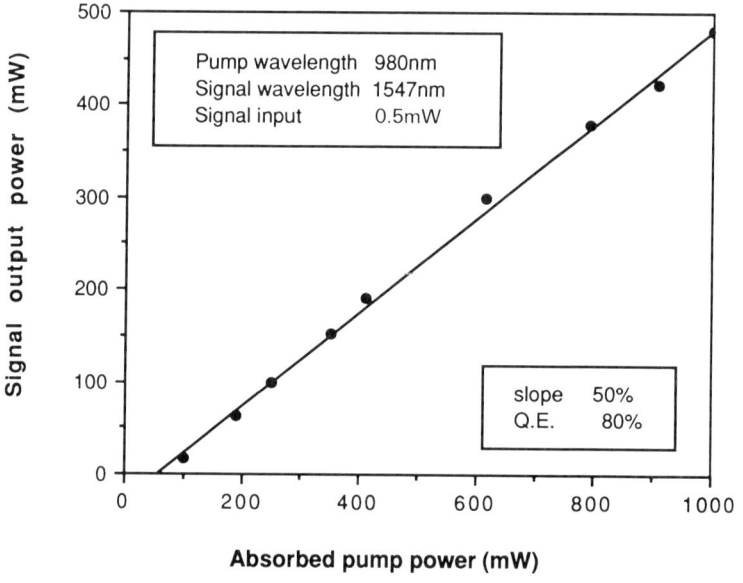

Figure 10.3 Signal output power versus absorbed pump power for the fibre power amplifier of Figure 10.2 (after [13]).

Ti:sapphire laser. Figure 10.2 shows the fibre power amplifier arrangement. Figure 10.3 plots the signal output power against the launched pump power for a fixed 0.5 mW input signal level at 1532 nm wavelength and a pump wavelength of 980 nm. The linear energy extraction efficiency can be clearly seen, with a slope of 50% (80% in quantum efficiency terms). If the input signal is removed from the amplifier, the device runs into oscillation from its own spontaneous emission and the distributed backscatter along the fibre or minute residual end-reflections. However, the gain recovery time is long (hundreds of microseconds) compared with the likely modulation rates encountered in a practical ASK-modulated system. Thus, modulated signals can be injected into this amplifier without concern about free-running lasing during the zero bit-period. Because the gain recovery time is long relative to the likely modulation rate, cross-talk in the heavily saturated condition is negligible, and the amplifier maintains a linear transfer characteristic.

No other optical amplifier has such advantageous signal power output capabilities.

Operating in heavy saturation conditions would in principle increase the noise penalty because the average population inversion is depleted. However, the spontaneous emission–spontaneous emission beat noise in fact diminishes because the homogeneously broadened nature of the line ensures that the saturating field depletes the spontaneous noise over the entire bandwidth of the amplifier. When a power amplifier is used in a communications system, the dominant quantum noise source is therefore the signal-spontaneous beat noise at the receiver.

However, the entire system loss lies between the noise source and the receiver, and the consequent degradation in overall performance due to amplifier noise is negligible. From the viewpoint of noise degradation, a power amplifier following the transmitter is the best option for improving the power budget of an optical communications system.

The large absolute levels of modulated optical signal power ($> 500 \text{ mW}$) provided by the power amplifier can be used in two ways:

(a) as a source of high peak-power pulses to increase the span or capacity of optical transmission systems; and
(b) as a source of optical energy that can be distributed by passive optical power splitters to a multiport optical distribution system (i.e. a broadcast system).

10.2.1.1 *Application of power amplifiers in transmission systems.* Bryant et al. [14] have demonstrated the largest unrepeatered direct-detection communication system using a single erbium–silica fibre power amplifier pumped using a Ti:sapphire laser at 980 nm in a 1.2 Gbit/s link over 250 km (see Figure 10.4). There was no observable degradation in the operation over the back-to-back (that is with the transmitter and receiver directly coupled without the fibre system present) BER performance of the optical system (Figure 10.5). Indeed, compensation of the residual frequency chirp in the laser source by the non-linear propagation of the high-power pulses provided by the power amplifier over the fibre (which was *not* dispersion shifted at the signal wavelength) resulted in record performance *and an apparent improvement* in the optical receiver sensitivity when the fibre system was placed between the amplifier and the receiver! This remarkable result indicates that not only can power amplifiers be used without introducing noise penalties in themselves, but also that the power levels generated at standard data rates over existing system fibres will introduce non-linear effects into the system. In this case, the influence was unexpectedly beneficial.

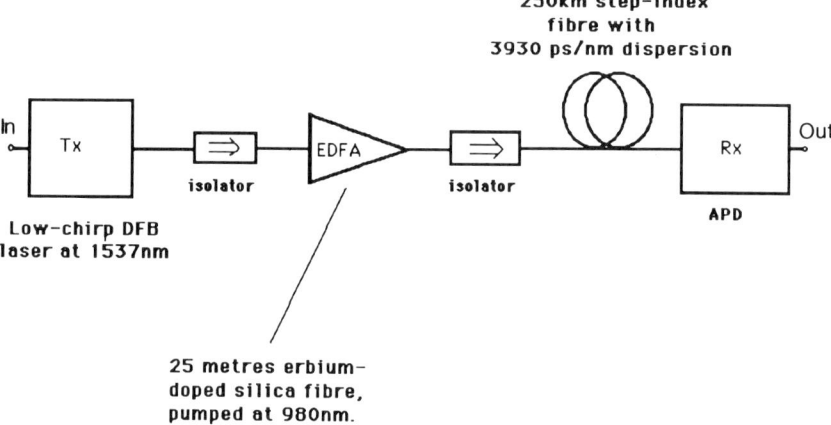

Figure 10.4 Representation of the longest unrepeatered direct-detection communications system, using 1300 nm optimised fibre, a low-chirp DFB laser transmitter, and an erbium-doped fibre power amplifier. (From [14].)

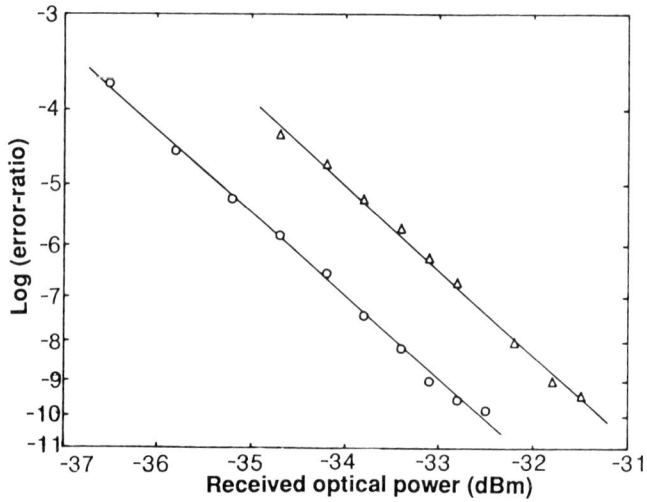

Figure 10.5 Bit error rate diagram for 2.4 G bit/s system shown in Figure 10.4. (From [14].) Triangles show 'back-to-back' error response; circles the error performance with fibre system *in situ*.

It is predicted that when power amplifiers become more commonly available, they will have a major impact in creating and upgrading communications systems that have no 'embedded' electronics or electrical power feeding. The implications of this realisation are significant. Increasing the system span by additions and improvements to the *terminal* equipment, as in the case of a power amplifier at the transmitter station, opens up windows of opportunity for longer unrepeatered links with no opto-electronic equipment in the transmission span itself.

Unlike terrestrial cable routes where buildings can conveniently house and power-feed repeaters, regenerators in undersea systems are extremely costly to install and repair. Two million US dollars is not unrealistic for the total lifetime cost of a *single* state-of-the-art suboceanic cable repeater. Reliability in a 25-year lifespan of a submarine cable system must therefore be exceptionally high, requiring major development and engineering costs for the cable and the repeaters. The use of fibre optical power amplifiers at the transmit end, as well as sensitive low-noise fibre pre-amplifiers at the receive end, will not only increase unrepeatered cable span but also dramatically reduce the capital and maintenance costs of undersea optical communications systems. This scheme (sketched in Figure 10.6) is likely to be the design objective of system researchers and planners over the next few years. For some medium-span systems common in Europe, Japan and the Caribbean (for example, around 300 km long), the cable costs could also reduce as the need for electrical power feeding in the cable may be completely eliminated.

Allowing major trunk routes to be 'all-optical' is also attractive for future uprading, as the technology allows. For example, it is well-known that the optical amplifiers are essentially transparent to bit rate, and could operate as ASK streams of many tens of gigahertz modulation rate. Also, the fibre amplifier's immunity to interchannel cross-talk allows wavelength or frequency multiplexed signals to be sent to line. Opto-electronic repeaters buried out in the system use non-reconfigurable electronics, and they cannot easily accommodate system upgrades in the time, frequency or wavelength domains. Embedded electronics become a bottleneck to upgradability. Although there will be limits to the extent to which a given link could be upgraded, determined (in the case of linear optical propagation) by the loss budget and the dispersion behaviour of the fibre, it is clear that the cable

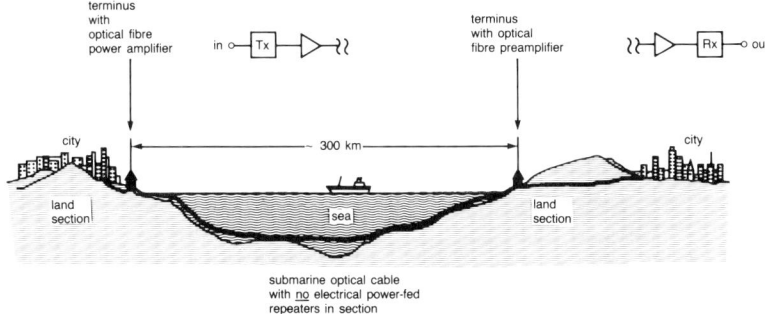

Figure 10.6 An all-optical submarine optical communications system, using shore-based optical amplifiers. Elimination of electrical power feeding could greatly reduce the cost of a large proportion of island-hopping links.

system using optical amplifiers at the terminal ends becomes effectively transparent to economic upgrading by TDM, WDM or FDM techniques, as and when the traffic demand requires.

When more information capacity is needed, electrical and optical multiplex/demultiplex equipment is added *only* at the accessible termini. This is probably one of the first examples of photonics bypassing a possible electronics communications bottleneck, although the scenario demands greater sophistication, speed and complexity at the information-processing nodes (i.e. at the terminal switching stations), which is of course where electronics excels.

10.2.1.2 *Application of power amplifiers in distribution networks.* The second exciting possibility is of multicustomer optical distribution networks employing power amplifiers to provide high initial levels of optical power at the source or 'head-end'. Figure 10.7 shows a passive optical network topology. The optical power is distributed using conventional fused-fibre directional couplers. With a fibre power amplifier placed at the head-end (dashed in Figure 10.7), the higher level of optical power sent to line is subsequently divided by passive fibre power splitters to many terminals, giving network designers increased scope and flexibility in their art.

In an ideal situation, if X is the number of times the amplifier doubles the optical power level sent to line, then every $3X$ dB increase in the system margin brought about by the introduction of a power amplifier allows 2^X as many

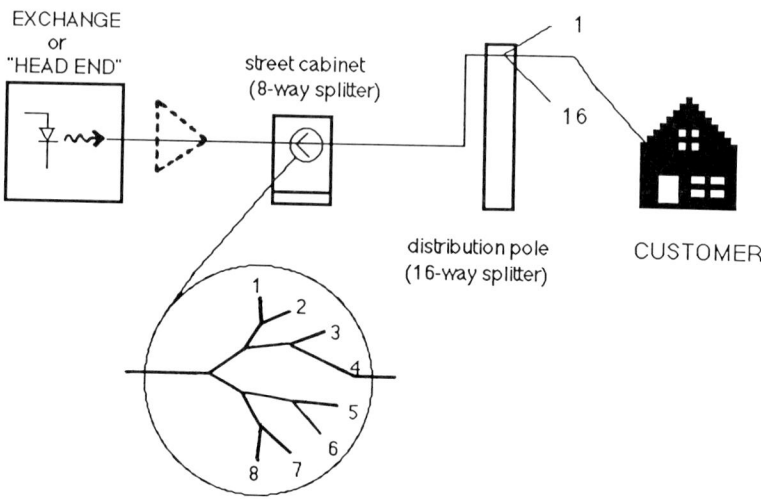

Figure 10.7 Schematic representation of a passive optical network (PON). Power amplification at or near the 'head-end' could result in a dramatic increase in customer numbers connected to a single source of high-bandwidth information.

customers to be connected to a given tree-and-branch network. Essentially, doubling the optical source power level doubles the number of receivers that can be served from that source. This ideal case ignores excess losses incurred by extending the network (i.e. splitter and fibre losses) and any noise penalties.

As an example of the enormous potential of fibre power amplifiers in this context, assume that a current unamplified broadband broadcast network may allow between 100 and 200 customers to receive a detectable signal up to 10 km from the 'head-end'. Broadband Passive Optical Networks (BPON) of this type are currently under intensive examination. A field trial in Bishops Stortford in England has been a recent 'showcase' for this new technology to demonstrate the feasibility of bringing fibres into homes and business premises [15], providing telephony, data, and larger bandwidth facilities such as multichannel entertainment or interactive video services. The laser at the common head-end would launch typically -3 dBm into the fibre exiting the exchange building. However, using a single $+23$ dB gain power amplifier in that building, the network size could expand to 128×2^7, i.e. 16 384 customers for the same maximum span and the same bit-rate.

This extrapolation has been verified in laboratory simulations. Hill and coworkers [16] have shown that this sort of progressive network expansion can be achieved in practice, when they used a single erbium–silica fibre power amplifier to extend a modest tree-and-branch passive fibre network to over 7000 terminals. The cost of a power amplifier may not be inconsiderable because, as previously explained, large amounts of pump energy are required to provide the required signal output power. With today's technology this requires either Ti:sapphire lasers, F-centre lasers or dye lasers, none of which is inexpensive or particularly 'wall-plug' efficient. However, the increased cost to the network-provider of the power amplifier is shared by an increased number of customers, so the additional cost per customer is small.

At British Telecom Research Laboratories the process of enlarging broadband optical distribution networks in this manner has continued. Over 39 million potential customer-size networks fed from a single head-end have been demonstrated, requiring, in practice, about one power amplifier per 10 000 customers. It is difficult to judge the impact of the amplifier technology on the future development of broadband multiport fibre networks – indeed, more modest discrete amplifiers located at strategic power-splitting points may be sufficient in the early stages of implementation. At the moment, the real value of these extraordinary laboratory demonstrations is in underlining the immense potential that amplifiers allow, by removing the constraint of propagation and splitting loss from the communication system budget.

The unique properties of the fibre power amplifier applied in the examples given above lead to the conclusion that 'hard-wired' optical communication of the point–point or distributed-station nature is poised to make significant technical advances in the 1990s.

10.2.2 Fibre amplifiers as signal repeaters

Power amplifers operate under the special conditions of signal saturation. In the small-signal regime it is well known that fibre amplifiers behave as efficient high-gain linear devices, with predictable noise characteristics. If this is so, how far can a point-to-point optical link be extended by cascading many small-signal amplifiers in series? Clearly, the noise penalties will accumulate by the 3–5 dB level experienced per amplifier, but there are other limitations that come to bear in such a cascaded amplifier chain (see Chapter 6). The fibre amplifier has a finite bandwidth, but, if they are all identical (note: the replicative nature of the gain spectrum of the co-doped erbium ion in silica is of major benefit here) then the system bandwidth becomes compressed as the number of amplifiers in the chain increases. After many optical repeaters, the

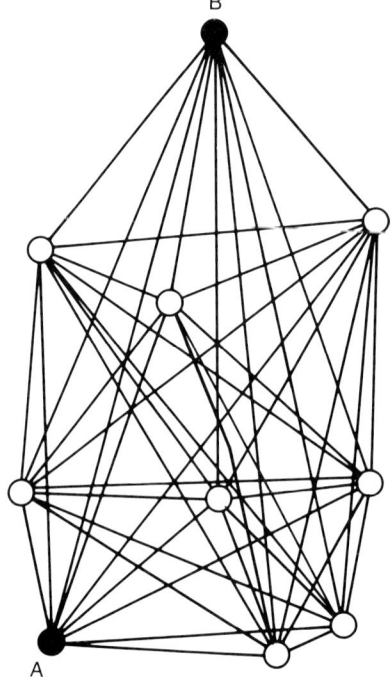

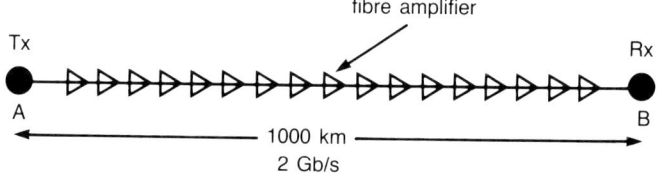

Figure 10.8 The fully-interconnected all-optical 'supertrunk' concept.

system bandwidth may only be a few nanometres wide, even although the 3 dB bandwidth of a single amplifier may be 30 nm. Despite these initial concerns, there is still scope for designing a much-enhanced communication system employing FDM or TDM multiplexing techniques.

Edagawa [18] demonstrated a 1.2 Gbit/s system over 906 km of fibre using 12 erbium–silica fibre amplifiers as repeaters in a series chain. No degradation in the BER performance was observed, although the system bandwidth was reduced to about 1 nm.

Let us assume that, on the basis of Edagawa's pioneering work, a 2 Gbit/s system over 1000 km was a realistic engineering target in the near future, using cascaded fibre amplifier repeater. Bearing in mind that the resulting link would have no signal electronics in the route, it could be imagined that there would be considerable interest in an 'optical supertrunk' overlayed on top of the existing fixed-lined telecommunication network. This would be particularly true in the UK and in European countries where demographic conditions are suitable for this network technology. A 500 km radius would be adequate to circumscribe most European States, and in any one country there may be of the order of 10 major conurbations. These cities could be fully interconnected for network security and routeing diversity, providing an all-optical web (Figure 10.8) carrying high-capacity trunk information. The amplifiers in this strategic network would be located in buildings and the optical reliability would be determined mainly by the continued availability of pump power and the integrity of the fibres in the network.

Note that this proposal results in all-optical communication highways, where the switching nodes of the network are pushed further apart in distance. Greater amounts of large-bandwidth information will descend upon and exit from these strategic nodes, making new demands for optical processing power. As we can see, improving the transmission technology through the use of amplifiers immediately provides new network opportunities, but simultaneously highlights the need to improve the optical signal processing capabilities (and indeed the network control aspects) at the switching nodes.

10.2.3 *Fibre amplifiers at 1320 nm wavelength*

The silica-based amplifiers that have attracted the most research interest for telecommunications purposes operate in the wavelength region around 1550 nm. Of great importance is the application of fibre amplifiers in different 'windows' of the optical spectrum, for example at 1320 and 2550 nm. Indeed, many installed optical systems operate exclusively in the 1320 nm wavelength region where the linear fibre dispersion is zero, with the option (as in the UK) to exploit the 1550 nm window as traffic growth expands. The exciting prospects for power and repeater fibre amplifiers discussed so far belie the reality that many current systems do *not* operate at 1550 nm, and some are only open for upgrading at around 1300 nm.

To avoid parasitic signal excited-state absorption (ESA) prevalent in silica-based fibres, amplifying single-mode fibres made from fluorozirconate glass have been doped with the trivalent neodymium ion (see Chapter 8), which exhibit signal gain in the range 1310–1360 nm. Gains of 7.5–10 dB have been reported [19, 20] around 1340 nm, and the advantageous saturation characteristics measured [21] with encouraging initial results. There are reports of prototype amplifiers being used in early laboratory demonstrations [22].

Larger gains for less pump power derived from diode lasers can be predicted as the obvious future developments in the search for practical fibre amplifiers for the 1320 nm window, along with increasing numbers of systems demonstrations.

10.2.4 Fibre pre-amplifiers

Coherent detection techniques were introduced in the early 1980s partly to increase receiver sensitivity in order to extend the optical system span. Heterodyne mixing on the receiver of the incoming light and a local oscillator affords possibilities of 'fine grain' demultiplexing of channels closely spaced in wavelength (in this case the channel separation is measured in gigahertz), but does so at the cost of engineering complexity involved in polarisation control and local oscillator frequency stability and referencing. However, detection sensitivities approaching the quantum limit of 25 photons/bit (at 140 Mb/s have been achieved by coherent techniques [23].

The more common direct-detection receivers rely solely on received power levels being detectable above a threshold level, this limit being set by receiver shot noise, thermal noise, bandwidth and quantum efficiency of the receiver photodiode. When a fibre amplifier is placed immediately prior to a receiver as an optical pre-amplifier, the signal-to-noise ratio at the receiver is dominated by the quantum behaviour of the optical gain medium. The dominant noise terms in the pre-amplifier/receiver combination are the unavoidable shot noise, the signal-spontaneous beat noise and the spontaneous-spontaneous beat noise, these characteristics swamping the thermal noise component from the receiver. We have already seen in Chapter 6 that for a signal-spontaneous beat noise limited case employing an out-of-band filter for the majority of the spontaneous emission, the noise penalty can be as low as 3.2 dB in a three-level laser system such as trivalent erbium in silica fibre. Optimum direct-detection pre-amplification should therefore provide a sensitivity of about 53 photons/bit at 1 Gbit/s. Figure 10.9 shows the improvement in receiver sensitivities for direct detection (on a log scale) against date in years. Clearly, the 151 photons/bit achieved in 1990 [24] (translated to 309 average photons per bit), obtained using an erbium fibre pre-amplifier and a PINFET receiver, is getting very close to the maximum sensitivity predicted for this configuration. A temporal extrapolation indicates that the fundamental limit may be achieved by 1993!

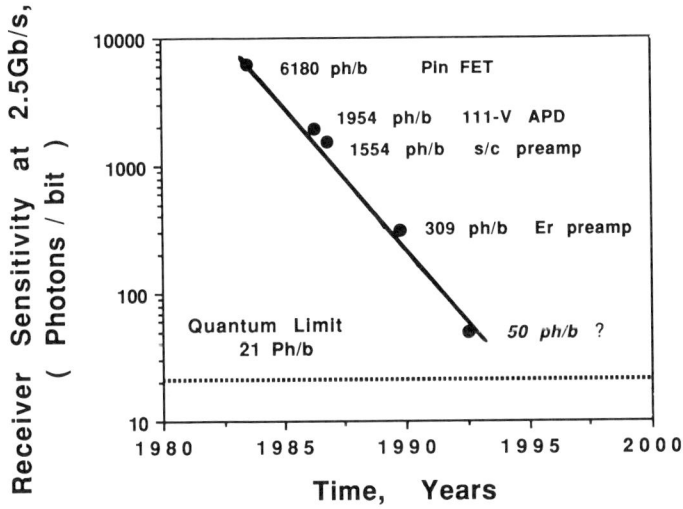

Figure 10.9 Reported receiver sensitivities for AM detection systems against date of report. Note that the number of photons per bit has been averaged over a 1–0 bit period.

The attainment of close to maximum receiver sensitivities for direct detection of only 3 dB away from the quantum limit indicates that the additional complexities of coherent transmission techniques and receivers will only increase the overall system budget by about 3 dB. Erbium fibre pre-amplifiers render a rationale for coherent technology based on increasing system power budgets rather questionable. However, demultiplexing techniques using 'fine grain' channel spacing and a tuneable local oscillator to demultiplex the channels at the receiver are extremely important, and it should be noted that fibre amplifiers operate very successfully with such multiplexing schemes [25], particularly as power amplifiers or repeaters.

10.2.5 Distributed fibre amplifiers

Fibre amplifiers are very different from discrete gain blocks, in that the gain medium can be distributed over very long propagation distances using very light doping of the active ion in the core of the fibre. Background loss (that is the loss not due to the ionic absorption but due to intrinsic, scatter and bending loss of the fibre) is not significantly higher than undoped fibre, which at the pump wavelength (at 1490 nm) and the signal wavelength (at about 1540 nm) is a few tenths of a decibel per kilometre. With this scheme, called *distributed amplification*, the entire propagation distance can exhibit gain. Indeed, the dopant ion concentration can be tailored to just overcome the background loss, giving a long-distance optical transmission system which is 'ether-like' at certain wavelengths. Craig-Ryan [26], Simpson [27] and

Bocko [28] have shown that low dopant levels can be achieved, and many kilometres of transparent fibre made. Williams et al. [29] have also shown that the pump power requirement for the transparency condition to be met is modest over 10 km.

The opportunities for such a distributed amplifier have yet to be fully explored, but two applications of potential significance are apparent – in soliton transmission systems [30] and in optical BUS distribution architectures [31].

10.2.5.1 *Soliton propagation in a distributed fibre amplifier.* A soliton is a high-intensity pulse of light which generates a non-linear self-phase shift as it propagates down a fibre, this non-linearity ameliorating the temporal spreading of the pulses by positive linear fibre dispersion. This non-linear behaviour of the soliton compensates for the influence of linear dispersion, allowing transmission of ultra-short pulses without temporal dispersion. By design, the pulses emerging from a long length of fibre can have exactly the same duration as the input pulses, and this is known as *soliton transmission.* For a soliton to be maintained it must not decrease in amplitude along the length of the fibre, i.e. it must not suffer from significant attenuation. Periodic amplification by discrete amplifiers (of the fibre or semiconductor variety) can effectively reinstate the soliton, as shown most dramatically by Mollenauer [32], who has demonstrated that periodic soliton amplification and propagation over 10 000 km in a recirculating ring configuration is possible. A transmission fibre which continuously maintains the pulse intensity over a long length – as in the case of a distributed fibre amplifier – would appear a logical step in the advancement of the subject. As Mollenauer also points out [33], the process of continuous amplification to either give gain or provide an 'ether-like' environment for the solitons would be a preferred route in WDM soliton systems.

In this discussion about soliton transmission systems in the context of amplification, either distributed or discrete, a further important point should be recorded. If amplification is used to extend the propagation distance over which an optical wave of almost constant intensity travels, then the average power level at which the onset of non-linear (soliton) behaviour occurs decreases dramatically. This is because the non-linear phase shift accumulates linearly with intensity *and distance.* Over a 1000 km optical span, pulses that are maintained with average power levels of only a few milliwatts will begin to show temporal behaviour that departs from the predictions for linear dispersion. Therefore, the increase in electronic-free transmission over long propagation distances using optical amplifiers will lead inevitably to soliton-like optical communication systems. Solitons should no longer be regarded as esoteric means of exploiting an intrinsic non-linearity in a communication fibre using optical pulses that are difficult to generate, encode, decode and detect. On the contrary, although the engineering aspects are still in their infancy, we have seen in this book how mode-locked fibre lasers can be used to

FUTURE DIRECTIONS

generate solitons, possibly even from semiconductor laser diode pumps. Detectors operating at microwave frequencies are available [34] and integrated optical signal processing offers scope for mux/demux operations or optical code generation and recognition [35].

Above all, by extending the capability of optics using optical amplifiers to avoid electronic regeneration, we rapidly reach the limits of linear optical transmission over long distances. In the future we may, by necessity, require that long, ultrafast optical systems operate in a soliton-like way. This is probably the first opportunity for non-linear optics in a mainstream communication application. In ten years' time, a commercial transatlantic optical submarine link may operate in such a manner.

10.2.5.2 Distributed fibre amplifier in a BUS network architecture. The distributed amplification obtained by the unique structure of a lightly doped fibre amplifier can be used to overcome a distributed loss. This loss may be the fibre background loss, but it may also include the loss experienced by a signal as power is tapped off at regular intervals along an optical information highway. This architecture, used to allow communication to and from many ports in a network, is known as a BUS system (Figure 10.10).

BUS architectures, and the derivatives of ring, taurus and ladder networks, have *not* become established in optical information networks because the number of tapping points from a passive BUS highway is strictly limited. Only ten or twenty taps are possible before the optical power travelling along the information highway reduces to unusable levels. One solution would be to provide optical amplification at each tap point to reconstitute the signal power, but a large number of discrete amplifiers is clearly impractical. In the distributed amplifier BUS concept, the fibre of the BUS is doped with an appropriate rare earth and conveniently pumped from either or both ends of the BUS. The amplification is continuous and compensates for the tapping loss, and is powered only from the ends of the BUS.

Urquhart [36] predicted extension of the concept to many hundreds of ports on the BUS, while Whitley [37] demonstrated the first experimental

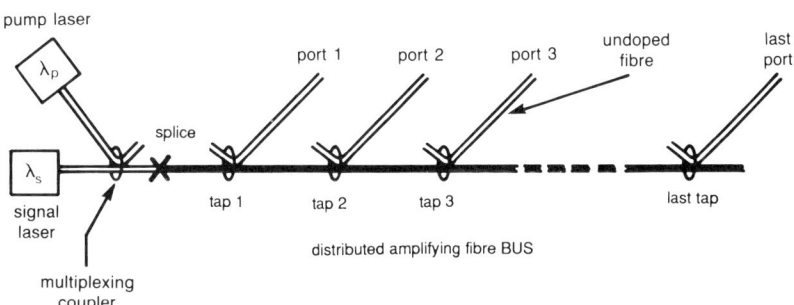

Figure 10.10 Optical BUS distribution network concept, using distributed fibre amplification between tap points.

realisation, both in 1990. The major application of this technology is thought to be in expanding the number of input/output ports in a multidimensional switching machine [38], which allows, in principle, gigantic switch throughputs by allocating moderate amounts of time, wavelength (or frequency) and space switching to an input/output array formed by the ports on a BUS communication highway. Whatever the outcome of this research, there will undoubtedly be optical switching networks that will use amplifiers to extend the switch size by increasing the fan-out potential from switching elements, and this is new territory in the field of optical signal processing and switching.

Clearly improved switching capabilities – brought about by greater exploitation of fibre amplifiers – will be needed in a scenario where improved span and/or capacity in trunk optical networks push switching and processing nodes to locations not necessarily governed by demographic considerations. Greater switch throughput will be required when amplifiers increase the amount and rate at which information arrives and leaves a processing node. Where amplifiers reshape the networks, optical communications and the way in which it is likely to develop in future will force change upon the way in which optical information is generated, processed, switched, stored and retrieved. The parallel between electronic communication network signal processing and electronic computers may find a mirror image in optical communication distribution and switching networks, where 'intelligence' will likely be required to control and manage photonic communications. The need for an optical computer may therefore arise from the demand to perform fast serial and/or slow parallel optical switching in this developing context.

10.3 Fibre lasers

As discussed in Chapter 2, fibre lasers have specific advantages that make them an attractive subset of laser technology. Condensing their key features into four main categories, it is clear that fibre lasers offer:

  (i) a wide range of operating wavelengths;
  (ii) broad tuneability (with corollary of short-pulse operation);
  (iii) high quantum efficiency and large output power handling;
  (iv) new transition opportunities, such as in up-conversion fibre lasers.

Their future impact is therefore likely to be found in applications that exploit these four features.

10.3.1 Short-wavelength fibre lasers

Short-wavelength lasers are important in applications where any form of diffraction-limited optical reading or writing process is involved, whether it be in writing short-pitch holographic gratings or reading digital information on CD players. Obtaining intense sources of radiation in the UV-blue region that are both practical and inexpensive would provide many technical and commercial opportunities.

Silica fibres incorporating rare-earth ions as the active medium are not particularly good for visible and short-wavelength lasers, and there has only been one report of a visible fibre laser in a silica host glass [39]. A probable drawback with the relatively insoluble rare-earth ions in a germania co-doped silica host seems to originate from the location of the ion at network defect centres, giving rise to enhanced photochromic behaviour. Pumping a silica fibre laser in the UV-blue-green region of the spectrum can give rise to strong photodarkening behaviour [40] at wavelengths up to about 900 nm, which results in cavity losses and an inability to sustain lasing.

Fibres made from a base of zirconium fluoride ('fluoride' or 'ZBLAN' fibres) do not appear to suffer similar photochromic sensitivity. In this glass the lanthanide ion is incorporated as a network former, and lattice defect centres seem less dense. Coupled with improved radiative transition probabilities due to poor phonon coupling in the glass, it is the fluoride glasses that offer the greatest potential for fibre lasers in the visible and short-wavelength region.

Recently, Allain and coworkers have shown that blue and green fibre lasers can be constructed using this material [41, 42]. Since it is rather fuzzy logic to strive to make a short-wavelength fibre laser pumped using a laser at an even shorter wavelength, a highly desirable and much sought-after device is one in which the pump wavelength is *longer* than the lasing wavelength.

From simple energy conservation considerations, this can be achieved by pump multiphoton absorption processes, commonly known as up-conversion lasing. As shown diagrammatically in Figure 10.11, rare-earth ions in fluoride glasses exhibit strong ESA of pump photons to higher energy levels of the ion.

This multiphoton absorption process populates energy levels above that of the pump band, which then allows spontaneous and stimulated emission between transitions at shorter wavelengths than that of the pump. Using this useful property of the fluoride glass fibre host and further exploiting the advantageous thermal dissipation properties of the fibre structure, Allain produced the first thulium-doped fluoride fibre laser pumped at 640 nm which was CW at 77 K. The laser operated at 455 and 480 nm. The same group have also produced a tuneable CW fibre laser operating between 540 and 553 nm at room temperature using holmium-doped fluoride fibre pumped at 647 nm. The up-conversion efficiency was 12%.

These first results clearly point the way towards achieving a highly desirable goal–that of producing the first diode-laser pumped up-conversion laser, the most likely pump source being GaAlAs semiconductor lasers operating around 800 nm. The up-conversion fibre laser structure offers the possibility of reasonably efficient, tuneable and moderate cost short-wavelength lasers that will have major significance in such applications as optical memory storage

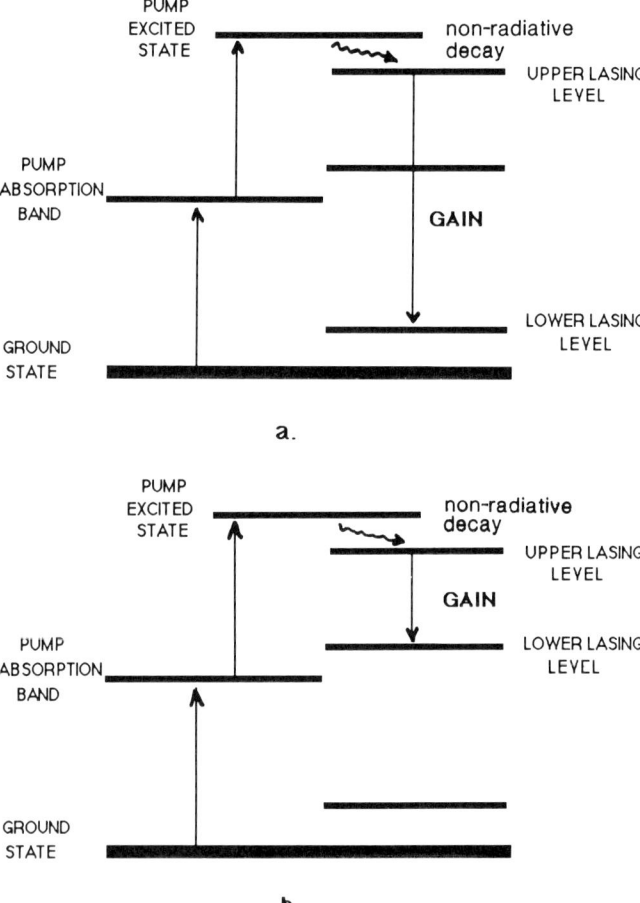

Figure 10.11 (a) General representation of an up-conversion pumped fibre laser scheme, where the emission wavelength is shorter than the pump wavelength. (b) Up-conversion pumped fibre laser scheme, where the emission wavelength is longer than the pump wavelength. Note that in both cases depopulation of the lower lasing level could be via radiative, non-radiative or excited-state absorption processes.

and information retrieval, printers, and specialist applications such as line-of-sight underwater communications.

10.3.2 *Long-wavelength fibre lasers*

A distinctive difference between silica fibres and fluoride fibres, brought about by the different phonon coupling energies, is the shift to longer wavelengths of the phonon edge in the latter material. Unlike silica-based fibres, fluoride fibres are transparent above 2000 nm wavelength. Indeed, the technology for manufacturing high-quality undoped fluoride fibres was developed for the prospect of optical communications in the low-loss window around 2550 nm, where ZBLAN fibres have a theoretical loss minimum of 0.01 dB/km. Currently, losses of about 0.65 dB/km are achieved in short fibre lengths [43]. For fluoride fibre laser oscillators using rare-earth dopants as the gain medium, these propagation losses are negligible in devices that are a few metres in length. Lasing in the 2000–3000 nm wavelength region has been observed in erbium-, holmium- and thulium-doped ZBLAN fibres [44–46], and indeed a simple communication system has been demonstrated using an erbium-doped fluoride fibre laser at 2700 nm [47]. A most impressive result comes from Wetenkamp [48] who achieved a 9.2% slope-efficient erbium–ZBLAN fibre laser at 2750 nm pumped using a wavelength of 795 nm, which is compatible with available GaAlAs semiconductor laser diodes. A threshold pump power of only 0.1 mW launched into a multimode fibre was recorded. Broad tuneability of these lasers are also reported [49].

The applications of fibre lasers in the 2000–3000 nm wavelength region are many and varied. Many of the spectroscopic signatures of important organic liquids and gases occur in this spectral region, so tuneable lasers in this range are useful for chemical sensing, analysis, and pollution monitoring. There is a line-of-site communications window close to 3000 nm. The peak OH^- absorption occurs at 2800 nm, and there is great interest in the use of lasers at that wavelength for laser surgery. Many of these non-telecommunications applications require optical powers that exceed those currently reported in fluoride fibre lasers, and there is a question mark over the limits of optical power that a fluoride fibre can tolerate without damage. Experience shows that high CW power densities of at least 3×10^6 W/cm^2 can be sustained without destroying a single-mode fluoride fibre, but that care must be taken to avoid hot spots occurring at the entrance and exit facets of the fibre, causing the fibre to melt.

10.3.3 *Fibre lasers for short-pulse sources*

The gain bandwidth of an erbium-doped fibre amplifier is in excess of 35 nm in the 1540 nm wavelength region. Transforming this optical bandwidth to the time domain reveals that pulses of picosecond duration should be possible when suitable intracavity mode-locking techniques are applied in an erbium fibre laser. Following the first reports by Kafka *et al.* [50], Smith *et al.* [51] obtained 2–3 ps duration pulses at 1560 nm from a 10 m long Fabry–Perot erbium-doped fibre cavity, producing peak powers of about 7 W (Figure 10.12). The pulses were close to the ideal sech^2 shape necessary for practical soliton transmission systems.

This demonstrates that soliton fibre lasers can be made in the correct wavelength region for soliton transmission systems, the importance of which was described in Section 10.2.5.1. Pumping soliton fibre lasers with CW

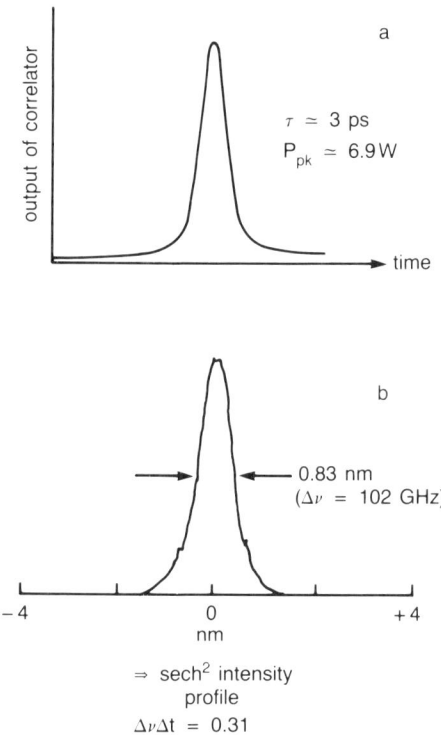

Figure 10.12 Transform-limited pulses from an erbium fibre laser. The width of the autocorrelation pulses was approximately 3 ps, and the peak power was 6.9 W. Spectral bandwidth was 102 GHz, indicating pedestal-free sech-squared pulse shapes required for soliton propagation studies in fibre systems.

semiconductor lasers looks a very likely possibility, and the future should see practical low-cost soliton fibre laser sources being developed for this application.

10.4 Fibre amplifiers at different wavelength regions

If communication systems are not constrained by the fibre loss budgets to specific wavelength windows such as 1320, 1550 or 2550 nm, one may ask what prevents operation of, say, a communication system operating at 1050–1100 nm using neodymium-doped fibre amplifiers and fibre lasers? Does the opportunity to open up a wider region of the electromagnetic spectrum to communications offer alternative optical strategies, components or materials systems? Another example may be in the relative attractiveness of the original communications window at 800–900 nm, where inexpensive laser sources and

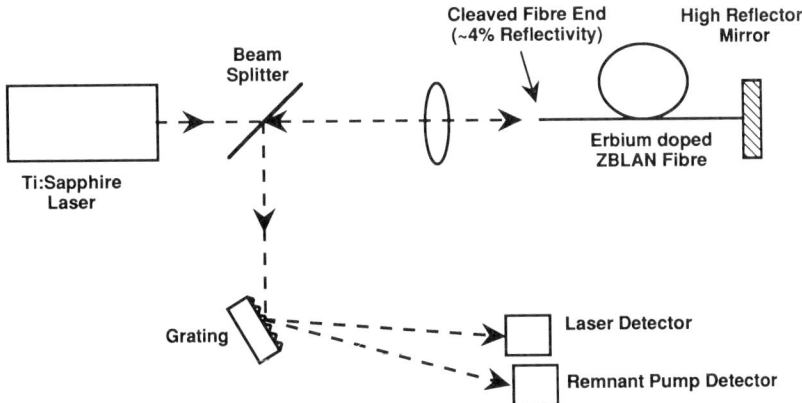

Figure 10.13 Cavity configuration for an up-conversion pumped erbium-ZBLAN fibre laser at 850 nm. The pump wavelength was 801 nm.

silicon detectors may provide economic options with erbium-, thulium- or neodymium-doped fibre amplifiers.

Fibre amplifiers operating at many different wavelength regions will be used as power amplifiers for short pulses. Operating windows of fibre amplifiers cover the most commonly used laser wavebands, for example the near-IR around 800–850 nm and the Nd:YAG region at 1064 and 1320 nm. We have yet to see the introduction of these devices in broader application areas of laser physics. The implications for optical fibre power amplifiers at different regions of the spectrum have not been realised or addressed, where the potential for research into new and exciting laser possibilities is considerable. For example, an up-conversion pumping scheme at 801 nm wavelength has produced a high power CW fibre laser at 850 nm [52], using erbium-doped ZBLAN as the active medium (Figure 10.13). The energy-level diagram for this unusual type of fibre laser is shown in Figure 10.14. The successful operation of this system would appear to suggest that a normally self-terminating transition can be made to operate CW using two-photon pump absorption processes.

The slope efficiency of this laser was 38% of absorbed pump power, which is quite remarkable for a normally self-terminating transition excited by a two-photon pump absorption process. From the low-Q cavity configuration shown in Figure 10.13, it is clear that the single-pass gain in the fibre must be substantial to overcome the power loss attributed to the 4% output coupler. Such a fibre amplifier might be applied to optical communications at 850 nm. For short-distance optical communication where fan-out capability is important, say in a low-cost data interconnect LAN, operating at 850 nm may lead to significant opto-electronic hardware cost reductions.

For optical processing and switching devices, we must ask if we are constrained to use material systems that are compatible with the currently

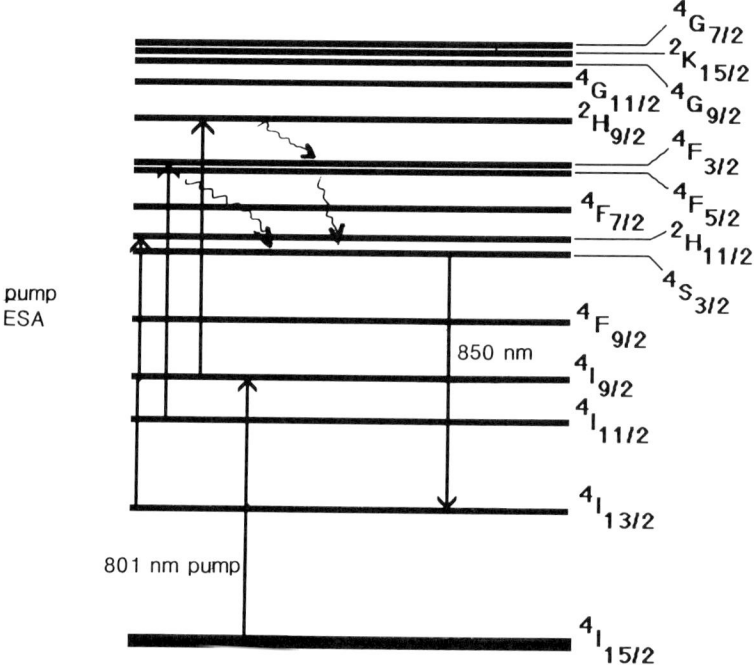

Figure 10.14 Energy level diagram associated with the laser of Figure 10.13. Note that the population of the $^2H_{11/2}$:$^4S_{3/2}$ levels and depopulation of the $^4I_{13/2}$ level take place via excited state transitions.

established communication windows. Some of the best resonantly enhanced non-linear materials, such as the semiconductor-doped glasses, operate in the visible and near-IR regions and the important fast low-energy switching functions that such a material provides in fibre form [53] are not readily applicable in current telecommunications windows of 1300 or 1550 nm. If, in the presence of amplification and large signal levels, attenuation of the optical signal in a communication system is of less importance, does this allow optical component technologies that were previously discounted to be brought back into consideration?

These questions arise from the likely availability of efficient fibre lasers and amplifiers operating across a large part of the optical spectrum. It is not yet possible to provide clear answers, but these may be the sort of questions currently being raised in research laboratories throughout the world.

10.5 Active fibre optical sensors

The work in optical fibre sensors often exploits the advances in technology brought about in other application areas such as telecommunications, creating new optical sensing opportunities. Because of competition from entrenched sensors of the electrical variety, the subject tends to be technically conservative and cautious in the mass-market, low-cost sensor area, but very quick to exploit the unique features of fibre sensors in specialist low-volume, high-cost markets. Thus, there is thriving activity in 'older' technology areas such as bend-loss induced multimode fibre collision sensors for assembly plant robots and passenger lifts. Simultaneously, advanced optical sensor technology appears in specialist areas such as miniature multiaxis optical gyroscopes for determining the location of oil-well exploration drill heads. (When considering the topic of fibre gyroscopes, it is worth recalling that a neodymium silica fibre, operating as an optically pumped superfluorescent laser, has produced the brightest incoherent source yet known [54], a device of considerable importance in this field.)

Optical fibres have the unique ability to distribute the sensing faculty to different physical locations, either by novel signal distribution and multiplexing techniques for extrinsic sensors similar to the BUS architectures previously described, or in distributing an intrinsic sensing feature. In the latter case, the distributed sensors are interrogated by optical pulse-echo techniques. One of the first rare-earth doped fibre sensors to be reported used the spectral shift of the ionic absorption with local ambient temperature at a fixed monitor wavelength [55]. Changes in the backscattered signature from the fibre identified the physical location and amount of the temperature change along the entire fibre length. This device used the redistribution of ground-state Boltzmann populations with temperature, and was not 'active' in the sense that it did not exploit excited-state transitions in a sensing mechanism. However, a subject for future study could be in the enhancement of optical sensing capabilities using the distributed gain available in an amplifying fibre. Any optical element which changes its absorption or reflectivity in the presence of a stimulus can be used with a fibre gain medium to create a change in a large signal output from an oscillator with the sensing element in, or part of, the cavity.

It is clear that the potential of 'active' fibres, either as enhanced optical sensors or to enable enhanced optical sensor networks, has not yet been fully exploited by the fibre sensor community. These concepts are the foundation for a new and fertile area of future study.

The ability of fluoride fibres to be transparent at wavelengths from the UV to greater than 3 μm can be combined with the successful incorporation of all of the rare-earths into ZBLAN fibres to make a very promising medium for optical sensors based on absorption or fluorescence spectroscopy. Techniques for the measurement of explosive or pollutant gases are well known, but not yet widely exploited using this technology. The use of active fibres in the optical sensor field is at a fledgling state, but interest and activity focused towards routine and niche applications should grow as doped fibres and pump sources becoming more commonly available to telecommunications users.

10.6 Conclusions

Research, development and application of fibre amplifiers and lasers in the past five years has brought about a dramatic reappraisal of optical telecommunications capabilities. The impact of these devices on the span, information capacity and distribution potential of optical systems will be considerable. These advances may make possible completely new network architectures, and may influence the requirements and design of future information switching and processing nodes.

Fibre amplifiers will provide a practical optical solution to an electronic bottleneck in optical transmission systems, opening up the possibility of the all-optical network. This prospect opens a window of opportunity for non-linear optical transmission, creating exciting prospects for ultra-long, ultra-fast communications. The need to control, switch and process information at rates and quantities surpassing the capacity of current electronic computers may provide a real driving force for the much-heralded 'optical computer', or more likely an optical/electronic hybrid.

Generation of short optical pulses for non-linear transmission studies seems to be practical with fibre lasers. New material systems will create new possibilities to exploit other wavelength 'windows' for communications. Lastly, the spin off from the communications research effort to other branches of laser physics has yet to take place, and the application of the technology to optical sensors and metrology is in its infancy.

As an exciting and important new technology, the optical fibre laser amplifier has come of age. It will be of immense future benefit through its pivotal role in improving global communications.

References

1. C.J. Koester and E. Snitzer, Amplification in a fibre laser. *App. Opt.* **3** 1182–1186 (1963).
2. V.S. Letokhov and B.D. Pavlik, Non-linear amplification of surface light wave in active optical fibre. *Sov. J. Tech. Phys.* **36** 2181–2187 (1966).
3. J. Stone and C.A. Burrus, Neodymium doped silica lasers in end-pumped geometry. *App. Phys. Lett.* **123** 388–389 (1966).
4. S.R. Nagel, R. McChessney and K.L. Walker, An overview of the modified chemical vapour deposition (MCVD) process and performance. *IEEE J. Quant. Elect.* **QE-18** 459–476 (1982).
5. C.J. Koester, Laser action by enhanced total internal reflection, *IEEE J. Quant. Elect.* **QE-2** 580–584 (1966).
6. K.C. Kao and G.A. Hockham, Dielectric-fibre surface waveguides for optical frequencies. Reprinted in *IEE Proc. J.* **133** 191–198 (1986).
7. S.B. Poole et al. Fabrication of low-loss optical fibres containing rare-earth ions. *Elect. Lett.* **21** 737–738 (1985).
8. R.J. Mears et al. Low noise erbium-doped fibre amplifier operating at 1.54 μm. *Elect. Lett.* **23** (19) 1026–1028 (1987).
9. W.I. Way et al., Multi-channel AM-VSB TV signal transmission using an erbium-doped optical fibre power amplifier. *Proc. IOOC 89*, Kobe, Japan, Paper 20PDA-10 (1989).
10. J.R. Armitage, Three-level fibre laser amplifier – a theoretical model. *App. Opt.* **27** (23) 4831–4836 (1988).

11. E. Desurvire, J.R. Simpson and P.C. Becker, High gain erbium doped travelling wave amplifier. *Opt. Lett.* **12** 888–890 (1987).
12. A. Lidgard, J.R. Simpson and P.C. Becker, Output saturation characteristics of erbium doped fibre amplifiers pumped at 975 nm. *App. Phys. Lett.* **56** (26) 2607–2609 (1990).
13. J.F. Massicott, R. Wyatt, B.J. Ainslie and S.P. Craig-Ryan, Efficient high power high gain erbium doped silica fibre amplifier. *Elect. Lett.* **26** (14) 1038–1039 (1990).
14. E.G. Bryant et al. Unrepeatered 2.4 Gb/s transmission experiment over 250 km of step-index fibre using erbium power amplifier. *Elect. Lett.* **26** (8) 528–529 (1990).
15. J.R. Stern and R. Wood, Broadband local networks. *Proc. Intl. Conf. on Integrated Broadband Services and Networks*, IEE, London, 15–18 Oct 1990.
16. A.M. Hill et al. 7203-user WDM broadcast network employing one erbium doped power amplifier. *Elect. Lett.* **26** (9) 605–607 (1990).
17. A.M. Hill et al. 39.5-million way WDM broadcast network employing two stages of erbium doped fibre amplifiers. *ibid*, (22) 1882–1884 (1990).
18. N. Edagawa et al. 904 km, 1.2 Gb/s non-regenerative optical transmission experiment using 12 erbium doped fibre amplifiers. *Proc. ECOC 89*, PDA-8, 33–36 (1989).
19. J.E. Pedersen, M.C. Brierley, S. Carter and P. France, Amplification in the 1300 nm telecommunications window in a neodymium doped fluoride fibre. *Elect. Lett.* **26** 329–330 (1990).
20. Y. Miyajima, T. Komukai, T. Sugawa and Y. Katsuyama, Neodymium doped fluorozirconate fibre amplifier operated around 1.3 μm. *Digest of OFC 90*, paper PD-16, Opt. Soc. Am., (1990).
21. J.E. Pedersen and M.C. Brierley, High saturation output power from a neodymium doped fluoride fibre amplifier operating in the 1300 nm telecommunications window. *Elect. Lett.* **26** (12) 819–820 (1990).
22. J.E. Pedersen, M.C. Brierley and R.A. Lobbett, Noise characterisation of a neodymium doped fluoride fibre amplifier and its performance in a 2.4 Gb/s system. *Photon Tech. Lett.* PTL-2, Oct 1990.
23. D. Malyon et al. PSK homodyne receiver sensitivity measurement at 1.5 microns. *Elect. Lett.* **19** 144–146 (1983).
24. P. Smyth et al. 152 photons per bit detection at 622 Mb/s to 2.5 Gb/s using an erbium fibre amplifier. *Elect. Lett.* **26** (19) 1604–1605 (1990).
25. M.C. Brain, Coherent optical networks. *Brit. Telecom Tech. J.* **7** (1) 50–57 (1989).
26. S.P. Craig-Ryan, B.J. Ainslie and C.A. Millar, Fabrication of long lengths of low excess-loss erbium doped optical fibre. *Elect. Lett.* **26** (3) 185–186 (1990).
27. J.R. Simpson et al. A distributed erbium doped fibre amplifier. *Proc. OFC 90*, paper PD19 Opt. Soc. Am., San Fransisco (1990).
28. P.L. Bocko, Rare-earth fibres by the outside vapour deposition process. *Proc. OFC 89*, paper TuG2 Opt. Soc. Am., Houston, Texas (1989).
29. D.L. Williams et al. Transmission over 10 km of erbium doped fibre with ultralow signal power excursion. *Elect. Lett.* **26** (18) 1517–1518 (1990).
30. L.F. Mollenauer and K. Smith, Demonstration of soliton transmission over 4000 km in fibre with loss periodically compensated by Raman gain. *Opt. Lett.* **13** (8) 675–677 (1988).
31. T.J. Whitley, C.A. Millar, S.P. Craig-Ryan and P. Urquhart, Demonstration of a distributed optical fibre BUS network. *Topical Meeting on Opt. Amplifiers and Appls.*, Monterey, Ca (1990).
32. L.F. Mollenauer et al. Experimental study of soliton transmission over more than 10,000 km in dispersion-shifted fibre. *Opt. Lett.* **15** (21) 1023–1025 (1990).
33. L.F. Mollenauer and S.G. Evangelides, Wavelength division multiplexing with solitons in ultralong distance transmission using lumped amplifiers. *Proc. CLEO 90*, paper CFI1, Anaheim, California, 21–25 May 1990.
34. D. Wake et al. Monolithic integration of 1.5 micron optical preamplifier and PIN photodetector with a gain of 20 dB and a bandwidth of 35 GHz. *Elect. Lett.* **26** (15) 1166–1167 (1990).
35. B.H. Verbeek et al. Integrated 4-channel Mach-Zehnder multi/demultiplexer fabricated with phosphorus doped SiO_2 waveguides on silicon. *J. Lightwave Tech.* **6** 1011–1015 (1988).
36. P. Urquhart, Optical fibre BUS network incorporating a long span fibre amplifier. Submitted to *Optics. Comms.* (June 1990).
37. S.A. Cassidy et al. Amplifying fibre LANS. *IEEE/LEOS Summer Topical Meeting on Opt. Multiple Access Networks*, Monterey, Ca (July 1990).

38. D.W. Smith, P. Healey and S.A. Cassidy Multidimensional optical interconnection networks. *Proc. of OE-Lase 90*, Los Angeles, (1990).
39. M.C. Farries, P.R. Morkel and J.E. Townsend, Samarium 3 + doped glass laser operating at 651 nm. *Elect. Lett.* **24** 709–711 (1988).
40. C.A. Millar, S.R. Mallinson, B.J. Ainslie and S.P. Craig, Photochromic behaviour of thulium-doped silica optical fibres. *Elect. Lett.* **24** (10) 590–591 (1988).
41. J.Y. Allain, M. Monerie and H. Poignant, Blue up-conversion fluorozirconate fibre laser. *Elect. Lett.* **26** (3) 166–168 (1990).
42. J.Y. Allain, M. Monerie and H. Poignant, Room temperature CW tuneable green up-conversion holmium fibre laser. *Elect. Lett.* **26**(4) 261–263 (1990).
43. S.F. Carter *et al.* Low loss fluoride fibre by reduced pressure casting. *Elect. Lett.* **26**(25) 2115–2116 (1990).
44. M.C. Brierley and P.W. France, CW lasing at 2.7 microns in an erbium doped fluorozirconate fibre. *Elect. Lett.* **24** (15) 935–937 (1988).
45. L. Wetenkamp, Efficient CW operation of a 2.9 micron holmium doped fluorozirconate fibre laser pumped at 640 nm. *Elect. Lett.* **26** (13) 883–884 (1990).
46. R. Allen and L. Esterowitz, CW diode-pumped 2.3 micron fibre laser. *Appl. Phys. Lett.* **55**(8) 935–937 (1989).
47. M.C. Brierley *et al.* Long wavelength fluoride fibre systems using a 2.7 micron fluoride fibre laser. *Proc. OFC 89*, paper PD 14, Houston, Texas, Feb 6–9 (1989).
48. L. Wetenkamp and U.B. Unrau, Optical fibre lasers made from rare-earth doped heavy metal fluoride glasses emitting in the 2.5–3 μm wavelength region. *15th Int. Conf. on infrared and millimeter waves, SPIE* **1514**, Paper Th 7.8, Orlando, Florida Dec. 10–14 (1990).
49. J.Y. Allain, M. Monerie and H. Poignant, Tuneable CW lasing around 0.82, 1.48, 1.88 and 2.35 microns in thulium doped fluorozirconate fibre. *Elect. Lett.* **25** (24) 1660–1662 (1989).
50. J.D. Kafka, T. Baer and D.W. Hall, Mode locked erbium doped fibre laser with soliton pulse shaping. *Opt. Lett.* **14** (22) 1269–1271 (1989).
51. K. Smith, J.R. Armitage, R. Wyatt, N.J. Doran and S.M.J. Kelly, Erbium fibre soliton laser. *Elect. Lett.* **26** (15) 1149–1151 (1990).
52. C.A. Millar, M.C. Brierley, M.H. Hunt and S.F. Carter, Efficient up-conversion pumping at 801 nm of an erbium doped fluoride fibre laser operating at 850 nm. *Elect. Lett.* **26**(22) 1871–1872 (1990).
53. D. Cotter, C. Ironside, B.J. Ainslie and H. Girdlestone, Picosecond pump-probe interferometric measurement of optical non-linearity in semiconductor doped glass fibre. *Opt. Lett.* **14** 317 (1989).
54. K. Liu, M. Digonnet, K. Fesler, B.Y. Kim and H.J. Shaw, Superfluorescent single-mode neodymium fibre source at 1060 nm. *Tech. Digest Optical Fibre Sensors Conf*, OFS 88, 462, (1988).
55. M.C. Farries and M.E. Fermann, Temperature sensing by a thermally-induced absorption in a neodymium doped optical fibre. *SPIE*, **798** 46–48 (1987).

Index

absorption 40
  excited state 92
  measurement 79
  multiphoton 186, 247
absorption bands, erbium 107
absorption spectra 70, 187
aerosol doping 63
all-optical supertrunk 240
alumina 54, 72, 162, 178
  coordination with 75
American Optical Company 1
amplified spontaneous emission (ASE) 136, 156
amplifier, fibre
  comparison with other types 151
  double pumped 137
  efficiency 140, 233
  gain bandwidth 131
  gain saturation 142, 208
  gain spectra 141, 206
  linearity 235
  noise 143
  optimum length 138
  polarization properties 143
  saturation power 131, 142, 232
amplifier, non-linear fibre 146
amplifier, semiconductor laser 149
  crosstalk in 150
  gain 149
  gain ripple 149
  optical processing in 150, 153
  polarization sensitivity 149
  saturated output power 149
automatic gain control (AGC) 143

background loss 38, 71, 77, 243
beat noise 143
beating (in multiplexer) 115
beatlength 116
Bellcore 154
Bishops Stortford 239
bit rate transparency 237
bonding in glasses 53
bottlenecks 238
Bragg grating reflector 168
branching ratios 23
Brillouin fibre amplifier 148
  shift 148

British Telecom Research Laboratory (BTRL) 118, 239
Broadband Passive Optical Networks (BPON) 239
broadcast system 235, 238
BT&D 131
BUS networks 245

cable TV systems 156
cascades 37
casting 198
CD players 205, 247
Ce^{3+} doped fibre 64
chemical analysis 249
chirp 135, 216
cladding pumping 10, 171
cleavage planes 110
cleaved fibre ends (as reflectors) 213, 219
clustering 35, 68, 99
co- and counter-propagation 46, 144
coatings 199
co-doping 32, 62, 210
  Yb^{3+}/Er^{3+} 6, 174
coherence length 2
concentration quenching 9, 57
confined fibre 75
couplers 10, 112
coupling loss, fibre to laser 151
Cr^{3+} 208
cross-sections
  absorption 23, 81, 140, 195, 197
  emission 80, 165, 197, 207
  ESA 86, 92, 207
  measurement of 80, 81
crosstalk 143, 152, 237
  Raman induced 152
Czochralski growth 107

defects 10, 178
devitrification 65
Dieke diagram 164
dielectric filters 119
dispersion 157
  in fibre laser 214
distributed gain 144, 243
doping density 17, 45, 74
double clad fibres 10, 171
dysprosium 187

effective core area 43
efficiency 87
energy levels 18, 20, 164
  labelling of thulium 194
energy transfer 69
erbium 6, 57, 166, 172, 187, 249, 251
europium 187
excited state absorption (ESA) 9, 86, 91, 139, 173, 175
  measurement of 88
external modulators 135

facet coating 110
facet reflectivities 111
fan-out 246
fibre amplifier *see* amplifier
  structure 107
fibre couplers
  (*see also* multiplexers) 112
  fused type 114
  fused type, packaging of 115
  polarization sensitivity 116
  polished type 112
fibre drawing 59, 83, 199
fibre gratings 127
  reflectivity and bandwidth 128
fibre gyro 176
fibre laser
  damage threshold 109
  effect of cooling 167
  efficiency 95, 174, 109
  external cavity 93
  fluoride 200
  high power 169, 213
  linewidth of 168
  loss modulation in
    (*see also* Q-switching) 214
  mode-locked Yb/Er 220
  phase modulation in
    (*see also* mode-locking) 215
  single longitudinal mode 128, 168
  soliton 250
  structure 106
  transitions in silica 162
  tuning of 127, 128, 164, 166, 174, 247
  with grating reflector 127
  with low QE 179
fibre ring laser 129
fibre sensors 252

INDEX

filters
 high pass 120
 pump rejection 119, 155
Finesse 129
flashlamp pumping 1, 169
fluorescence
 decay, for erbium 102
 decay, for neodymium 99
 excitation dependent 25
 lifetimes 7, 81
 line narrowing 9
 measurement of 80
 non-exponential decay 97
 quenching 29
 spectra 71, 187
 time resolved 67, 71, 96
fluoride fibre lasers
 Er^{3+} 204
 Ho^{3+} 202
 No^{3+} 200
 Tm^{3+} 202
fractional inversion 83
free spectral range (FSR) 129

gain
 modulation of 142, 206
 recovery time 83, 234
 saturation see amplifier
 spectra 84, 206
gas phase doping 66
glasses (see also host glasses)
 crystallization rate 52
 germania 54, 146
 network former 185, 247
 network modifier 54, 57, 185
 substituted 54
gratings
 fibre 127
 for feedback 127
 for multiplexers 123
 photoresist 128
group velocity dispersion 217

half coupler blocks 113
head-end 238
HfF^{4+} 184, 198
high peak-power pulses 235
hole burning 9, 25, 169
holmium 9, 179, 185, 187, 194, 247
host glasses see glasses
 borate 55
 fluoroberyllate 175
 fluorophosphate 175
 fluorozirconate 36, 184
 phosphate 55
 silica 36
hypersensitive transitions 8

intermodulation distortion
 see crosstalk 143
inversion parameter, μ 85
ion–ion interactions 7, 16, 32, 35, 67, 96
isolators 155

Judd–Ofelt parameters 8, 26, 196

KDD 154
Kerr non-linearity 157
Kramers–Kronig in gain medium 218

LaF 184
laser rangefinder 12
lasers
 distributed feedback (DFB) 135
 Fabry–Perot 109
 high power semiconductor 110, 112
lasing criteria 30
lifetime
 non-radiative 28, 30
 radiative 27, 195
 shortening 164
ligand field 7, 57
line broadening
 homogeneous 22, 26
 inhomogeneous 22, 185
linear gain stage
 (see also repeater amplifier) 135
linestrength 23
local broadcasting 156, 235, 238
loop reflectors
 see Sagnac interferometer 126
loss barriers 230
LP modes 44

MCVD 161
memory storage 247
mirror damage 247
mode field mismatch in splices 130
mode-locked semiconductor lasers 157
 dynamics of 220
mode locking 4, 213, 249
molecular stuffing 65
monomode fluoride fibre 199
multicustomer networks 239
multimode fibre laser 171
multiple quantum well lasers 109
multiplexers
 (see also fibre couplers) 115
 comparison of 124
 dielectric filter type 121
 fused type 116
 grating type 123
 planar wave guide type 122
 twisted/fused type 118

Nd:YAG laser 3, 10
neodymium 3, 9, 25, 29, 38, 57, 58, 67, 161, 166, 187, 194
net amplifier gain 151
network management 246
networks 238
nodes 241
noise 47
noise accumulation 155
noise figure 143
non-linear transmission 157
non-linearities in fibres 215
non-radiative transitions 8
Nova laser 5
NTT 154

optical signal processing 245
optical time domain reflectometer (OTDR) 158
optimum pump wavelengths 89
output power of a laser 47
overlaps (of pump, signal and ion fields) 45
oxide glasses 56

packaging
 amplifiers 130
 fibre couplers 115
 filters 120
phonon coupling 8, 247
phonon edge 56, 248
phonon energies 185
phonons 148
photodarkening 180, 247
photorefractive effect 10
photoresist 128
Pirelli 154
plane-wave approximation 42
polarization controller (fibre) 166
population dynamics 28
populations 93
power amplifier 1, 135, 152, 233
 noise penalties in 235
praseodymium 8, 180, 187
preamplifier 2, 136, 153, 156
preamplifier receiver sensitivity 156
preform making (fluorides) 198
propagation equations 42, 46
pulse shaping 143, 150
pump/probe measurements 92
pump sources 107
 Alexandrite at 786 nm 202
 argon ion laser 93, 172
 colour centre 88
 dye laser 172
 early diode laser 2, 10
 flashlamp 1, 169
 GaAlAs 34, 173, 249
 InGaAs/GaAs strain layer at 980 nm 81, 108, 173
 InGaAs/InP at 1480 nm 108, 173, 197
 injection-locked array 139, 174
 intensities 16, 32
 Nd:YAG 11, 171, 172
 Nd:YLF 174